Insect, Disease and Weed Control

E.R. DE ONG
D. BISHOP
J.L. BISHOP, JR.

CHEMICAL PUBLISHING CO., INC. 1972

Insect, Disease and Weed Control

ISBN: 978-0-8206-0006-2

Chemical Publishing Company:
www.chemical-publishing.com
www.chemicalpublishing.net

First Edition:

© **Chemical Publishing Company, Inc.** - New York 1972

Second Impression:

Chemical Publishing Company, Inc. - 2011

Printed in the United States of America

FOREWORD

To provide a reasonably comprehensive guide to agricultural pests and their control, we attempt to briefly describe the most important pests and agricultural products. In describing the control for each crop, we try to provide you with some information to aid your understanding of both the specific problem and its relation to other crops and pests. Through this understanding, the first stage in the intelligent application of control procedures is achieved. Thus, an over-application of pesticides to overcome the effect of unknowingly harboring a pest reservoir in or at the edge of the field/yard can be avoided.

The second stage of intelligent (safe, economic) pest control is gained through thorough familiarity with the use of the chosen method of pest control. The effort expended by academic, industrial, and governmental experts in determining the best dosage levels and methods of application is wasted unless the product label is read and followed.

Your interest, hopefully aroused by our work, can be further developed by contacting the appropriate experts in government, academia, or industry. To aid you in this activity, we try to provide references in each section.

Finally, this effort is dedicated, with yours, to the production and consumption of agricultural products of the best quality.

CONTENTS

I

Introduction

The use of chemicals in the control of pests dates back hundreds of years. Not until the early part of the twentieth century, however, was any thought given to the safety of those applying the material or to the possibility of dangerous amounts of the pesticide* (residue) remaining on the raw fruit or vegetable.

The first general use of insecticides was that of Paris green in the control of the Colorado potato beetle which was sweeping the country. Later it was used against grasshoppers so successfully that orchardists adopted it in their fight to control codling moths on apples and pears. The eggs of the moth are laid on the young fruit and when hatched, the young larvae burrow into the fruit to develop. Homemade calcium arsenite and the commercial compound, Paris green, were commonly used against this insect. Both of these compounds are effective insecticides, but at that time contained so much uncombined arsenic as to be dangerous to foliage. Later with the newly developed lead arsenate, the applications in apple orchards became frequent and heavy.

Arsenical injury in the orchard is not confined to the foliage and fruit, but after years of heavy spraying the trees themselves may be injured and the soil poisoned. In some apple districts, this effect became so general that large numbers of orchard trees were pulled out and the land planted with resistant crops such as barley.

---

* The term "pesticide" is applied to insecticides, fungicides, rodenticides, and herbicides (weed-killers). The term "insecticide" means any substance or mixtures of substances intended for destroying, controlling or repelling any insect and their relatives the mites and ticks. The term "fungicide" means any substance or mixture of substances intended for preventing, destroying or repelling any fungi or bacteria attacking plants and also includes agricultural disinfectants. The term "rodenticide" means any substance or mixture of substances intended for destroying, controlling or repelling rodents or other vertebrate animals which the Secretary of Agriculture shall declare to be a pest. The term "herbicide" means any substance or mixture of substances intended for preventing, destroying or repelling any weed. The term "weed" means any plant which grows where it is not wanted and includes shrubs and trees.

Early Inorganic Insecticides.

The first insecticides consisted principally of inorganic chemicals including arsenic, calcium, lead, barium, mercury, sodium fluoride, and phosphorous pastes for controlling ants and cockroaches. Thallium was used in ant baits and for rodent poisoning, while selenium was recommended for the control of plant-feeding mites (spider mites). Mercury in the form of calomel acted as an insect repellent and as a fungicide. Lime-sulfur both in the liquid and dry form was the usual control agent for scale insects on deciduous trees. Citrus trees were fumigated under canvas tents with hydrocyanic gas for checking the ravages of scale insects. The gas was generated from either sodium or calcium cyanide in pots placed under the tent and later in fumigating machines with the delivery hose for the gas laid under the tree. During the early part of the twentieth century, however, resistance of the scale insects to the action of the gas became so great that heavier dosages of gas were required than the tree would tolerate. Recourse was the use of highly refined petroleum fractions applied as a spray to the citrus tree as a protection against insects and mites.

Early Organic Insecticides.

The early history of organic insecticides is very meager. Kerosene emulsion has a limited use in the garden and on the citrus tree but was found ineffective at dosages that were tolerated. Various crude distillates of petroleum oil were found to be more effective as insecticides but were limited in use because of plant injury. Several years later the "white" oils were developed that could be used on foliage as well as on dormant trees. Kerosene continued to act as the base for fly sprays, as a control agent for flies in the home, and on livestock.

Creosote, derived form coal tar, has been used extensively on timbers and on utility poles as a wood protectant.

The introduction of certain plant derivatives as insecticides was a decided contribution to safe practice as they break down quickly into harmless compounds on exposure.

Ground pyrethrum flowers have been marketed for many years as "insect powder" for use against household insects. Results varied, however, with the content of the active principals known as "Pyrethrins I" and "Pyrethrins II". In commercial practice the active principals are extracted, standardized and marketed alone or in combination with organic compounds such as thiocyanate.

To increase the activity of the pyrethrins and to lower their cost, they are usually combined with a synergist such as piperonyl butoxide or sulfoxide. Sprays for use in the home or on livestock are made with a kerosene base. For garden use a base is employed that will be safe on foliage.

The principal value of the pyrethrum sprays in the home is against flies, mosquitoes, clothes moths and other household insects. They are also of value in controlling stored grain pests. On livestock they control flies, lice and ticks. In the home garden they are valuable in the control of aphids and the younger stages of caterpillars.

Rotenone, an extract of the tropical plants *Derris* and *Lonchocarpus*, has wide use on livestock as a protection against lice, fleas, and the cattle grub of dairy and beef cattle. Rotenone is also a control agent for the Mexican bean beetle and for various insects of the home garden. The ground powder and the extract of rotenone are marketed in standardized form to permit exact dosages. As rotenone is an active fish poison, care must be taken to avoid drift and barn drainage where cattle have been treated.

The value of a nicotine extract of tobacco as an insecticide was one of the first to be recognized. However the only source of supply was a water extract of tobacco stems, and the results varied with the nicotine content of the tobacco. It was many years later that the factory standardized product of nicotine sulfate became available and could be used accurately both as a spray and a dust.

Nicotine is especially valuable in the home garden and for certain vegetable crops as it volatilizes in two or three days with little danger to beneficial insects. Its principal value is in the control of aphids, thrips and the young stages of certain caterpillars.

Nicotine is marketed as a dust and a 40 percent concentrate. The former should carry 3 to 4 percent of nicotine except for susceptible species. The manufacturers directions should be followed for sprays.

Sabadilla, an extract of the seeds of the plant *Schoenocaulon officinale*, is a general purpose insecticide for gardens and livestock.

Ryania is from the plant *Ryania speciosa*. It has had extensive use against the sugarcane borer. More recently it has been recommended for use in the control of the codling moth on apples.

CLASSIFICATION OF INSECTICIDES

Contact Insecticides.

This term refers to substances which, in lethal concentration, kill by intimate contact with the body of the insect. It may either act when newly applied as a spray or dust, or by residual deposit with which the insect comes in contact. It is most frequently used against sucking insects such as aphids, scale insects, thrips and sawfly larvae. Examples are phrethrins, petroleum oil sprays and many of the synthetic organic insecticides.

Stomach Insecticides.

This term refers to substances which when applied in lethal concentration to a plant surface will destroy or repel insects feeding upon the leaves or fruit.

Examples are calcium arsenate, Paris green and the fluorosilicates. Many of the new types of organic insecticides act both as contact and stomach poisons, as DDT in its effect on the gypsy moth larvae. Chlordane and lindane may also act as fumigants when in confined spaces. Stomach insecticides are used principally against defoliators and in the control of the codling moth on apples and pears.

Fumigants.

These are volatile substances which release lethal concentrations of vapor in confined spaces at room temperatures or above. Examples are hydrocyanic acid, methyl bromide and carbon tetrachloride. Fumigants are used in warehouses and mills for destroying insects in stored grain, dried fruit and nuts.

Methyl bromide and carbon disulfide are used as soil fumigants in the control of nematodes and wireworms.

Systemic Insecticides.

These substances, when applied to one area of the organism that is to be protected, may be translocated by that organism to the area of itself needing protection from insect pests. The application sites of a plant may be as diverse as the leaf or the seed. Demeton (Systox) or schradan (OMPA) are examples of substances used for the control of mites and aphids on plants. The cattle grub may be controlled by ronnel (Dow ET-57) and coumaphos (Co-Ral, Bayer/ 21/199).

Irradiation as an Insecticide.

Sterilization of insects in the larval or pupal stage is accomplished by treating them with X-rays or gamma rays. The irradiated insects are then released to mate with wild insects of the same species. An example is sterilization of male screw-worm flies which attack livestock.

Diseases of Insects.

The bacteria, fungi and viruses which cause diseases among certain forms of insects are cultured commercially and distributed to areas infested with susceptible insects. As examples, diseases may be given to the Japanese beetle and the alfalfa caterpillar.

Chemosterilants of Insects.

This term refers to substances which, when applied as a spray to the body of the adult insect or when fed to the adult or the larva, retard the development of the reproductive organs or produce genetic changes that cause sterility or death. Examples are the affect of Apholate on the development of the ovaries of house flies, and Imidazole for fabric proofing against insect attack (Knipling, 1960).

Desiccating dusts.

These are substances which when applied at lethal concentrations to the body of the insect absorb or disrupt the protective waxy layer (lipoid) covering the insect's body thus causing a rapid loss of water. The dusts are very finely divided ranging in size from 3 to 10 microns. Examples are a silicogel powder known as SG 67 (Dri-di) (Wagner and Ebeling, 1959), and diatomaceous earth for applying to stored grain.

FEDERAL AND STATE REGULATIONS FOR PESTICIDES

All pesticides offered for sale in interstate commerce must be registered with the United States Department of Agriculture at Washington D.C. together with a fully warranted statement of active and inert ingredients, or with State officers if in intrastate commerce alone. Shipments which, after sampling and analysis, do not show conformity with the filed statement are removed from the market. Materials which have years of safe and effective usage behind them are registered with little question. New materials must furnish satisfactory evidence of effectiveness for the purpose sold, and if there is any question as to safety in its use, confirmatory data are required. Such tests are varied in nature owing to the nature of the material and the amount of its residue.

The Federal Food, Drug and Cosmetic Act, as amended in 1954, provides methods for controlling the amount of residue on raw agricultural products, such tolerances to be established within 90 days after the Secretary of Agriculture has filed a certificate of usefulness for the product.

The Food and Drug Administration has the authority to determine the amount of chemical residue from the preharvest or preslaughter application of pesticide to any food crop and the tolerance permitted. There is thus a dual system of control on the use of pesticides.

The amount of residue remaining on the raw (unprocessed) product of vegetable or fruit at harvest time must conform to the tolerance (amount considered harmless) as established by the Federal Food and Drug Administration. This is expressed as parts per million (p.p.m.). Food products found on the market which exceed the established tolerance are subject to seizure. A list of pesticides showing those requiring no exemptions and those having zero tolerance is given in the Appendix. Since this list is subject to revision and additions, an agricultural officer should be consulted for the latest information. For their own protection, growers should be informed of any established tolerance for crops they expect to market.

In addition to federal and state laws regulating the marketing of pesticides, a number of the states and other municipalities have official regulations covering the sale and use of the more injurious forms of materials. Complaints of damage from dusts and sprays which drifted onto nearby property have been largely directed against aircraft applications of calcium arsenate and the herbicide 2,4-D and related compounds. Damage has also resulted from the drift arising from power operated ground machines. However, weed control by the use of chemicals has become so necessary for many growers of rice and small grains that profitable crops cannot be grown without them. Adjustment of the conflicting needs is being met by the use of granules, instead of dusts and sprays, as a carrier of pesticides. A number of states have regulations restricting the use of injurious pesticides to areas where there is the least possibility of damage from their use. Permits are obtained from an agricultural officer before other than small amounts of injurious materials may be purchased. Appropriate warnings of the proposed application of an injurious material shall be given to all persons working in the area and to all owners of any animals, and honey bees, known to be present or on nearby properties.

Labels and Brands.

Attached to every pesticide container is a printed label which, together with the accompanying circulars and leaflets, make up the "labeling" required by law. The following information is included:

1. Common or manufacturer's name of the active principal e.g. (malathion, Sevin)
2. Chemical name, used for identification
3. Net content of active ingredient stated as percentage e.g. (rotenone, 5 percent dust)
4. Directions for use including the name of the pest(s) for which it is recommended
5. Warnings of hazard, if any
6. Name and address of manufacturer
7. Materials highly toxic to man should show the word "Poison", and the skull and cross-bones in red as well as an antidote statement.

Brands.

In addition to the label there is usually a "brand" name which serves to identify the manufacturer's or processor's product. This name may or may not indicate the name of the active ingredient. For example, "Black Leaf 40" shows the strength of the active ingredient (nicotine) but does not name it. Since most processors and even many dealers have their own brand for each pesticide which they carry, there are many thousands of brand names. For this reason, purchasers should be guided by the use of the common or manufacturer's name of the pesticide.

THE ESTABLISHMENT OF A PEST CONTROL PROGRAM AND THE USE OF PESTICIDES

The use of chemical pesticides has been a significant factor in the production of vegetable crops of both high quality and quantity. The use of these materials has improved the standards of the vegetables that are marketed and used. The success that has been achieved in the past may lead to the use of higher dosages, more toxic chemicals, and more frequent applications than are necessary for the control of the pests. When these excesses occur, residues of the pesticides in the environment may approach or exceed the levels of the chemicals that are established as reasonable maximums. The reduction in dosages, use of less toxic or less persistant chemical agents, and fewer applications are recommended by entomologists and ecologists when this problem persists. Wherever practical, nicotine, rotenone, and malathion may be substituted for DDT and parathion. Nicotine is especially useful in the home garden since this agent disappears within two or three days to be followed by the natural means of control of the pests. The home gardener is in a particularly advantageous position to forego the use of the more toxic and persistant chemicals since he can better tolerate a partial loss in yield rather than risk the excess-use of the chemical agents.

The value of pesticides is well established and their use is often necessary to produce saleable, profitable crops. However, many factors must be considered in the choice of a control technique in addition to the immediate cost.

1. The manufacturer is required by law in the United States to include on the label a list of insects or plant diseases for which the material is recommended. Any experimentation outside of the recommendations must be undertaken at the risk of the user. The Federal Government and many of the State governments publish recommendations periodically to aid the grower in his choice of the most suitable means of pest control given the current state of the development of the various control techniques and the needs of society.
2. The insecticidal effect of a pesticide that is considered for use is a primary concern in the evaluation of alternative control techniques. Chemical pesticides can be classified as either temporary or persistant in their effect. Temporary insecticidal effect includes the protection of seedlings against the attack of cutworms and earwigs, for example, while persistant effect is descriptive of the attack of the cabbage aphid throughout the growth period of the crop.
3. The degree of control is another important variable in the choice of a control technique. "Economic control" generally refers to less than complete control (or annihilation) of the pest and acceptance of partial impairment of the crop quality or quantity by the pest as "cheaper" than the complete control of the pest. Thus the pickers may be required to discard a very small portion of the crop rather than the application of an amount of insecticide that would be

necessary for complete annihilation of the pest. The comparative toxicity of groups of insecticides is occasionally the subject of agricultural research. For instance, Young and Ditman (1959) evaluated the relative toxicity of several insecticides under existing conditions in Maryland. All of the insecticides were used in the form of emulsions with the following concentrations in pounds of active ingredient per acre being used in the program: endosulfan (Thiodan) 0.28, ethion (Nialate) 0.28, carbaryl (Sevin) 0.44, malathion 0.56, naled (Dibrom) 0.14, carbophenothion (Trithion) 0.26. Carbophenothion and endosulfan, alone, resulted in the reduction of more than seventy-five-percent of the potato aphids on both eggplant and tomato. All of the pesticides, with the exception of carbaryl and naled, resulted in a reduction of the population of Cabbage Aphid of at least seventy-five percent. Seventy-five percent reduction, in the numbers of Melon Aphids on the cucumber plants were reported from the applications of carbaryl, malathion, endosulfan, and carbophenothion. Only the use of endosulfan resulted in a significant reduction in the small population of Flea Beetles, whereas all chemicals caused a large reduction in the population of the Potato Leaf Hopper, Cabbage Looper, and the Imported Cabbage Worm. No "off-flavor" was detected in any of the crops as reported by the investigators.

4. The amount of active ingredient of the pesticide that remains on the plant after the application of the spray, dust, etc. is termed the "residue". The amount of residue that is "tolerated" (permitted) on the marketable portion of the crop is generally stated in parts of pesticide per one million parts of harvested crop (parts per million or ppm). Crops that are marketed and contain residues that are in excess of the legal tolerance are subject to condemnation.
5. Since many processors of foodstuffs have established their own regulations regarding the type and amount of the residue allowable on the crops that they purchase, they should be consulted prior to the use of pesticides.
6. The form in which the pesticide is to be used, whether emulsifiable concentrate (EC), wettable powder (WP), dust, granule, dip, or bait, should be considered in relation to the equipment that is available for application. The residues from pesticides that are applied in sprays from emulsifiable concentrates frequently disappear more quickly than those that are applied from wettable powder bases or as granules. Granules frequently are preferred if the pesticide must be applied from an aircraft both to avoid the drift of the chemical and to penetrate dense foliage.

   Baits are made from bulky materials such as wheat bran, that is mixed with an active ingredient, such as sodium fluorosilicate in the case of control of Earwigs. Baits are generally applied by hand around seedlings and under the boards where the pests hide during the day.

7. Careful handling of the pesticides is always necessary. They must *always* be stored in a location that is inaccessible to unauthorized users, such as irresponsible persons and children. The user of pesticides must always remember that the purpose of pesticides are generally to kill the pest, and the chemicals are not capable of distinguishing between pests and pets, children, or desired wildlife. For maximum effectiveness of chemical compounds, storage in cool, dry locations is generally recommended.
8. The frequent and heavy applications of pesticides may result in dangerous accumulations of their residues in the soil. Recent reactions to the use of DDT in the United States and other countries are a response to the deleterious affects that such residues have on fish and other wildlife.
9. The persistence of pesticides in silt loam soil ranging from two weeks to three years was reported by Lichenstein, et al., 1959. In this study, ten to one hundred pounds of lindane and two, twenty, and two hundred pounds of aldrin per six inch acre were evaluated. Within two weeks after its application, lindane decomposed to a nontoxic compound that was still detectable by the colorimetric assay technique for the determination of lindane, but was not detectable by bioassay. Two weeks after the treatment, the quantity that was estimated by bioassay amounted to sixty-six percent of that obtained by chemical analysis. After three years, the same ratio was found between bioassay and chemical analysis. Aldrin was converted to dieldrin, the presence of which was shown by chemical analysis. The peak of the formation of dieldrin was at fifty-six days after treatment. This peak was, naturally, followed by a decrease in the formation of dieldrin. Dieldrin was found to control the Alfalfa Snout Beetle to a satisfactory extent, and, for this reason, has been used extensively in New York to prevent further infestations of this pest. Dieldrin applied as granules at 3 to 5 pounds per acre to alfalfa seedlings produced detectable residues of dieldrin on alfalfa as late as 32 months after application when forage samples were analyzed by gas chromatography. Similar analysis of soil samples taken in 1-inch increments to a depth of 6 inches from treated fields showed a majority of the dieldrin residue in the first inch of soil, suggesting that splashing of dieldrin-treated soil onto the alfalfa was the mechanism of contamination (Hardee, et al., 1964). An apparent buildup of zonal accumulation of dieldrin at the 6-inch soil layer was noted and is probably attributable to concentration of the insecticide following leaching downward to a natural or plow-made hardpan. Dieldrin residues in soil were readily detectable as long as 44 months after application (Hardee, et al., 1964).
10. The flavor of vegetables and fruits may be affected either favorably or unfavorably by the applications of pesticides. The latter effect is illustrated by the unpleasant flavor that is imparted to root crops by the application of lindane to the soil.

The recommendations for the application of pesticides will vary in different locations throughout the United States and the world. For this reason, the dosages and recommended pesticides are presented as a guide and are not to be construed as being an unconditional recommendation for, and assurance of safety and effectiveness in use. The expert advice of local agricultural agents, such as those employed by the Land Grant colleges in the United States as well as officials of the local and federal governments should be consulted in the preparation of a pest control program.

BIBLIOGRAPHY

Decker, Geo. C., N.A.C. News 20(6):3 (1962).
Knipling, E.F., J. Econ. Ent. 53:415 (1960).
Koehler, James W., Calif. Dept. Agr. Bull. 52(2):133 (1963).
Nichols, James O., Penn. Mis Bull. 4404 (1962).
Staff, U.S. Pub. Health Service, Dept. Health, Education & Welfare (1949, 1952).
Staff, N.A.C. News 22(3):3 (1964).
Storer, Tracy I., Calif. Agr. Exper. Sta. Cir's. 410;434 (1952, 1953).
Wagner, R. E. and Ebeling, W., J. Econ. Ent. 52:208 (1959).

II

Nontoxic Methods and Materials as Insecticides

The use of toxic chemicals (pesticides) as the sole control of agricultural and forestry pests has long been recognized by many biologists as being unsatisfactory. This belief is based on the failure of many such applications to give sufficient protection to growing crops, or the applied chemical may be of a hazardous nature to those applying it, or the residue on the fruit or vegetable may be above the permitted tolerance, or there may be a harmful drift of the spray or dust to adjoining fields.

The uncertainty and risk of chemical control has in recent years led to serious investigations of other means of insect control as follows:

1. The release of insect species made sexually sterile by gamma radiation (the only method that has already been employed successfully)
2. The use of chemicals that produce sexual sterility in the natural population of insects
3. The release of adult insects infected with disease organisms that would destroy larval progeny by transmitting the disease to other insects and contaminating the environment
4. The development and release of insect strains that carry deficient genetic characteristics. "Such genetic deficiencies should be a type that would not prevent rearing the insects under controlled conditions or seriously interfere with the ability of the released males to compete with normal males in mating" (Knipling, 1960).

IRRADIATION

A field test was made in 1955 on the island of Curacao 40 miles off the coast of Venezuela. Screw-worm flies were abundant at the beginning of the experiment and the large flocks of sheep and goats were suffering severely from their attack. Sterilized fly pupae were prepared at Orlando, Florida and sent by air-mail to the island. The adults emerged two days after irradiation. Sterilized males were distributed over the island by airplane at the rate of about 100 per square mile. Within 8 weeks the percentage of sterile egg masses increased from 69 to 100 and the population declined more than 99 percent. No fly activity was found after the eighth week of release (Smith, 1960).

Following eradication of the screw-worm fly on the island of Curacao, plans were made for similar work in Florida and the southeastern states. The annual damage to livestock from the fly attack in the region is estimated at 30 million dollars. Immense numbers of flies were reared on meat, sterilized, and released in the fly infested areas. More than 3 billion of fly pupae were produced, sterilized, and distributed in the southeastern states in the eradication program. Livestock losses have been eliminated in the states concerned. An inspection force is maintained to protect against livestock with fly infested wounds entering the treated areas (Smith, 1960).

Irradiation of the malaria mosquito is thought to be a practical method to use in the control of this insect. Such practice might not be suitable when used alone but may be useful as a supplement for an anti-malaria campaign (Davis, et al., 1959).

Codling moth pupae, exposed to gamma radiation, have shown as high as 98 percent lethality in the male sperm. The females were much more radio-sensitive than the males. Field experiments were then conducted in the irradiation of male codling moth pupae in a cage over a living apple tree. The male moths from the irradiated pupae were then allowed to mate with normal female moths confined with them. "The presence of irradiated males had no measurable effect on the total number of eggs laid but the developing embryos usually died at, or before, the red ring stage of the egg".

"The next step . . . will be to release irradiated male codling moths (exposed as fully developed pupae . . .) in a small isolated apple orchard . . . The aim will be to introduce the irradiated individuals in sufficient numbers as to approximate a ratio of 25 irradiated males to one normal male (Proverbs and Newton, 1962)".

A problem in quarantine practice is to prevent any living stage of the fruit flies, found in the Hawaiian Islands, from being introduced into continental United States. The flies in the larval stage feed on a large variety of fruit and vegetables and would be extremely injurious if established in California or other fruit and vegetable growing states where the crops grown are suitable as hosts for the flies. The species concerned include the Oriental fruit fly, the melon fly and the Mediterranean fruit fly. Since much of the life of the fruit fly is spent within the fruit itself, the experiments were designed to simulate the natural medium (Balock et al., 1963).

Young fruit fly eggs (2 to 4 hours old) were found to be susceptible to low dosages (600 to 1200 roentgens) of radiation. Much higher dosages, however, were required to prevent the hatching of older eggs. "Low dosages of gamma radiation prevented the emergence of adult flies from irradiated eggs, larvae, or young pupae. Much higher dosages of gamma radiation were required to prevent fly emergence from older pupae than from eggs, larvae or young pupae", (Balock et al., 1963).

"These studies have served to provide much basic information on the effect of gamma radiation on three species of fruit flies. Before the use of gamma radiation could be considered for quarantine treatments, it would be necessary to determine . . . whether the various fruits and vegetables in commercial cartons and packages would be adversely affected by such treatments at the dosages required", (Balock et al., 1963).

Experiments with the European corn borer showed that when virgin females were mated to male moths treated with X-rays, only 1 percent of the eggs hatched. Exposure of pupae of both sexes resulted in a reduction of egg hatch as the dosage was increased. The percentage of egg hatch varied with the age of the pupae at the time of treatment (Walker and Brindley, 1963).

The pink bollworm treated with gamma radiation in the pupal stage has been found to be effectively sterilized. "When 1 and 3-day old pupae were irradiated, dosages inducing complete sterility also caused some pupal mortality and appreciable moth damage; however, irradiation of 5 or 7-day old pupae induced almost complete sterility without pupal mortality or adult malformation". Longevity of males from 7-day old pupae was shorter than that of the control. Such reduced longevity of sterile males is not necessarily a disadvantage in mating experiments. It has been found that most of the matings of untreated moths occur well within the period of the shortened life span of the sterile males. The dosages in the effective area of the cobalt-60 source ranged from 530 to 590 roentgens (r) per minute (Ouye et al., 1964).

CHEMOSTERILANTS*

The use of chemicals as sterilizing agents rather than for immediate killing bears some promise for solution of the problem. One of the reasons for substituting chemicals is that gamma radiation is somewhat injurious to the screwworm. It reduced longevity and perhaps vigor and mating ability. It has therefore

*A chemosterilant is a chemical capable of causing sexual sterility, that is, a failure to reproduce. Insect sterilants may act in several ways. They may prevent production of ova or sperm, may kill ova or sperm after they have been formed, or they may damage chromosome material in the sperm or ova so they cannot function (Weidhass and McDuffie, 1963).

been necessary to release correspondingly more males in eradication programs than normally would be required to destroy wild screw-worm populations. Sterility without injuring the insect is desirable. For this purpose a screening program has been operating for several years (Knipling, 1962). During this time more than 300 antimetabolites** and alkylating agents*** have been evaluated as chemosterilants (Weidhaas and McDuffie, 1963).

LaBreque et al., (1961, 1962, and 1963) have conducted a large number of preliminary experiments with these hazardous materials in the control of flies and mosquitoes by baiting with chemosterilants. A summary of results (1961) stated that the compounds apholate, aphoxide and aphomide merit field evaluation". In 1963 LaBreque et al. released a list of compounds tested as chemosterilants from which 20 were found to affect the fertility of house flies at concentrations ranging from 0.1 to 5 percent. Egg hatch varies with the compounds used from zero to almost normal. "The results of the experiments, although not conclusive in themselves, indicated that chemosterilants might be used to eliminate an insect from an extensive area in eradication programs".

A field experiment in the control of house flies with chemosterilant baits was made by LaBreque et al., (1962). The location was on a Florida key sufficiently isolated to prevent migration and on which a dump consisting of trash and garbage was located. The chemosterilant, aphoxide, was applied once a week for 9 consecutive weeks in a bait consisting of cornmeal 67 percent, sugar 15 percent, powdered milk 15 percent, powdered egg 2.5 percent, and aphoxide 0.5 percent. Counts were made of flies lighting on a wooden grid in 1 minute.

House fly populations were reduced from 47 per grid to 0 within 4 weeks. The proportion of egg masses (from females collected at the dump) containing at least 1 viable egg was reduced from 100 percent to 10 percent within 4 weeks and the percent hatch among all eggs laid was reduced to 1 percent within 5 weeks (LaBreque et al., 1962).

Experiments in the control of house flies with the chemosterilant, methaphoxide (metepa), were also conducted by LaBreque et al., (1963b). Applications were made at weekly intervals for 9 weeks, then semi-weekly. Granular baits with cornmeal as a carrier were the most effective, granulated sugar was less effective, and vermiculite was unsatisfactory. The bait was applied to the droppings beneath the poultry cages. A dosage of 0.5 percent of metepa was used in the baits. Fly population remained high until the applications were changed from once to twice a week.

Similar results were obtained by Gouck et al., (1963) in house fly control on a garbage dump on Pine Island, Florida. A cornmeal bait containing 0.75 percent

**An antimetabolite is a compound wherein a metabolite essential to cell development has been changed in one of several ways which, when introduced into an animal, will show signs associated with a lack of the metabolite (LaBreque, 1961).

***Alkylating agents . . . replace hydrogen with an alkyl group in fundamental genetic material with an effect similar to that of irradiation (LaBreque, 1961).

of the chemosterilant apholate was applied once a week for 7 consecutive weeks then five times each week for 5 weeks.

"The fly population decreased from 68 per grid to 5 to 20 during the first 7 weeks and remained between 0 and 3 per grid the following 5 weeks. The egg hatch ranged from 81 percent before treatment to 12 to 49 percent while applications were made once each week (7 weeks) and 2 to 26 thereafter. The male fertility decreased very little with weekly bait applications but averaged only 22 percent when food was available at all times" (Gouck et al., 1963).

Sexual sterility of the screw-worm fly by the use of various antimetabolites and alkylating agents has been studied by Crystal (1963). It was concluded "that the alkylating agents appear to offer the greatest promise of success as chemosterilants against screw-worm flies. Although the aziridinyl compounds are outstanding in this respect, others need not be neglected".

Harris (1962) has studied the effect on stable flies of the chemosterilants apholate, aphoxide and methaphoxide. The materials were applied topically to the dorsal wall of the thorax and by exposure of the flies to a residual film. The chemosterilants were dissolved in methanol and applied with a graduated applicator. Both males and females were treated in all the tests except those in which treated flies were mated with untreated flies.

It was found that "those chemosterilants sterilized the stable fly within the dosage range in which DDT produces mortality in the insect". Apholate was found to be more effective as a sterilant of stable flies than of screw-worm flies when both sexes were treated topically. The principal effect of the three chemosterilants on stable fly production was the failure of the eggs to hatch. Egg production was not reduced except at a high dosage of apholate and methaphoxide. A chemical need not reduce egg production to be an effective sterilant provided it reduces egg hatch to a very low level. Male stable flies were more sensitive to apholate and aphoxide than females (Harris, 1962).

The practical value of chemosterilants as a control of the house fly must be studied from a number of angles. A knowledge of mating behavior dispersal is vital. "Not only must the concentration of the chemicals needed to sterilize be known but also what other physiological effects these chemicals may have on the organisms" (Murvosh et al., 1964). Under laboratory conditions both apholate and metepa had a bad effect on the longevity of both male and female flies. A 1 percent concentration of the chemosterilant in the food both affected the longevity of the males similarly with but slight difference with the females. At 0.5 percent concentration the results were comparable. "Although both sterilants appeared to shorten the life span . . . a period of 10 days or more was required for an appreciable drop to occur in the male population, since at least 80 percent of the males and females survived the first 10 days. It seems probable that sterile males in the field would have an opportunity to mate with most of the females that emerged within the same week before the chemicals exerted any toxic effects other than sterility of the males" (Murvosh et al., 1964). The average life span of the house fly is thought to be about 73 days.

A series of fifteen compounds were tested for chemosterilant effect on house flies by feeding the chemicals at levels up to 1 percent to adult flies. Under the conditions of the test, 9 compounds had little or no effect on sterility, and 2 compounds caused sterility with no oviposition; 3 were sterile temporarily, and 2 compounds (apholate and thiotepa) caused permanent sterility with oviposition. Apholate at a food level as low as 0.1 percent reduced the viability significantly. Food levels of the sterilant as low as 0.25 to 1.0 percent resulted in fewer eggs as well as causing almost complete sterility (Painter and Kilgore, 1964).

The compound imidazole has been found to be an active antimetabolite and contact insecticide. Because of rapid evaporation it has little residual value except when present in wool fibers. Imidazole has a low mammalian toxicity and is quite free from odor and irritating qualities. These factors make it especially useful in working with textile insects such as clothes moths and carpet beetles. Preliminary experiments also show promise for use against a number of plant feeding insects. It has been shown that the toxicity of imidazole may be increased by combining it with the synergists formaldehyde, phenol, 2-aminopyridine and pyrethrins (Pence, 1963).

DESICCATING DUSTS

Dust coatings of insects with chemically inert dusts of very small particles is known to be of limited value as a control measure. Resistance to insecticidal effect does not develop as it does for certain chemical insecticides. Of all the materials tested against the rice weevil and grain weevil, silicate aerogels were found to be the most effective. Such dusts are usually applied as a dust stream from a blower.

Aerogels are formed when sulfuric acid is added to sodium silicate to produce a firm "hydrogel". It is then washed and dried to form a granular dehydrated silicate aerogel that can be reduced to a very small porous particle size. In commercial practice the silica gel is impregnated with ammonium silico fluoride at a concentration of 4.7 percent. This prevents the formation of a hard cake when the gel dries.

The powder or dust is marketed under the brand of SG 77 and SG 67. The addition of the fluoride compound makes the powder less susceptible to variations in the humidity.

The aerogel dusts have been used successfully in the control of dry wood termites, cockroaches, fleas and the two-spotted mite (which attacks roses) and a number of other types of insects. (Wagner and Ebeling, 1959), (Ebeling and Wagner, 1959).

DISEASES OF INSECTS

Insect diseases are an important factor in reducing the size of insect populations although this value is seldom recognized. The organisms causing disease among insects include viruses, bacteria, fungi, protozoa and nematodes. These are very rarely the same as those attacking man or plants. Several attempts have been made in recent years to utilize disease producing organisms as a control of injurious insects. Unfortunately the first attempts were made with fungi which are quite dependent upon favorable atmospheric and temperature conditions for development. These restrictions rule out a large part of the United States and account for the failure of culturing chinch bug diseases in Kansas. Natural conditions for the growth of fungi are almost ideal in much of Florida where parasitic fungi are recognized as a valuable aid in the control of certain pest insects.

Infection by bacteria, viruses and protozoa are due, just as in man and the higher plants, to accidental contamination by wind, flowing water, sucking insects and cultivation tools. Favorable weather conditions are, however, necessary for building up populations sufficiently dense to disseminate disease rapidly.*

Commercial development of cultures of the organisms causing insect diseases is comparatively recent but the future is very promising for use against certain insects. As yet research has been principally among insects of agriculture and forestry but will be further extended.

Japanese Beetle.

This insect in the adult stage feeds on the fruit and foliage of orchard trees and on many ornamentals and in the grub (larval) stage on the roots of lawn grasses. The grubs are subject to infection by two bacterial diseases. Infected grubs die and disintegrate in the soil to infect other grubs. The spores are very resistant and may live for years in the soil. Cultures of the bacteria are developed by the U.S. Bureau of Entomology and Plant Quarantine and combined with a dust base ready for distribution in infested areas. The powder may be mixed with fertilizer or certain of the insecticides, scattered over the lawn and washed in by sprinklers. The disease develops more rapidly in lawns that have a large population of grubs (Hawley, 1952).

Alfalfa Caterpillar.

The alfalfa caterpillar is widely distributed in western areas damaging the alfalfa both for hay and for the seed crop. Two forms of disease organisms

---

*de Ong, E.R., Chemical and Natural Control of Pests, New York, Reinhold Publishing Corp. 1960.

attack the caterpillar, a virus and a bacterium (*Bacillus thuringiensis* Berliner). Damage is caused not only by feeding, but where caterpillars have been killed by the virus, the hay becomes unpalatable to cattle. The economic value of the virus attack is dependent upon the density of the insect population for infection spreads slowly in light infestation (Steinhaus, 1951).

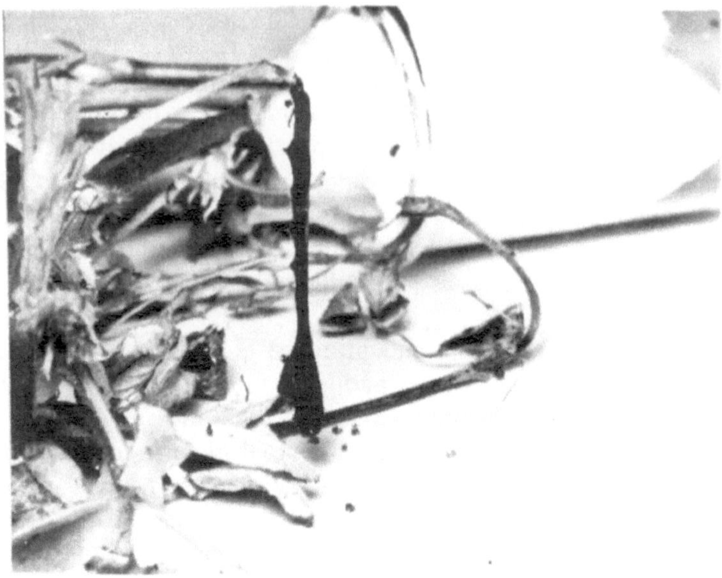

Figure II-1. Larva of alfalfa butterfly, dead of polyhedrosia used in sprays. Note how fluid contents of body fluids have gravitated to anterior end of the insect.

A comparative study has been made of the control of the larvae of the alfalfa caterpillar by applications of various commercial preparations of the bacillus and two chemical insecticides, Phosdrin and Dylox. The dosages of the former was 2 ounces per acre, of the latter 5 ounces per acre, while the bacillus preparations of wettable powders ranged from 3 to 6 ounces per acre. The mortality of the large caterpillars with bacillus preparations at dosages ranging from 1.4 to 3.6 ounces per acre was from 72 to 93 percent. The mortality from Phosdrin was 92 percent, from Dylox 98 percent, and with a second trial 95 percent. Counts on all materials were made at 3 or 5 days interval.

The conclusions were that the "tests show that most of the small larvae surviving the first exposure to bacillus preparations are unable to continue to develop to the later instars. Thus, when properly used, the microbial materials are as capable as chemical insecticides in bringing about effective control of population of the alfalfa caterpillar". "When applied at the rate of 6 ounces per acre satisfactory control can be obtained with commercial preparations (of bacillus) comparable to those tested" (Hall and Stern, 1962).

The value has been shown of a commercial culture (Thuricide) of the bacterium, *B. thuringiensis*, as a control of the imported cabbage worm but its efficacy is limited by the waxy surface of the cabbage leaf and related forms. Laboratory tests have shown that the addition of small amounts of amine-stearate (mixture of a volatile amine and stearic acid) reduced evaporation, increased suspension and made the deposit resistant to heavy rainfall. Dosages of 0.25 and 0.5 pound of wettable powder per acre were applied by spraying without difficulty, but heavier dosages were impractical. No loss of insecticidal value was noted from the addition of the amine-stearate (Geering and Lloyd, 1962).

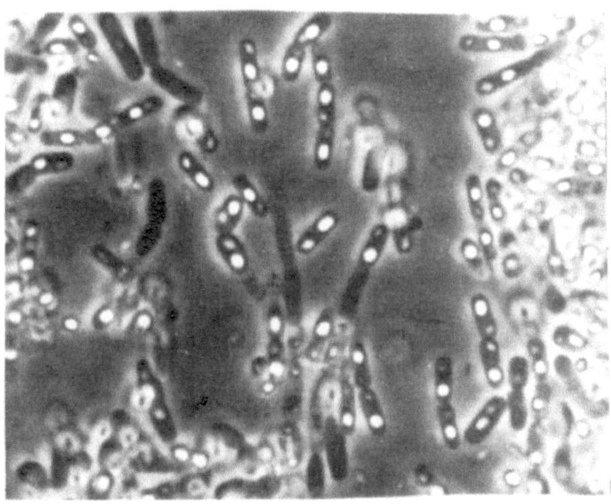

Figure II-2. Crystals in *Bacillus thuringiensis* Berliner, taken by light microscope.

European corn borer.

To control the borer, under Quebec conditions, requires 4 applications of DDT or other toxic insecticide thus making the corn fodder unfit for livestock feeding. Ryania is also used but is expensive and requires heavy dosages. For these reasons trials were made with 3 commercial preparations of *B. thuringiensis*, the borer being known to be susceptible to the bacillus. Combination spray formulations were also made of DDT and with Sevin (carbaryl) and the bacillus.

"The results of the 2-year study indicate that the European corn borer's susceptibility to *B. thuringiensis* in the form of wettable powder appears to be rather low and inconsistent, and that 4 pounds per acre do not give effective control of a heavy infestation" (Hudon, Marcelm, 1962).

INSECT PARASITES AND PREDATORS*

Plant feeding insects and a number of those that annoy our person are subject to attack by either parasitic insects or predators or both. They are very effective in reducing insect populations because of their great number and their ability to attack at varying stages of development of the pest insect from egg to adult.

Parasites feed in or on the body of the insect they are preying upon, but without causing immediate death. Predators kill their prey at the time of attack by sucking out the body juices, or like the hornet, storing the paralyzed prey for their larva to feed upon. These helpful insects are working all about us but so quietly that we do not recognize them. Parasites or the predators are usually found where their host insects abound. Dragonflies, which feed on mosquitoes and midges both in the larval and adult stages, are found about swamps. The lady beetle and the syrphid fly may be seen where there are colonies of aphids, mites or other suitable food insects.

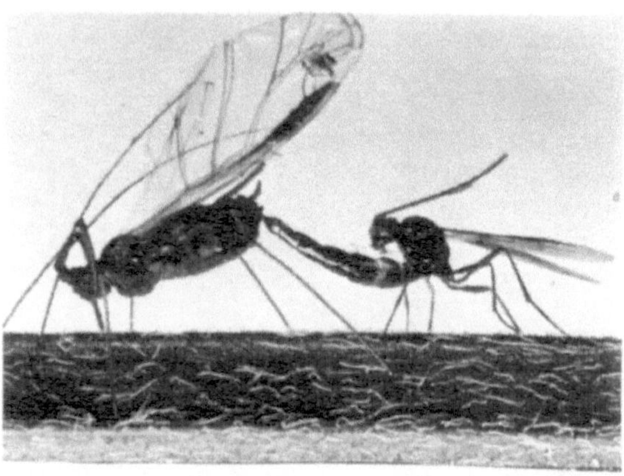

Figure II-3. Parasitic wasp ovipositing in aphid. (Photo Doten and F. E. Skinner.) Figure II-4.

Parasites.

The wasps that prey on other insects range in size from 0.3 to 2 mm. (millimeter) in length to several mm. long. Many of these attack only one species of food insect or closely related species. Others attack a large variety of hosts and for that reason are better able to maintain themselves throughout the year. They attack principally aphids, scale insects, caterpillars and the eggs of certain larger insects as the katydid.

Scale insects are frequently parasitized by the wasps, and this attack together with that of predators, insect diseases and unfavorable weather may hold the scale below the economic level thus making control with chemicals unnecessary.

*de Ong, E.R., Chemical and Natural Control of Pests, New York, Reinhold Publishing Corp., 1960.

To be effective in the control of scale insects, the parasite must be able to find the necessary type of food throughout the year or at least during the warmer seasons and pass the winter in suitable stages of the scale insect.

Aphids are one of the chief sources of food for parasitic wasps and only because of their very rapid rate of increase are they able to sustain the attacks. However the aphid has one advantage in that they emerge from winter quarters at a slightly lower temperature than does the wasp and become well established before the attack begins. The parasite builds up slowly at first so that it is usually late spring or early summer before the combined attacks of parasites and predators hold the aphids in check. It is for this reason that chemical control is frequently necessary early in the season.

Figure II-4. Aphid showing exit hole of parasitic wasp. (Photo F.E. Skinner.)

The tiny wasps become very active on sunny days searching for colonies of aphids, tapping an aphid with its antenna and then with a quick motion plunging its ovipositor into the aphid's body and depositing its egg. For a very few days the aphid remains apparently normal, except for a darker color, but as the parasite feeds reproduction ceases and the aphid is soon dead.

Many of the caterpillars (larvae of moths and butterflies) are also fed upon. Several destructive species such as the imported cabbage worm and the Oriental fruit moth have entered this country but their parasites were left behind. This permitted the pest insect to build up rapidly since there was no natural control until the parasites were imported and bred up in large numbers for release. In some instances native parasites have attacked the foreign pest and been helpful in its control. Tachinid flies also attack many of the caterpillars, certain beetles and grasshoppers.

The tiniest parasites are those developing in the eggs of the host insect. The latter include a number of our most injurious moths such as the bollworm, the sugarcane borer and also certain of the leafhoppers.

Parasitic wasps develop at a very rapid rate. The life span from egg to adult may not be more than 6 to 10 days. This makes possible a number of generations in one season. If, however, the later generations develop within a period of food shortage they may die without an opportunity of ovipositing.

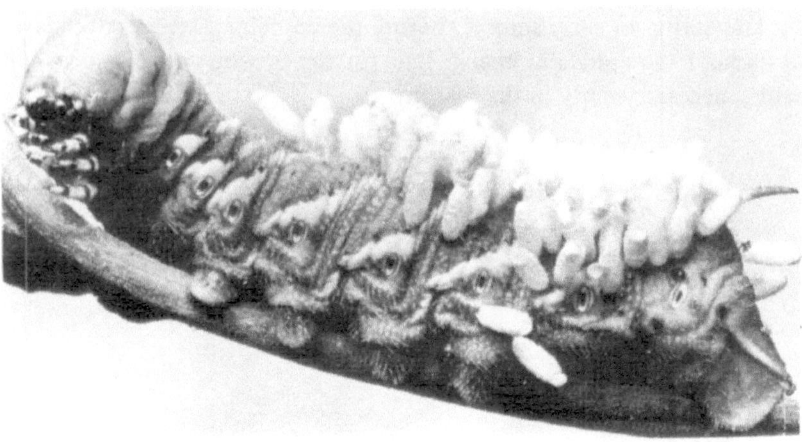

Figure II-5. Cocoons of a parasitic wasp that attacks Sphinx moth larvae. (Photo F. E. Skinner.)

Predators.

Wandering among the rose leaves may be found a brightly colored beetle, about one-quarter of an inch long, shiny black with red spots on the wing covers and white marked thorax. This is one of the most common of the lady beetles. They are predaceous both in the adult and larval stage. The latter are sluggish, gray with pale colored spots. They also are seeking colonies of aphids, scale insects, mealybugs and mites. In addition to the lady beetles, there are many forms of predators; in fact, most groups of related insects contain one or more predators. There are large fierce flies taking their prey on the wing just as the dragonfly does. In another large family, the adult fly feeds on pollen and nectar, while the larva attacks every aphid it can reach. Hiding under rocks and boards during the day but active at night, are the ground beetles and the tiger beetles, feeding on caterpillars and almost every type of insect they meet. Even the wood-borers are not safe for they are attacked by predaceous beetles and large parasitic wasps. Because they are more conspicuous, the predators are better known to the public and should be encouraged whenever practical.

BIOLOGICAL CONTROL OF INSECTS

In 1888-89 the predaceous beetle vedalia was imported from Australia, established in California, and within two years was clearing the citrus groves of the cottony-cushion scale insect that was threatening their existence. This success was followed a few years later with the collection and importation into California of the predaceous beetle, *Cryptolaemus*, which feeds on the citrus mealybug. This beetle is being bred in great numbers and distributed wherever the

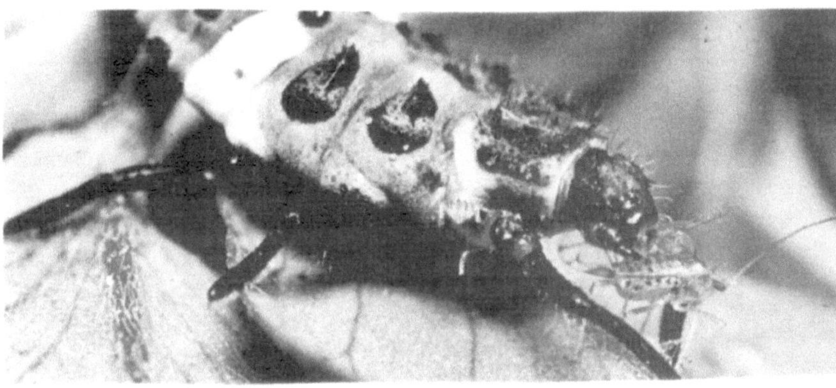

Figure II-6. Larva of ladybeetle, feeding on winged adult spotted alfalfa aphid. (Photos, F. E. Skinner.)

mealybug is found. This practice makes available to the citrus growers of the state ample supplies of the beetle whenever needed. These two examples are among the most successful introductions of beneficial insects to feed on crop pests. (Clausen, 1956).

Figure II-7. Syrphid fly larva feeding on aphids. (Photo Doten & Skinner.)

Biological control has in recent years become an important part of insect control especially in citrus groves and in protecting the hardwood forests of the northeastern states from forest insects. A number of parasites have also been introduced for work on the European corn borer. Among the 95 species that were established in this country by 1953, 81 were parasitic in habit and 14 were predaceous. The parasites are principally wasps but there are also a number of flies, the tachinids, feeding mostly on forest and shade-tree insects (Clausen, 1956).

Figure II-8. Green lacewing *(Chrysopa)* the larva of which feeds on aphids. (Photo F. E. Skinner.)

A large number of the introduced insects have come from continental Europe while others have been brought from Australia and South Africa (Clausen, 1956).

Biological control, where established, is more economical and less hazardous than the use of chemicals. It is a slow process and must be carried on with great care to avoid the possibility of introducing insects that would develop dangerous feeding habits such as attacking other parasites or predators.

WEATHER

Unfavorable weather is another vital factor in preventing the development of insects. The egg and pupal stages of insects frequently offer means of passing severe winters. Very low temperatures may kill a large proportion of over-wintering insects. This is especially true in variable temperatures of very low and moderate warm weather. In the citrus districts of the southwest high temperatures may kill a large proportion of the newly hatched scale insects especially on the sunny side of the tree. Hard dashing rain will wash away aphids and other small insects on exposed parts of trees and bushes.

CULTIVATION PRACTICES

Good farm and garden practices are helpful in controlling many pest insects. They include tillage, crop rotation and adjusted planting dates. Corn that is planted late in the season may escape severe injury by the corn rootworm in the southeast and suffer less from attacks by the European corn borer in the northeastern and central states. Corn planted at midseason in southeastern Texas is less subject to injury by borers than are earlier or later plantings.

Chinch bug damage to sorghums in southwestern Oklahoma is less severe to crops planted in April than to crops planted in early June. In western Nebraska the Colorado potato beetle is controlled (with fewer insecticide sprays) on potatoes when planted in June than on those planted earlier (Baker and Mathews, 1952).

Wireworm damage to crops in the Pacific northwest can be reduced by crop rotation and good cultivation methods. "Flooding infested fields for 6 to 7 days with an inch or two of water during extremely hot weather will kill nearly all the wireworms. Soil temperatures under the water must remain above 68°F for the entire period for the best results". Another method for killing most of the wireworms, is to allow the top 15 inches of soil to remain very dry for several weeks in the summer. The latter treatment may not have to be repeated oftener than once every 5 or 6 years. Plowing infested fields between July 15 and August 15 will reduce the number of adult wireworms that will lay eggs the following spring (Lane and Stone, 1959).

One of the most destructive insects of the cotton belt is the pink bollworm. In addition to chemical treatments there is much that can be done by sanitation to control its ravages. The eggs of the moth are usually placed at the base of maturing green bolls or in heavy infestation of the leaves. When the crop if harvested in August and the stalks destroyed, there is no place for the late emerging moths to deposit their eggs and the population is greatly reduced. In many districts local regulations require not only early harvesting and the destruction of cotton stalks in the field but also the destruction of all volunteer cotton in adjoining fields.

The destruction of the stalks is done with special machines which shred the stalks. This kills many of the larvae in the crop residue and leaves the trash ready for plowing under. The larval kill from this process ranges from 50 to 98 percent (Chapman et al., 1961).

The above references will serve to illustrate the practical value of cultivation control measures which in many instances can be employed at a very nominal cost and without hazard.

QUARANTINE

Our first and most effective weapon against further introduction of injurious insects from foreign fields are quarantine regulations. These measures either forbid the importation of any materials that might harbor injurious insects not known to be present in this country or require a very thorough inspection of such material. Such regulations are applied only to districts where it is known that such undesirable insects are present. Quarantine regulations may be established and often are set up to prevent insect infested material being moved from one country to another or even introduced from any district where a local infestation of injurious insects is established. An example of the latter is the various quarantines set up in the cotton belt in the attempt to prevent the spread of the injurious pink bollworm. The enforcement of regulations prohibiting the transportation of materials containing dangerous insects and plant diseases requires a large force of inspectors stationed at ports of entry.

Unfortunately many injurious insects such as the San Jose scale were widely established before the danger was recognized. It resulted in heavy expense to orchardists throughout the land to protect apple and peach trees. More recent invasions are those of the gypsy moth in the northeastern states; the Japanese beetle, and the khapra beetle, a warehouse pest. The cost of control of these invaders amounts to millions of dollars annually and will probably continue.

BIBLIOGRAPHY

Baker, W. A., and Mathews, O. R., "1952 Yearbook", Washington, D.C., U.S.D.A., 1952.
Balock, J. W., et al., J. Econ. Entom., 56:42:(1963).
Chapman, A. J., et al., *ibid.*, 54:972 (1961).
Clausen, C. P., U.S.D.A. Technical Bull. No. 1139 (1956).
Crystal, Maxwell M. J. Econ. Entom., 56:468 (1963).
Davis, A. N. *ibid.*, 52:868 (1959).
de Ong, E. R., Chemical and Natural Control of Pests, Reinhold Publishing Corp., New York (1960).
Ebeling, W., and Wagner, R. E., J. Econ. Entom., 52:190 (1959).
Geering, Q. A., and Lloyd, J. H., J. Econ. Entom. 55:786 (1962).
Gouck, H. K., et al., *ibid.*, 56:445 (1963).
Houdon, Marcel, *ibid.*, 55:115 (1962).
Hall, I. M., and Stern, V. M., *ibid.*, 55:862 (1962), 55:782 (1962).
Hawley, Ira M., "1952 Yearbook", Washington D.C., U.S.D.A. (1952).
Harris, Robert L., J. Econ. Entom., 55:882 (1962).

Knipling, E. F., 53:415 (1960); *ibid.*, 782. (1960).
LaBreque, G. C., *ibid.*, 54:684 (1961); *ibid.*, 55:449 (1962); 56:150; 476b (1963).
Murvosh, Chas. M., et al., *ibid.*, 57:1964 (1964).
Ouye, Milton T., et al., *ibid.*, 57:387 (1964).
Painter, Ruth R., and Kilgore, W. W., *ibid.*, 57:154 (1964).
Pence, Roy J., *ibid.*, 56:1 (1963).
Proverbs, M. D., and Newton, Newton, J. R., *ibid.*, 55:934 (1962).
Smith, Charles D., *ibid.*, 53:1110 (1960).
Steinhaus, Edward A. "1952 Yearbook", Washington D.C., U.S.D.A. (1952)
Wagner, R. E., and Ebeling, W., J. Econ. Entom., (1959). JEE. 52:208 (1959).
Walker, John R., and Brindley, T. A., J. Econ. Entom. 56:522 (1963).
Weidhass, D. E., and McDuffie, W. C., Entom. Soc. Amer. Bull. 9:268 (1963).

III

Weeds

Any vegetation which grows in an undesired location may be classified as a weed. Such plants usually exhibit rampant growth, if only in the eyes of their potential subjugator. Moreover, they have occupied the attention of those associated with crop and stock production since prehistoric times.

Some weeds are so classified because they are intrinsically noxious and their elimination is desired to protect the health and safety of man or beast. Other plants, which may be of value themselves, become "weeds" by growing where, for aesthetic or economic reasons, they are not wanted. The desired crop must share the available moisture and nutrients in the soil with the unwanted plant growth. Sunlight, essential to maximum plant development, may be obscured by tall weed growth. Weeds whose growth is characterized by long vines may strangle the development of the desired plant growth. The presence of green weeds in a crop can measurably increase the difficulty of harvesting. In addition, green weeds in stored grain (or other crops) are undesirable since they release both heat and moisture, which are detrimental to the storage of the crop.

Frequently these intruders are eliminated manually by stooping over and pulling them out by their roots. The author and, doubtless, the reader too can attest personally that this is a wearisome task. A more evolved technique for weed control involves the use of crude hand tools to "chop out" this volunteer growth. Mechanization of farm work enabled the grower to control weed growth by replacing "the sweat of his brow" with animal or mechanical power. These techniques all involve either the removal of the weed or curtailment of its growth by mechanical action. In recent times the basic approach to weed control changed with the popularization of chemical weed control. Chemicals were developed that selectively kill some forms of plant growth and leave others unaffected.

Most families of plant life consist of species of economically useful plants, and species of plants that are classified as weeds. A plant's classification as a weed is dependent upon location, (corn in a bean field is a weed, unless succotash is considered a crop at the field level) and culture. (Burdock is a weed in America, a food in Japan), etc. However, knowledge of the horticultural classifications (genus-species-family, etc.) is helpful in recognizing common characteristics of weeds, their growth, and propagation. In addition, knowledge of the genetic characteristics of both weed and crop are helpful in determining effective chemical or genetic control of the weed.

Anyone interested in weeds should be able to recognize at least in a general way a wide variety of plants, since most any native or naturalized plant can become a "weed". In order to achieve effective weed control, one must be able to identify various plants so that the appropriate herbicides and cultural practices are employed to control weed growth. Without a great deal of special training and experience, it is difficult to identify a great number of plants, although most people are familiar with quite a few weeds by their common names. It is not our intent to present a botanical dissertation but simply to acquaint the reader with some of the more common and troublesome plant growth which occurs in the temperate zones of North America. Any number of complete botany books are available for more minute and detailed descriptions of plant names and growth habits. Most state Agricultural Departments provide extensive literature regarding local weed problems and detailed advice for their control. The general outline which follows lists the plant families and may be an aid in gaining familiarity with the characteristics of weeds. The various plant classifications are established on the basis of seed production, degree of evolution of flower petals, method of propagation, etc.

Most botanical classifications are based upon the following outline:

- DIVISION
  - Class
    - Order
      - Family
        - Genus
          - Species

Most plants are then characterized by the combination of the two lowest level classifications–the familiar genus-species (Latin or Latinized) name.

Some of the most primitive forms of plant life *e.g.* the mosses, figworts, lichens, ferns, etc. constitute a large amount of natural (voluntary) plant growth covering a great portion of the earth and so, strictly speaking, are "weeds" However, these seldom constitute a major problem for the person involved with agricultural production, gardens and lawns, so they will not be discussed extensively.

SYNOPSIS OF THE ORDERS AND FAMILIES OF VASCULAR PLANTS

Division I. Peteridophyta (Vascular Cryptogams)

The Vascular Cryptogams consist of fern-like, moss-like, rush-like and quill-leaved plants which have no true flowers and reproduce without embryos (by spores).

Class I Articulatae
- Order I Equisetales
  - Fam. 1. Equisetaceae

Class II Lycopodiinae
- Order II Lycopodiales
  - Fam. 2. Lycopodiales
- Order III Selaginellales
  - Fam. 3. Selaginellaceae

Class III Isoetinae
- Order IV Isoetales
  - Fam. 4. Isoetaceae

Class IV Filicinae (Ferns and Fern-Allies)

Subclass I. Eusporangiatae
- Order V Ophioglossales
  - Fam. 5. Ophioglossaceae

Subclass II. Leptosporangiatae
- Order VI Eufilicales
  - Fam. 6. Osmundaceae
  - Fam. 7. Schizaeceae
  - Fam. 8. Hymenophyllaceae
  - Fam. 9. Polypodiaceae
- Order VII Hydropteridales
  - Fam. 10. Marsileaceae
  - Fam. 11. Salviniaceae

Division II. Spermatophyta (Seed-Plants, Flowering Plants

These are plants both with and without green leaves, they produce true flowers which bear stamens or pistils or both. They normally reproduce by seeds containing an embryo or a minute plant.

Subdividion I Gymnospermae

Gymnosperms are trees or shrubs with needle-like, scale-like or firm broad leaves, its wood is often gummy, containing pitch, and its flowers are mostly unisexual.

Order VIII Coniferae
Fam. 12. Taxaceae
Fam. 13. Pinaceae

Subdivision II Angiospermae

Angiosperms are trees, shrubs, herbs, and flowers with leaves of a variety of forms and textures.

Class I Monocotyledoneae
Order IX Pandanales
Fam. 14. Typhaceae
Fam. 15. Sparganiaceae
Order X Helobiae (Najadales)
Fam. 16. Zosteraceae (Potamogetonaceae)
Fam. 17. Najadaceae
Fam. 18. Juncaginaceae
Fam. 19. Alismataceae
Fam. 20. Butomaceae
Fam. 21. Hydrocharitaceae
Order XI Glumiflorae (Graminales)
Fam. 22. Gramineae (grasses)
Fam. 23. Cyperaceae (Sedges)
Order XII Spathiflorae (Arales)
Fam. 24. Araceae
Fam. 25. Lemnaceae
Order XIII Farinosae (Xyridales)
Fam. 26. Xyridaceae
Fam. 27. Eriocaulaceae
Fam. 28. Bromeliaceae
Fam. 29. Commelinaceae
Fam. 30. Pontederiaceae
Order XIV Liliiflorae (Liliales)
Fam. 31. Juncaceae
Fam. 32. Liliaceae
Fam. 33. Haemodoraceae
Fam. 34. Dioscoreaceae
Fam. 35. Amaryllidaceae
Fam. 36. Iridaceae
Order XV Scitamineae
Fam. 37. Marantaceae
Order XVI Microspermae (Orchidales)
Fam. 38. Burmanniaceae
Fam. 39. Orchidaceae
Class II Dicotyledoneae

Subclass I. Archichlamydeae (Choripetalae)
- Order XVII Piperales
  - Fam. 40. Saururaceae
- Order XVIII Salicales
  - Fam. 41. Salicaceae
- Order XIX Muricales
  - Fam. 42. Myricaceae
- Order XX Leitneriales
  - Fam. 43. Leitneriaceae
- Order XXI Juglandales
  - Fam. 44. Juglandaceae
- Order XXII Fagales
  - Fam. 45. Corylaceae (Betulaceae)
  - Fam. 46. Fagaceae
- Order XXIII Urticales
  - Fam. 47. Ulmaceae
  - Fam. 48. Moraceae
  - Fam. 49. Cannabinaceae
  - Fam. 50. Urticaceae
- Order XXIV Santalales
  - Fam. 51. Santalaceae
  - Fam. 52. Loranthaceae
- Order XXV Aristolochiales
  - Fam. 53. Aristolochiaceae
- Order XXVI Polygonales
  - Fam. 54. Polygonaceae
- Order XXVIIa Centrospermae (Chenopodiales)
  - Fam. 55. Chenopodiaceae
  - Fam. 56. Amaranthaceae
  - Fam. 57. Nyctaginaceae
  - Fam. 58. Phytolaccaceae
  - Fam. 59. Aizoaceae
- Order XXVIIb Centrospermae (Caryophyllales)
  - Fam. 60. Portulacaceae
  - Fam. 61. Caryophyllaceae
- Order XXVIII Ranales (Ranunculales)
  - Fam. 62. Ceratopyllaceae
  - Fam. 63. Nymphaeaceae
  - Fam. 64. Ranunculaceae
  - Fam. 65. Berberidaceae
  - Fam. 66. Menispermaceae
  - Fam. 67. Magnoliaceae
  - Fam. 68. Calycantraceae

Fam. 69. Annonaceae
Fam. 70. Lauraceae
Order XXIX Rhoeadales (Papaverales)
Fam. 71. Papaveraceae
Fam. 72. Capparidaceae
Fam. 73. Cruciferae
Fam. 74. Resedaceae
Order XXX Sarraceniales
Fam. 75. Sarraceniaceae
Fam. 76. Droseraceae
Order XXXI Podostemonales
Fam. 77. Podostemaceae
Order XXXII Rosales
Fam. 78. Crassulaceae
Fam. 79. Saxifragaceae
Fam. 80. Hamamelidaceae
Fam. 81. Platanaceae
Fam. 82. Rosaceae
Fam. 83. Leguminosae
Order XXXIII Geraniales
Fam. 84. Linaceae
Fam. 85. Oxalidaceae
Fam. 86. Geraniaceae
Fam. 87. Zygophyllaceae
Fam. 88. Geraniaceae
Fam. 89. Simaroubaceae
Fam. 90. Meliaceae
Fam. 91. Polygalaceae
Fam. 92. Euphordiaceae
Fam. 93. Callitrichaceae
Order XXXIV Sapindales
Fam. 94. Buxaceae
Fam. 95. Empetraceae
Fam. 96. Limnanthaceae
Fam. 97. Anacardiaceae
Fam. 98. Cyrillaceae
Fam. 99. Aquifoliaceae
Fam. 100. Celastraceae
Fam. 101. Staphyleaceae
Fam. 102. Aceraceae
Fam. 103. Hippocastanaceae
Fam. 104. Sapindaceae

Fam. 105. Balsaminaceae
Order XXXV Rhamnales
Fam. 106. Rhamnaceae
Fam. 107. Vitaceae
Order XXXVI Malvales
Fam. 108. Tiliaceae
Fam. 109. Malvaceae
Order XXXVII Parietales (Violales)
Fam. 110. Theaceae (Ternstroemiaceae)
Fam. 111. Guttiferae (incl. Hypericaceae)
Fam. 112. Elatinaceae
Fam. 113. Tamaricaceae
Fam. 114. Cistaceae
Fam. 115. Violaceae
Fam. 116. Passifloraceae
Fam. 117. Loasaceae
Order XXXVIII Opuntiales
Fam. 118. Cactaceae
Order XXXIX Myrtiflorae (Myratales)
Fam. 119. Thymelaeaceae
Fam. 120. Elaegnaceae
Fam. 121. Lythraceae
Fam. 122. Nyssaceae
Fam. 123. Melastomataceae
Fam. 124. Hydrocaryaceae
Fam. 125. Onagraceae (Oenotheraceae)
Fam. 126. Haloragaceae
Fam. 127. Hippuridaceae
Order XL Umbelliflorae (Umbellales)
Fam. 128. Araliaceae
Fam. 129. Umbelliferae
Fam. 130. Clethraceae
Subclass II Methachlamydeae (Gamopetalae or Sympetalae)
Order XLI Ericales
Fam. 131. Clethraceae
Fam. 132. Pyrolaceae
Fam. 133. Ericaceae
Order XLII Diapensiales
Fam. 134. Diapensiaceae
Order XLIII Primulales
Fam. 135. Primulaceae

Order XLIV Plumbaginales
- Fam. 136. Plumbaginaceae

Order XLV Ebenales
- Fam. 137. Sapotaceae
- Fam. 138. Ebenaceae
- Fam. 139. Symplocaceae
- Fam. 140. Styracaceae

Order XLVI Oleales
- Fam. 141. Oleaceae

Order XLVII Contortae (Gentianales)
- Fam. 142. Loganiaceae
- Fam. 143. Gentianaceae
- Fam. 144. Apocynaceae
- Fam. 145. Asclepiadaceae

Order XLVIII Tubiflorae (polemoniales)
- Fam. 146. Convolvulaceae
- Fam. 147. Polemoniaceae
- Fam. 148. Hydrophyllaceae
- Fam. 149. Boraginaceae
- Fam. 150. Verbenaceae
- Fam. 151. Labiatae
- Fam. 152. Solanaceae
- Fam. 153. Scrophulariaceae
- Fam. 154. Bignoniaceae
- Fam. 155. Martyniaceae
- Fam. 156. Orobanchaceae
- Fam. 157. Lentibulariaceae
- Fam. 158. Acanthaceae
- Fam. 159. Phrymaceae

Order XLIX Plantaginales
- Fam. 160. Plantaginaceae

Order L Rubiales
- Fam. 161. Rubiaceae
- Fam. 162. Caprifoliaceae
- Fam. 163. Adoxaceae
- Fam. 164. Valerianaceae
- Fam. 165. Dipsacaceae

Order LI Cucurbitales
- Fam. 166. Cucurbitaceae

Order LII Campanulatae (Campanulales)
- Fam. 167. Campanulaceae
- Fam. 168. Compositae

The previous orders circumscribe those plants which are generally recognized as the higher plants. Of those families which have been mentioned, only a few are of commercial significance at the present time. For those few families further brief discussion is warranted.

Fam. 1 Equisetaceae (Horsetail Family)

The horsetail is also known as snake-grass, horse-pipes, scouring-rush, pine-grass, or snake-grass. It may be found in sandy or gravelly soils, primarily in uncultivated areas. The deep-creeping rootstocks of this plant are perennial. The reproductive stems are composed of quite brittle jointed sections 1-2 inches long, without branches, and grow 4 to 10 inches high. They bear seeds (spores) at the tips of the stems. These fertile shoots appear early in the spring and then disappear. These are followed by the sterile shoots which remain until fall. The sterile shoots are 3-6 inches high, branched and resemble miniature pine trees. The horsetail plant contains a poisonous substance which causes a disease called "equisetosis" when eaten by animals. The affected animals exhibit muscular weakness, nervousness, diarrhea and in some cases paralysis, coma and death. Sheep and cattle are susceptible but the plant is especially poisonous to horses.

Fam. 14. Typhaceae (Cat-tail Family)

This family includes some of the simplest of the Monocotyledons. Their flowers are borne in long, brown fuzzy spikes at the stem tips. They have creeping rhizomes and members of the family are generally found in shallow water or in marshes.

Fam. 22 Gramineae (Grass Family)

The representatives of this family are widely found and include many valuable plants, e.g. Corn, Wheat, Oats, Rye, Barley, Sugarcane, Millet, Rice, Sorghum, Blue-Grass, and Bamboo. Many of these grasses are also useful for grazing and hay, as well as basic foods and for lawns. However, as black sheep may be found in any family, the "weeds" in this family include: Foxtails, Chess, Wild Oats, Quack Grass, Sand Bur, Johnson Grass, and Darnel (which produces poisonous seeds).

Quack Grass (Agropyron repens)

This is a perennial with pale yellowish-white rootstocks that spread 2-12 inches below the ground surface. The rootstocks are so tough that they can be pulled from the soil without severe breakage, so hardy that they can be plowed to the surface, partially dried, and still grow when they are covered again with soil. The most vigorous rootstock development occurs in cultivated fields, whereas in undisturbed soil the root development is more shallow. New plants may be produced at each joint of the rootstock (approximately every two inches). Quack grass can reach one to three feet high and has dull green leaves. The plant produces many viable seeds in the seed head which grows atop a long, slender spike. Quack grass was introduced to the United States for grazing cattle, but it has become a very noxious weed in many areas.

Figure III-1. Quack Grass. (Iowa Department of Agriculture Photo 1965, E. P. Sylwester.)

Wild Oat (Avena fatua)

This is an annual which reproduces by seeds only. The seeds may retain their vitality for several years when buried in the soil. They may sprout in a depth as great as 12 inches.

Cheat Chess (Bromus secalinus L.)

This is an annual or winter annual which is propagated by seed. It produces a large crop of seeds which retain their vitality for years when buried in the soil.

Darnel (Lolium temulentum L.)

This is an annual which is sometimes called poison ryegrass. It has flat leaves (slightly rough on the upper side), smooth leaf sheaths, and 1½-2 ft. stems.

Little Barley (Hordeum pusillum Nutt.)

This is an annual borne on stems 4-14 inches tall. It is found in wasteland and along roadsides.

Squirrel-tail Grass (Hordeum jabatum L.)

This is a perennial which grows in clumps. It has smooth leaf sheaths, 1-2 foot stems and dense fibrous roots. It is found in waste areas.

Figure III-2. Wild Oat. (Courtesy of Adolf Coors Co., Golden, Colo.)

Bermudagrass (Cynodon dactylon (L.) Pers.)

This is a tender perennial which is sometimes called Wire Grass or Joint Grass. It propagates by seed, underground rootstocks, and surface-creeping stems. Its stems are flattened, have two leaves at each joint, and a ring of white hairs at the base of each leaf blade.

Crowfoot Grass (dactyloctenium aegyptium)

This has culms (stems) which are spreading and branching. It has seedheads with two or more short, thick spikes which spread finger-like from the top culm.

Barnyard Grass (Echinochloa crusgalli (L.) Beauv.)

This is a coarse annual with thick stems which often grow to 4½ feet tall. It has smooth leaf sheaths and blades, fibrous roots and may be found in moist places.

Crabgrass (Large and Small) . . . (Digitaria sanguinalis (L.) Scop. and Digitaria ischaemum (Schreb.) Muhl.)

These are annuals with hairy sheaths and leaf blades. The reddish, spreading stems often root at lower joints and 4-7 flowering branches radiate outward from the flattened stem. Crabgrass is common in cultivated fields and is an unsightly lawn pest.

Goose Grass (Eleusine indica (L.) Gaertn.)

This is a coarse annual bunch grass 1-2 feet tall. The leaf blades are often folded. The leaf sheaths are smooth or slightly hairy at the edges, and the base of its stems are somewhat flattened.

Texas Millet or Buffalo-Grass (panicum texanum Buck 1.)

This is an annual with erect culms. It often roots at the nodes. These plants are pale green and can be further identified by a velvety hairiness below their nodes.

Panic Grass or Fall Panicum (Panicum dichotomiflorum Mich X.)

This is an annual with smooth, spreading stems twenty to forty inches long. It is also characterized by smooth, loose, leaf sheaths, often purplish, and smooth leaf blades 1/8 to 3/4 inches wide. Panic Grass commonly roots at the lower joints and resembles Tickle Grass.

Tickle Grass (Panicum capillare L.)

This is an annual with hairy stems, leaf sheaths, and leaf blades. This plant has fibrous roots and is commonly found in abandoned farm fields.

Sandspur (Cenchrus tribuloides L.)

This is an annual with flattened stems, leaves which may be flattened or coiled; in either case fine hairs may be found at the top of the sheath.

Its seeds are borne in a spiny burr about 1/2 inch across with two seeds to the pod. Other plants similar to this one are also commonly known as Sandspur.

Green Foxtail (Setaria viridis (L.) Beauv.)

This is an annual with smooth leaf blades, and is commonly found in poorly drained soil.

Yellow Foxtail (Setaria lutescens (Weigel) Hubb.)

This is an annual with long (twenty to forty inch) stems branching from the base. The leaf sheaths are generally smooth and its flat leaf blades are broad and often covered with a waxy film. The leaves have long hairs on the upper surface where attached to the sheath. The Yellow Foxtail is found in many of the same locations as the Green Foxtail.

Wild Barley, Squirrel-tail Grass, (or Tickle-Grass) (Hordeum jabatum L.)

This grows in swampy lands, along roadsides, and in other such unused lands. Growing one to two feet high, this plant is distinguished by a fibrous, compact root which sends up many stems. The two to four inch long leaves resemble those of Bluegrass, but are paler. The flowering head takes the form of a dense spike two to four inches long. These flowering heads produce leaf-like projections (glumes) that can work their way into the mucous membranes of grazing cattle. These glumes can then cause inflammation, ulceration, and, in extreme cases, loss of teeth in the victim. The latter eventuality will usually result in the starvation of the victim.

Johnsongrass (Sorghum halepensex III-3 (L.) Pers.)

This is a perennial with long, scaly rootstocks; flat, smooth leaf blades 16 to

20 inches long, and stems two to six feet tall. Johnsongrass may be distinguished from the related cultivated species by its long, scaly rootstocks. It is found in cultivated fields and in bottomlands. This plant is reported to have been introduced to the United States in the mid- nineteenth century as a forage grass by a Colonel William Johnson, a "gentleman farmer." Col. Johnson's experiment has become a pernicious perennial weed.

Figure III-3. Johnson Grass. (Courtesy of Adolf Coors Co., Golden, Colorado)

Broomsedge (Andropogon virginicus L.)

This is a perennial bunch grass with dense, fibrous roots. Reaching three to five feet, it is propagated by its wind-borne light seed. As is common with such weeds, it is found chiefly in abandoned farm areas or unimproved areas.

Annual Blue Grass (Poa Annua)

This is emerald green and has soft slender leaves with loose sheaths. Its stems are flattened, tufted, and frequently root at the lower nodes. Similar in appearance to some perennial bluegrasses, *Poa annua* is lighter colored and forms seeds during most of its growing season. Turning brown in warm weather, this plant becomes even more undesirable in a lawn.

Fam. 30. Pontederiaceae (Pickerelweed Family)

This family consists of aquatic and marsh growing herbs and includes the Water-Hyacinth (*Eichornia* Kunth), the Mud-Plantain (*Heteranthera* R.&P.), and the Pickerelweed (*Pontederia* L.). Pickerelweed is an aquatic herb which grows along the edge of slowly moving water. It is a very hardy plant and produces a single stem bearing a spike of blue-violet flowers and long-petioled leaves. It prefers to grow in about a foot of water, but may grow in less.

Fam. 23. Cyperaceae (Sedge Family)

Meadow Rush (Scirpus atrovirens Willd.)

This is a perennial with scaly horizontal rootstocks and fibrous roots developed from a taproot. The plants can extend three to five feet tall on erect stems, resembling grass in appearance. The meadow rush grows best in wet swampy soil and produces greenish-brown flower clusters at the top of the plant.

Nutgrass (Cyperus rotundus L.)

This is a ½ to 1½ foot tall perennial propagated by seeds and tuber-bearing rootstocks. Its stems are smooth, slender, three-sided and leafless except for three or four bracts (a structure belonging to the stem which is a somewhat modified leaf) at the seed head.

Fam. 31. Juncaceae (Rush Family)

The Rush family consists of grass-like herbs that grow in water or wet soil. They commonly have hollow or pithy stems and small clustered flowers.

The Field Rush (Juncus macer S.F. Gray)

This is a perennial, grass-like plant with hollow, dark green stems and fibrous roots. It grows in poorly drained, uncultivated areas.

Fam. 32. Liliaceae (Lily Family)

The Liliaceae is considered a very important family, containing more than two thousand species. Many of these species (*e.g.* lilies, hyacinth, tuberose, lily-of-the-valley, and tulip) are cultivated because of their fragrance and beauty. Lily-of-the-valley is reported to be poisonous if eaten. Asparagus is one of the food plants of the Lily family; the *Allium* species (including the onion, leek, garlic, and chives) also contains edible representatives as well as several universally acclaimed weeds (wild onions and wild garlics). These various species are perennials with hollowish, grass-like leaves which arise from a whitish bulb, often covered with a papery coat. In addition to the underground bulbs, these weeds produce numerous small aerial bulblets in the flowering head. The species catagorized as wild onion and wild garlic include *Allium vineale, Allium canadense*, and other Allium species.

DICOTYLEDONS

Fam. 50. Urticaceae (Nettle Family)

Burning Nettle or Dog Nettle (*Urtica urens*) is a soft stemmed annual with many stinging bristles. It is found on waste ground and has somewhat toothed, greenish flowers.

Fam. 54. Polygonaceae (Buckwheat Family)

Pennsylvania Smartweed (Polygonum pennsylvanicum L.)

This is an 8 to 24 inch tall annual with smooth, mostly erect stems. The spear shaped leaves are smooth on the lower side. The flowers are rose colored and are borne in upright finger-like clusters. The smartweed is found in low wet areas, such as pastures and stream banks.

Spotted Smartweed is an annual closely resembling the Pennsylvania Smartweed in both appearance and habitat except that its leaves have a purplish triangular spot near the middle.

Prostrate Knotweed (Polygonum aviculare L.)

This is an annual reproduced from its seeds. The stems are highly branched, slender, bluish-green, wiry, and form a dense mat. The leaves are ½ to 1½ inches long, narrow, oblong, and bluish-green in color, while the flowers are small, white, yellow, or pinkish and are borne at the base of the leaves.

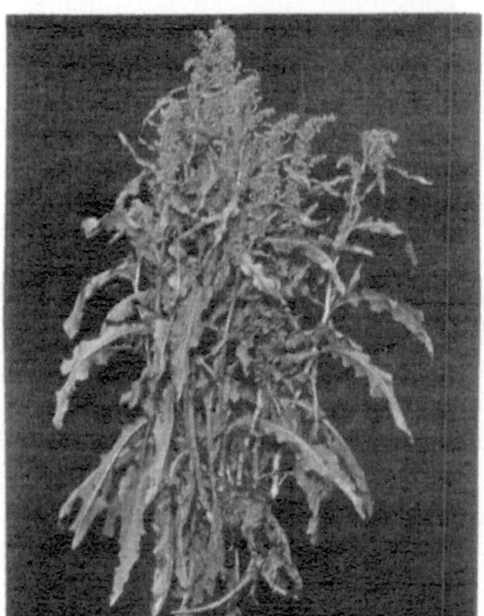

Figure III-4. Curly Dock. (Courtesy of Adolf Coors Co., Golden, Colorado.)

Water Smartweed or Water-Pepper (polygonum hydropiper L.)

This is an 8 to 24 inch annual found in poorly drained areas. The narrow, pointed leaves taste peppery. The flowers are borne in small, narrow, greenish clusters at the tip of the stem.

Curly Dock or Narrow Leaved Dock (Rumex crispus L.)

This is a perennial with large, deep, yellowish taproots. Its stems are erect, smooth and grow one to three feet tall. The plant may be recognized from the large, elongated leaves with with wavy edges, and the large, branched clusters of small greenish flowers. It is commonly found in waste spaces and pastures. This weed is also sometimes referred to as Yellow Dock.

Broad-Leaved Dock (Rumex obtusifolius L.)

This is sometimes called Bitter Dock or Red Leaved Dock, and is a perennial with a big, deep taproot. It resembles Curly Dock except that its stems are slightly rough and its lower leaves are broad and not very wavy at the edges.

Bindweed, or False Buckwheat (Polygonum scandens L.)

This sometimes behaves as either a perennial or an annual. Its twining stems reach from 6 to 40 inches. The Bindweed may be further recognized from heart shaped leaves and greenish flowers borne in small clusters. This weed is commonly found along field borders, in low woods, or in damp thickets.

Figure III-5. Sheep Sorrel. (Iowa Department of Agriculture Photo 1965, E. P. Sylwester.)

Sheep Sorrel, Red Sorrel, Field Sorrel, Sourgrass, or Sourweed (Rumex acetosella L.)

This is a perennial that grows in "sour" soil and is difficult to eradicate without "sweetening" the soil. The rootstocks in older plants spread horizontally. The 4 to 12 inch stems are smooth. The sour tasting leaves, more numerous at the base of the plant, vary from arrow-shaped at the bottom of the plant to spear-shaped at the top of the plant. Its numerous small flowers are found on small branching clusters at the stem end. Sheep sorrel propagates itself by underground rootstocks and seeds. This weed is commonly found in pastures or abandoned fields.

Fam. 55. Chenopodiaceae (Goosefoot Family)

This family is composed of succulent herbs characterized by greenish flowers that are minute to the point of being inconspicuous.

Lamb's Quarters (Chenopodium album L.)

This is a 1 to 6 foot annual with a deep taproot. The stems are erect and smooth except for vertical ridges with leaves growing alternately on the stem. The leaves have toothed edges and are grayish-green on the undersurfaces. Its flowers are very small, greenish, and appear in clusters developing from the ends and sides of the branches. Lamb's Quarters may be found in cultivated fields as well as in abandoned areas.

Fam. 56. Amaranthaceae (Amaranth Family)

This family consists of herbs which often have red stems and minute flowers. The genus Celosia includes common garden flowers and the other side of the family's desirability is represented by the genus Amaranthus which consists of several species of the troublesome pigweeds.

Prostrate Pigweed (Amaranthus graecizans L. or Amaranthus blitodes Wats.)

This is an annual with a short taproot, and smooth, reddish stems that usually spread over the ground except for the erect tips. The smooth, egg-shaped leaves are alternate on the stem with their broad end uppermost. The small, greenish flowers are formed in small dense clusters at the base of the leaf stalks. Prostrate pigweed is found in neglected fields and yards.

Pigweed, Wild Beet, Green Amaranth (Amaranthus hybridus L.)

This is an annual with a reddish taproot and a slightly rough, erect stem bearing deep green egg-shaped leaves.

Small flowers grow in dence finger-like clusters arising from the bases of the leaf stalks and the ends of the stems and branches.

Wild Beet or Rough Pigweed (Amaranthus retroflexus L.)

This is an annual which resembles Green Amaranth although coarser.

Fam. 57. Nyctaginaceae (Four-o'clock Family)

The Wild Four O'Clock or Umbrella Wort (Mirabilis nyctaginea (Michx.) MacM.

This is a bunchy perennial with a number of stems arising from a thick, fleshy, branched tap root. Its stems, which grow 1-3 feet high, have broadly forked branches. The stems are usually smooth and shiney, but sometimes are ridged, hairy and purplish. The leaves are opposite on the stem and resemble those of the lilac. The flowers are small, pale purple, and are 3-5 to a cluster. The Wild Four O'Clock, reproducing by both seeds and roots, may be found on roadsides, railroad embankments, along fence rows and on prairie land in Midwestern United States.

Figure III-6. Wild Four O'Clock. (Iowa State University Photo, E. P. Sylwester.)

Fam. 58. Phytolaccaceae (Pokeweed Family)

Pokeberry (Phytolacca americana L.)

This is a perennial with a large taproot and 3 to 9 foot stems which become woody at the end of the growing season. It has large and smooth leaves and produces small, white flowers, followed by dark purple fruits with crimson juice and shiny, small black seeds. The old leaves, berries, and roots are thought to be poisonous. Pokeberry is found along roadsides and the borders of fields.

Fam. 61. Caryophyllaceae (Pink Family)

This family consists of herbs with stems that may be swollen at the nodes or joints and have mostly opposite leaves. Flower lovers are indebted to this family for the Sweet Williams, Carnations, and Pinks while being less than happy with the Chickweeds, Corn Cockle, and other weed-classified members.

Corn Cockle (Agrostemma githago L.)

This is a "winter annual" with a straight taproot and erect, silky, hairy stems 1 to 3 feet tall. Its long, narrow leaves are attached at the joints of the stem. The plant bears large, purple flowers and is often found in grain fields. The seeds are reported to be poisonous, making the Corn Cockle especially objectionable in grain for feed or milling.

Mouse-Ear Chickweed (Cerastium vulgatum L.)

This is a perennial with short roots, propagated by seeds and its creeping stems. The leaves are small, oval, and hairy and the weed is usually found in lawns, yards, and gardens.

Common Chickweed (Stellaria media (L.) Cyrillo.)

This is an annual or winter annual with a branching root system and stems that form spreading clumps on the ground except for the erect tips. The lower joints of the stems often take root. The egg-shaped leaves join the stem in pairs while the stems support small, white flowers characterized by five petals. Its habitat is in cultivated fields, lawns, yards, and gardens.

Fam. 64. Ranunculaceae (Crowfoot Family)

A few of the more well known members of this family include: Anemone, Larkspur, Columbine, Wolfsbane, and Buttercups. The Larkspur (*Delphinium* spp.) is alleged to cause more cattle poisoning in the Western parts of the United States than any other plant. Sheep and horses may also be affected by ingestion of this weed. Trembling, depression, and constipation, followed by death from respiratory paralysis are the recognized symptoms of poisoning from the ingestion of Larkspur. This weed produces long stems topped with flower heads composed of small bluish flowers. The Larkspur resembles the lovely Delphiniums and Larkspurs of the flower gardens.

Fam. 73. Cruiferae (Mustard Family)

The family name, Cruiferae, was inspired by the cross-like arrangement of the flower petals of many plants in the family. The Mustard Family contributes: turnips, cabbage, water cress, radishes, and horse radish to the garden for their food value, and Sweet Alyssum and Candytuft for decorative value. The familiar condiment, mustard, is prepared from the seeds of one of the varieties of the mustard plant. Shepherd's Purse, Peppergrass, and Wild Mustard are generally viewed as the weeds of the family.

Wild Mustard (Brassica Kaber (DC) L. C. Wheeler)

This is an annual with erect stems branching near the top and hairy near the base. The upper leaves have rather smooth edges while the edges of the lower leaves are usually cut into tooth-like portions. Wild Mustard (Crunchweed) bears yellow flowers in finger-like clusters followed by long, smooth seed pods.

Black Mustard (Brassica nigra (L.) Koch)

This closely resembles Wild Mustard and is also an annual. Grain fields or uncultivated fields are favorite haunts of this weed. The seeds of many of the mustard plants retain their viability for many years even though buried in the ground.

Shepard's Purse or Shepard's Pouch (Capsella bursa-pastoris (L.) Medic.)

This is an annual with a short, whitish taproot. The 8 to 16 inch stems are smooth and branched with alternating arrow-shaped leaves having cuts along their edges. Very small white flowers borne in finger-like clusters on the ends of the stems and branches are the precursors of distinctive triangular, flat seed pods. This weed is sometimes found in cultivated fields, in yards, and abandoned places.

Figure III-7. Wild Mustard. (Iowa Department of Agriculture Photo 1965, E. P. Sylwester.)

Pepper Grass or Poor-man's Pepper (Lepidium virginicum L.)

This is either an annual or biennial with a short, whitish taproot, smooth branching stems and long, narrow dark-green leaves. The flowers are inconspicuous and grow in finger-like clusters at the ends of the stem and branches.

Fam. 82 Rosaceae (Rose Family)

Members of the Rose family provide mankind with many pleasantries, *e.g.* apples, pears, cherries, plums, and similar fruits as well as such ornamentals as roses, spireas, and hawthorns. However, some members of the Rose family are less than beneficial. For example, the Cokecherry (*Prunus virginiana* L.) is poisonous to livestock. Ordinarily disdained by grazing animals, it is often eaten when overgrazing or drought impoverishes the healthful forage. Various forms of chokecherry are widely distributed over continental United States.

The genus *Agrimonia* L. includes the weeds Agrimony and Harvest-lice; the genus *Potentilla* L. harbors Cinquefoil and Five-Finger. In California, at least, several species of wild Blackberry have become troublesome, forming dense thickets and invading fields, and orchards as well as being pests along ditches and fences with their pesky brambles.

Figure III-8. Perennial Peppergrass. (Iowa Department of Agriculture Photo 1965, E. P. Sylwester.)

Fam. 83. Leguminosae (Pulse Family)

Among the plants of economic importance in the Pulse Family are the edible beans, peas, peanuts, lentils, soy beans and clovers. Fine cabinet wood is furnished by Black Locust and the ornamentals from this family are: Wisteria, Judas Tree, Lupine, Mimosa, Sweet Peas, Licorice, and Senna. However, thirty or more of the leguminous plants are weeds. The Lupines and the Locoweeds are poisonous to live-stock. A member of the Trifolium species, White Dutch Clover, is a common lawn pest except to little children who while away many hours on warm summer days searching among stems of lawn clover for the lucky "four-leaf clover".

The plants collectively known as Crazyweed or Locoweed are members of the *Oxytropis* species. Crazyweed commonly grows in dry areas of western United States. All parts of the plant are poisonous. The toxic substance in the weeds is reported to be an alkaloid. Animals which have consumed an amount equal to approximately 90% of their body weight over a two-month period may be adversely affected. They become weak, dull, drag their feet, become shaggy, seek solitude, lose sensibility, and then stop eating and die.

The *Lupinus* species contains a number of members that are poisonous. The wild lupine species closely resemble the cultivated garden variety. The flowers are pea-like and usually blue, although they sometimes are pink, white, or yellow. The resultant seed pod has a characteristic appearance, resembling a soft wooly pea pod. The lupines, like the Crazyweeds and Locoweeds inhabit the cattle grazing areas of the western parts of the United States.

Fam. 86. Geraniaceae (Geranium Family)

Carolina Cranesbill, Cranesbill, or Wild Geranium Carolinianum (*Geranium Carolinianum* L.) has an annual or biennial nature. Its taproot is slender and whitish supporting erect, hairy stems which branch widely at the base. Small purple or pale pink flowers are produced at the tips of the stems and branches. The flowers are followed by long, pointed seed pods which resemble beak-like capsules. The Cranesbill is found predominately in poor soils in warm climates, such as southern United States.

Fam. 92. Euphorbiaceae (Spurge Family)

The Spurge Family contains many tropical species. The plants often contain a milky juice which is poisonous in many species. The Poinsettia and the Castor Bean are common ornamental species. The seeds of the beautiful Castor Bean are very poisonous.

Nodding Spurge (Euphorbia maculata L.)

This is an annual with a short taproot. Generally it is erect with smooth stems which contain a milky juice. Its leaves are oblong and have slightly toothed edges. It produces small greenish flower clusters borne at the ends of the stem and branches. These flowers yield smooth, three-sided seed pods.

Milk Purslane (Euphorbia supina Raf.)

This is an annual whose stems form flat mats on the surface of the ground. It has small, hairy leaves and seed pods.

Leafy Spurge (Euphorbia esula L.)

This is a perennial, herbaceous, noxious field weed. The plant is a long lived, deeply rooted herb which grows 8 inches to 3 feet tall. The simple or clustered stems support bluish green leaves that turn brownish-orange in late summer. All portions of the plant exude a milky juice when it is injured. The flowers are small and greenish-yellow. The plant is very responsive to soil moisture and will often cease growing or blossoming during extremely dry soil conditions.

Reproducing by seed, spreading roots, and rootstocks, the leafy Spurge occurs in patches in fields, in pastures, along roadsides, and in other uncultivated places. The plants commence blooming by mid-May and continue throughout the growing season. When the seed capsule matures, it is capable of throwing the seeds as far as fifteen feet from the plant. The abundant production of seed combined with this "throwing" ability and deep (up to fifteen foot) rootstocks suggest that this weed should not be allowed to go to seed unless its semi-permanent presence is desired.

Figure III-9. Leafy Spurge. (Iowa Department of Agriculture Photo 1965, E. P. Sylwester.)

Fam. 97. Anacardiaceae (Cashew Family)
Poison Ivy (Rhus radicans L.)

This is related to common sumac. It is poisonous the year around (even in its winter dormant condition). It is found in parks, pastures, alluvial woodlands, roadsides, fence rows and along rivers and streams. Birds probably are responsible for the wide dissemination of poison ivy seed. They utilize the berries in the winter when food is scarce. Poison ivy may appear as an erect shrub or as a vine climbing by aerial rootlets. The stems are usually woody. The leaves are alternate on the stem, and always borne in groups of three leaflets. They may be glossy or dull green; usually smooth, but occasionally hairy. The edges of the leaves vary from smooth through toothed to lobed. Flowering usually occurs in June and July and the resulting fruit is a small white berry cluster.

Contact with poison ivy frequently produces blistering and severe inflammation. Although immunity seems to vary with individuals, contact should be avoided by everyone. Poisoning occurs when a susceptible person contacts the plant or its volatile oils. After contact with it, a person should wash and rinse several times the exposed skin area as soon as possible with an abundance of warm water and strong alkali laundry soap. Severe cases should be treated by a physician.

Poison Oak (Rhus toxicodendron L.)

This is usually an erect shrub 2-6 feet tall. It may also trail along the ground or climb fences and trees. The leaves are 3-6 inches long, shiny and shaped like oak leaves. The plant flowers in April or May. The leaves turn red or yellow when the soil begins to dry out in late spring or early summer. It is not related to true oaks. Warning: DO NOT burn Poison Oak. Oils in the smoke, inhaled, can be more harmful than touching the plant.

Fam. 109. Malvaceae (Mallow Family).

This family is distinguished by the Cotton plant. Ornamentals include the Shrubby Althaea and Hollyhock. Weeds in this family are Flower-of-an-hour, Indian Mallow, and other mallows.

Common Mallow, Cheeses, (Malva neglecta Wallr.)

This is a biennial with a deep root. Its stems are trailing but not rooting. The leaves are heart-shaped and borne on long stalks. The small flowers are white or pale lilac. The seed pods are round and flat with the seeds arranged in a distinctive circular manner in the pod.

Roundleaved Mallow (Malva rotundifolia L.)

This is similar in appearance to the common Mallow. By nature it is either annual or biennial, reproduces by seeds, and has rounded leaves. Its stems may be smooth or faintly hairy, they are branched and may be nearly erect or spreading on the ground's surface. Its flowers are small and whitish.

Prickly Sida, Spiny Sida or Prickly Mallow (Sida spinosa L.)

This is an annual with a shallow taproot and softly hairy stems which grow 8 to 20 inches tall. The ovalish leaves have sawtooth edges and are ½ to ¾ inches long. The ½ inch diameter flowers have five petals and are followed by cup-like seed pods containing ten seeds in a circular arrangement.

Fam. 116. Passifloraceae (Passion-Flower Family)

Maypop, Passion Flower or Apricot Vine (*Passiflora incarnata* L.) is a perennial reproduced by root shoots and seeds. The plant climbs by tendrils or trails its stems along the ground. The flowers are showy, purplish, and fragrant with many frilly petals. The Passion Flower produces a grayish to brownish-black fruit that resembles a small melon.

Fam. 125. Onagraceae (Evening-Primrose Family)

Evening Primrose (Oenothera biennis L.), a biennial, has hairy stems that often branch. The long, narrow leaves have coarse cuts in their edges and are most deeply scored near their base. Yellow flowers are borne on long stalks. Cylindrical, ribbed seed pods bear many yellow or brown seeds.

Fam. 129. Umbelliferae (Parsley Family)

The Parsley family contains about 1300 species including carrots, parsnips, caraway, anise, celery, dill and fennel. Wild Carrot is one of the pesky weeds in this family. It also numbers the poison Hemlock and Water Hemlock.

Figure III-10. Wild Carrot. (Iowa Department of Agriculture Photo 1965, E. P. Sylwester.)

Wild Carrot or Queen Anne's-Lace (Daucus carota L.)

This is a biennial with a white, fleshy taproot that is as thick as the taproot of a garden carrot. The 1½ to 4 foot tall hollow stems have a bristly outer surface while the leaves are fernlike. The small white or pinkish flowers are borne in round, dense, flat, or concave clusters at the stem tips and branch tips. The wild carrot is frequently found in old pastures and along roadsides.

Water Hemlock (Cicuta maculata L.)

This is a 3 to 8 foot tall perennial or biennial with smooth, hollow, erect stems. The rootstock is thick and short, developing hollow horizontal chambers as the plant matures. If the rootstock is cut, drops of a yellowish, oily liquid, soon turning reddish-brown, gather on the cut surface. This aromatic liquid is reported to contain cicutoxin, the poison of the plant. The leaves are large and divided into many lace-shaped leaflets with finely toothed edges. The numerous white, small flowers are borne in clumps or clusters and the resulting seeds are small and roundish. Water hemlocks grow only in wet areas, such as stream borders, marshes, and swamps. Although the plant is toxic to both man and beast when ingested, cattle are usually the victims when they graze in infested areas and pull the rootstocks from the wet ground. Water hemlock is capable of causing such rapid death that the animals are often not observed during their illness. Poisoned animals exhibit muscle twitching followed by convulsions and death if sufficient quantity of the Water Hemlock has been ingested. Since no specific antidote has been found for *Cicuta* poisoning, prevention of contact with the Water Hemlock is the only safeguard for the health and life of both beast and man. The following plants have a similar appearance to the Water Hemlock, and are therefore often confused with it: Water Parsnip, Water Parsley, Cut-leaved Water Parsnip, and White Heads.

Figure III-11a. Water Hemlock, **11b.** Water Hemlock Roots (Iowa State University Photo, E. P. Sylwester.)

Poison Hemlock (Conium L.)

This is a branching biennial that commonly grows 4 to 6 feet tall but may occasionally reach 10 feet high. The taproot is long and white while the stem is smooth, hollow, stout, and has distinctive red or reddish-purple spots. The leaves are fern-like and the lower leaves may be more than two feet long. The small white flowers are arranged in clumps of small tufts similar to those of the wild carrot. Poison Hemlock may grow in either moist or fairly dry soil. The Poison Hemlock contains several toxic alkaloids. The alkaloids are found in the stems, green leaves, and fruits of the plant, but are sufficiently volatile that the dry plants are not considered dangerous (if mistakenly included in hay, for instance).

If cattle consume an amount equal to one to two percent of their body weight of young Poison Hemlock plants, the poisoning may be severe enough to result in death. *Conium* affects the animal's brain, causing weakness, the buckling of the hind legs, resulting in inability to stand. As the poisoning progresses the front legs become paralyzed, the victim becomes drowsy, finally slipping into a coma. With proper care, however, all but the most serious cases have a good chance of recovery. Poison Hemlock, according to tradition, furnished the "cup of death" from which Socrates died.

Fam. 135. Primulaceae (Primrose Family)

The perennial Moneywort (*Lysimachia nummularia* L.) derives its name (coin-like) from its characteristically round leaves. The Moneywort is a trailing or creeping plant that reproduces by seeds and creeping stamens which root at the joints. Moneywort has yellow flowers that appear at the base of the leaves.

Fam. 145. Asclepiadaceae (Milkweed Family)

Common Milkweed (*Asclepias syriaca* L.) This is a lovely perennial plant that grows in two to five foot tall clusters. The roots develop from long, horizontal rootstocks and support the erect stems. The leaves, green and smooth on their upper surface but whitish and hairy below, are large, long and oval shaped. Large ball-like clusters of sweet smelling pinkish-purple flowers, borne at the stem and branch tips, precede the greenish-gray, hairy seed pods that are 3 to 5 inches long. The white, flat, egg-shaped seeds become brown at maturity. Milkweeds grow in pastures, cultivated fields, and on abandoned ground.

Whorled Milkweed (Asclepias verticillata L.)

This is a perennial that propagates by seeds and underground roots. The slender roots support a slender stem that is either single or slightly branched and reaches 1 to 2½ feet. The plant contains a milky juice and has narrow leaves borne 3 to 6 in a whorl which give the plant a fern-like appearance. The flowers, greenish-white, are borne in umbrella-like clusters. The smooth, spindle-shaped seed pods are carried erectly on the stems, with the seeds attached to silky, fine hairs. The Whorled Milkweed contains a substance that is poisonous to sheep, cattle, and horses. All portions of the plant that grow above ground are toxic to these animals if ingested, whether the plant is fresh or dried. Affected animals exhibit a weak, high pulse, weakness of the hind quarters, listlessness, weight loss, and a general deterioration of muscular control. Poisoned animals stagger, fall, have violent spasms, and finally die of respiratory paralysis. No reliable antidote is reported.

Fam. 146. Convolvulaceae (Morning Glory Family)

The plants of this family are chiefly trailing or twining herbs, such as the sweet potato. Members of this family which are generally regarded as weeds include the Morning Glory, Bindweeds, and Dodders. The Morning Glory and Bindweeds twine around cultivated plants and are difficult to eradicate, especially the Bindweed whose spreading rootstocks propagate the plant rapidly. The Dodders are parasitic plants that twine around Clover, Alfalfa, and Flax, growing their roots into the stems of the cultivated crop and robbing the host of its nourishment.

Figure III-12. Whorled mildweed plant. (Iowa State University Photo, E. P. Sylwester.)

Dodder (Cuscuta L.)

This annual, resembles common plants in many respects, but is a parasite, lacking a conventional root system. The seeds produce shoots that twist around the host and send small, penetrating suckers into the stem of the host. Encouraged and sustained by its first success, the Dodder sends its stems from one host to another; soon the infested area can become a mass of thread-like, reddish stems, and weakened or destroyed "hosts". The yellowish-brown "leaves" are merely scales. The small, brownish-white flowers, borne in clusters along the stems, produce the brownish, rough-surfaced seed pods that contain the raw material for next year's infestation. Common Dodder hosts include alfalfa, clover, wild sunflowers, goldenrod, smartweed, asters, chrysanthemum, hydrangea, petunia, eggplant, carrot, and raspberries. It is reported that Dodder will not usually infest small grains, corn, sorghum, or most grasses.

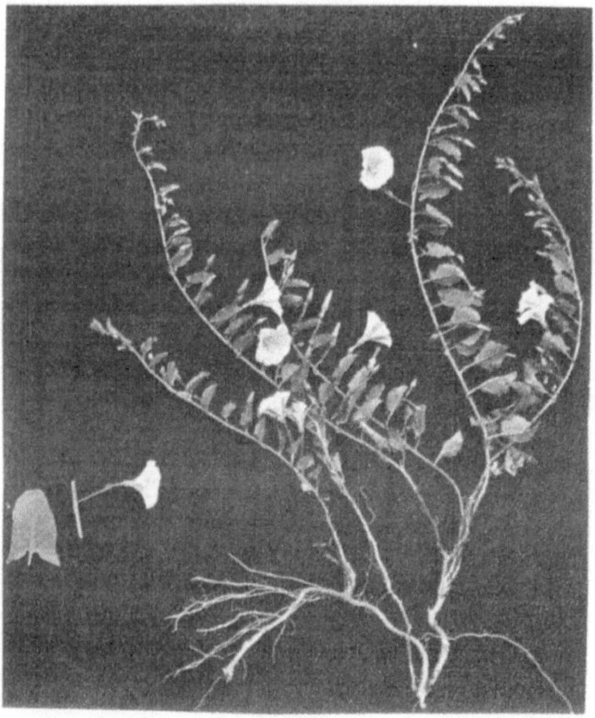

Figure III-13. Field Bindweed. (Eugene Heikes, Cooperative Extension Survey, Fort Collins, Colo.)

Field Bindweed or Creeping Jenny (Convolvulus arvensis L.)

This is a perennial whose roots develop from extensive horizontal rootstocks. The smooth stems twine while spreading over the surface and support oblong leaves ½ to 1 inch long, with two outward pointing lobes at their base. The pale pink or white, funnel-shaped flowers have purplish edges and are borne on long slender stalks. The egg-shaped seed pod splits into two sections when ripe. The ability of the seeds to lie dormant in the soil for several years before germinating makes control of this weed more difficult.

The Hedge Bindweed (Convolvulus sepium L.)

This perennial, closely resembles the Field Bindweed except for the very long stems, leaves pointed at the tip, and much larger flowers. Both Bindweeds are considered noxious weeds of the same category.

Ivy-Leaved Morning Glory (Ipomoea hederacea (L.) Jacq.)

This is an annual with a slender taproot and hairy, twining stems that spread on the ground. The three-lobed leaves grow alternately on the long stem. Large, funnel-shaped flowers vary from white through pale blue to purple and this Morning Glory is found in both cultivated and abandoned fields.

Common Morning Glory (Ipomoea purpurea (L.) Roth.)

This is also an annual like the Ivy-Leaved Morning Glory and has a similar appearance except that the former has heart-shaped leaves and its flowers are borne in clusters of three to five on the stem.

The Wild Sweet Potato or Wild Potato Vine (Ipomoea pandurata (L.) G.F.W.Mey)

This closely resembles the Common Morning Glory except for the formers' large fleshy roots.

Fam. 151. Labiatae (Mint Family)

The Family name *Labiatae* is derived from the Latin noun *labium* (lip) and refers to the irregular, two-lipped corolla (inner petals) that are characteristic of the flowers of plants belonging to this family. Members of this family are widely spread throughout the temperate regions and, because they are easily recognized, this family is one of the oldest known families. Many members of the Mint family are found along roadsides, stream beds, and in waste places. Although no important food plants are found among the Mints, they provide many flavorings, including spearmint, peppermint, thyme, sage, and sweet marjoram.

Henbit (Lamium amplexicaule L.)

This is a winter annual or biennial with a branching, spreading taproot and square, nearly smooth stems that grow 4 to 6 inches tall. The Henbit produces roundish, hairy leaves with deeply scalloped edges and purplish or reddish-purple flowers at the bases of the upper leaves.

Ground Ivy (Glechoma henderacea L.)

This is a shallow-rooted perennial that propagates by creeping stems and seeds. The 15 to 30 inch long stems will root if the joints touch the soil. Kidney-shaped leaves are bright green and have an aromatic, minty odor. The Ground Ivy produces purplish-blue flowers that are borne in small clusters.

Fam. 152. Solanaceae (Nightshade Family)

This family includes the economically important or aesthetically desired Irish Potato, Tobacco, Tomato, Egg Plant, and Petunia. The roster of weeds includes the Black Nightshade and Jimson Weed, both poisonous, the Horse Nettle, and the Buffalo Burr.

Jimson Weed (Datura stramonium L.)

This annual has a rank odor and may be found in dumping grounds, yards, and hog pens. The thick, short, branching taproot supports thick, erect, 1 to 5 foot tall stems that branch widely near the top. The leaves are smooth, large, long and somewhat arrow-shaped with irregular toothing at the edges. Whitish flowers are borne singly on short stalks. These large flowers grow from the base of leaf stalks or the forks of branches and are followed by large, egg-shaped seed pods that are covered with sharp, short spines. The seeds are considered to be very poisonous if ingested.

Figure III-14. Horse Nettle. (Iowa Department of Agriculture Photo 1965, E. P. Sylwester.)

Horse Nettle (Solanum carolinese L.)

This perennial has deep, horizontal rootstocks and erect stems covered with many short spines and hairs. These 6 to 20 inch tall stems are branched in the upper portion and support hairy, egg-shaped leaves. The middle rib of the leaf is covered with yellowish prickles on the underside while the leaf edges have large indentations. The branch tips bear clusters of white or purplish flowers that produce large, yellow-orange, smooth berries. The Horse Nettle may be found in both waste and cultivated land.

Buffalo-Burr (Solanum rostratum Dunal.)

This is an annual characterized by a deep taproot and bright yellow flowers. The leaves and stems are covered thickly with long spines similar to those of the Bull Nettle.

Black Nightshade (Solanum nigrum L.)

This is an annual with a slender taproot and smooth, erect stems, branched in the upper portion, that reach from ½ to 3 feet tall. The smooth, rather egg-shaped leaves have wavy or toothed edges. Small, white flowers hang in clusters on long stalks midway between the leaf pairs, and produce spherical, smooth, black berries. The green foliage and green, unripe berries are poisonous, although the berries are "said" to lose their poison when ripe and are used in pies in some parts of the country. The seeds are dull yellowish or dark brown, and have a pitted surface.

Figure III-15. Black Nightshade. (Eugene Heikes, Cooperative Extension Service Colorado State University.)

Fam. 153. Scrophulariaceae (Figwort Family)

The following members of the Figwort family are common representatives of our native flora: Mullein, Toadflax, Figwort, Beard-Tongue, Speedwell, Lousewort, and Painted Cup, as well as the ornamental plants Foxgloves and Snapdragons.

Tyme-leaved Speedwell (Veronica serpyllifolia L.)

This is a perennial whose stems form dense mats. The lower leaves are thick and rather smooth while the upper leaves are more narrow. The flowers are pale blue or bluish white with dark markings.

Corn Speedwell (Veronica arvensis L.)

This is an annual or winter annual having a short, slender taproot and erect, short, slightly hairy stems which branch widely from the base. The stems bear somewhat hairy leaves, usually in pairs at the joints, and very small, bright blue flowers. The seedpods are small and flat. The Corn Speedwell grows in pastures and in open woods.

Purslane Speedwell or Neckweed (Veronica peregrina L.)

This is an annual that is similar in appearance to the Corn Speedwell except that it is smooth instead of hairy. This rather fleshy plant produces whitish flowers and grows in damp open soil, fertile fields, and gardens.

Moth Mullein (Verbascum blattaria L.)

This is a biennial with a slender taproot and erect, unbranched 2 to 5 foot tall stems that are smooth except for sticky hairs at the top of the stem. The leaves are dark green and alternate on the stems. The plant's many pinkish-white or yellowish flowers are borne separately on short stalks on the upper end of the stem. Moth Mullein plants are found in pastures, along fence rows, and in abandoned areas.

Common Mullein or Flannel Plant (Verbascum thapsus L.)

This is a biennial with pale green, wooly leaves and yellow flowers that are crowded together into long, wooly, club-like clusters on short stalks. This plant is found in unimproved pastures and along roadsides.

Fam. 155. Martyniaceae (Martynia Family)

The Unicorn Plant, Devil's Claw, Ram's-Horn, or Proboscis-Flower (Martynia Louisiana Mill.) or (Proboscidea Louisianica (Mill) Thell.)

This is an annual of tropical origin that infests river banks, pasture land, waste places, and hog lots from southern Missouri, Illinois, and Indiana southward to Mexico. This weed has a large, purplish, creamy-white or spotted with yellow and purple flower. The leaves are hairy and resemble those of squash or muskmelon, but may occasionally be bright reddish-purple on the underside. The fruit resembles a small cucumber that has a long curved beak and is the troublesome aspect of this weed, especially so in the range lands of the western states. As the fruit matures, the outer fleshy part becomes dry and the inner seed pod becomes woody. The long "beak" splits lengthwise and becomes a sharp, curved hook or claw. These claws can become snagged in the wool of sheep or the legs of cattle and irritate the livestock. These seed pods may also interfere with grazing if they become attached to the livestock in the mouth region. In addition, the seed becomes widely dispersed and spread by its livestock carriers.

Fam. 156. Orobanchaceae (Broom-Rape Family)

Branched Broom-rape, Hemp Broom-rape, or
Tobacco Broom-rape (Orobanche ramosa L.)

This lacks chlorophyll and is a parasite to tomatoes, hemp, tobacco and, secondarily, to safflower, melon, lettuce, and grapes. The roots fuse with those of its host allowing the Broom-rape to draw its nourishment from the host, causing greatly reduced yield from the host crop. The yellow shoots of the Broom-rape may be smooth or hairy and arise like branches from a central shoot. The flowers may be white, pale yellow, blue, or violet, are borne in spikes, and resemble tiny snapdragons. This weed thrives best in a semi-arid climate and is a large economic problem to vegetable growers in parts of California.

Fam. 160. Plantaginaceae (Plantain Family)

The Plantain Family consists of mainly stemless herbs with multiple flowers carried on spikes.

Bracted Plantain or Rat-Tail Plantain (Plantago aristata Michx.)

This is an annual or winter annual with a fairly short, slender taproot and erect, hairy, unbranched stems 2 to 14 inches tall, leafless, and terminating in a flower cluster. The leaves are usually hairy, long, narrow and are attached at the base of the plant. It has small flowers borne in a dense, tapering cluster resembling a bottle brush. This plant is found in overgrazed pastures, wastelands, and roadsides but doesn't usually compete successfully with other plants on good soil.

Figure III-16. Buckhorn. (Iowa Department of Agriculture Photo 1965, E. P. Sylwester.)

Buckhorn Plantain (Plantago lanceolata L.)

This is a perennial with a thick taproot. Its leaves are longer and broader than the bracted Plantain and have ribs on the upper surface. Its long stems end in a denser, more cylindrical, leafless, flower cluster than the bracted plantain.

Broad-leaved Plantain (Plantago major L.)

This is a perennial with a short, thick taproot that supports stems terminated with dense, pencil-like flower clusters. The broad, egg-shaped, wavy-edged leaves are usually smooth and attached with long, thick stalks to the stem's base. Broad-leaved Plantain is found in lawns and waste spaces.

Hoary Plantain or Pale-Seed Plantain (Plantago virginica L.)

Also known as Dwarf Plantain, it is a small, broad-leaved plant distinguished by the conspicuous, white hairiness of the stems and leaves.

Fam. 165. Dipsacaceae (Teasel Family)

Slashed Teasel (Dipsacus laciniatus L.)

This is a biennial which grows as tall as six feet. Its stout stems are erect, prickly and branched; and its small, white flowers occur in large heads resembling those of the thistle. The leaves are prickly and lobed. Teasel is a serious roadside and pasture-land pest in the Great-Lakes area and the Mid-Eastern region of the United States. When the flower heads mature, they are dry, very sharp, and bristly. Because dry Teasel flower heads are used for bases for outdoor wreaths, cemetery refuse often contains the remains of Teasel seedheads.

Wild Teasel (Dipsacus sylvestris Huds.)

This is similar to slashed Teasel but is shorter (3 to 4 feet tall), has lavender blossoms and smaller flower and seed heads. The leaves of Wild Teasel are lance-shaped with toothed edges.

Fam. 168. Compositae (Composite Family)

Ox-Eye Daisy, (Chrysanthemum leucanthemum L.)

This is a perennial found in unimproved pastures and wastelands. Short rootstocks and shallow roots support smooth stems that branch at the base. The lower leaves are often deeply notched but the upper leaves are smooth with toothed edges. Its large flower heads (1½ or 2 inches across) have yellow centers surrounded by white petals.

Horseweed or Fleabane (Erigeron canadensis L.)

This is an annual having a short, slender taproot and erect 1-6 feet tall stems that branch bushily in the upper portion. Its numerous leaves are narrow and hairy. The flowers are produced in small flower heads composed of yellow centers and many short, greenish-white, petal-like flowers. Horseweed is often found in recently abandoned fields.

Jerusalem Artichoke (Helianthus tuberosus L.)

This is a perennial that is sometimes grown for its edible, white tubers. Fibrous roots develop from short rootstocks that produce the egg-shaped tubers. The erect, 3-9 foot stems are tall, rough and hairy and branched in the upper portion. Leaves are large and vary from spear-shaped to egg-shaped. The yellow flower heads resemble sunflowers. Jerusalem Artichoke may be found along roadsides and in old fields as well as in vegetable gardens.

Spanish Needles (Bidens bipinnata L.)

This annual has a branching taproot, and erect, smooth stems branching in their upper portions. Deeply cut leaves are borne in pairs at the joints of the stems and branches. Yellow flower heads, terminating sparsely leafed branches, precede narrow, long, sharp-pointed, dark-brown or black seeds. The seeds bear 3 or 4 stiff, bristled spines. Spanish Needles may be found in old fields, fence rows, open woods and other rocky, sandy soils.

Cudweed or Rabbit Tobacco (Gnaphalium purpureum L.)

This annual, has fibrous roots supporting velvety gray 6-10 inch stems that often branch at the base. On the underside, the leaves are velvety-gray, but are green and hairy on the topside. On the lower part of the plant the leaves are narrow at their base but broad and rounded at their tip while the leaves on the upper parts of the plant are narrow. Tannish-white flowers are borne in clusters at the bases of the upper leaves.

Black-Eyed Susan (Rudbeckia hirta L.)

This biennial or perennial grows a thick, short taproot covered with long, fibrous roots. Its 1-3 foot stems are rough-hairy, erect and slightly branched. The flower heads are large (2-4 inch diameter) with dark purplish-brown, cone-shaped centers surrounded by orange-yellow petal-like flowers. This hairy-leaved plant may be found in pasture land.

Blessed Thistle (Cnicus benedictus L.)

This annual, produces 1½-3 foot stems covered with fine, white hairs. The leaves are deeply cut, covered with sticky hairs, and have spines on their edges. The yellow flowers are borne in compact, egg-shaped heads with sharp barbs. The seed is 1/4-3/8 inch long, cylindrical, bristled and heavily ribbed.

Spiny-Leaved Sowthistle (Sonchus asper (L.) Hill)

This annual, with milky juice, and a deep, slender taproot, grows in grain fields, waste fields and cultivated fields. Its stems are erect, 1-6 feet tall, pale green, smooth, and branching in the upper portion. The lower leaves' edges are often deeply cut into irregular spiny lobes, while the upper leaves often have edges cut into small, spiny tooth-like portions. The flower heads, ½ inch in diameter, produce pale yellow, petal-like flowers.

Dog Fennel or Mayweed (Anthemis cotula L.)

This is an odoriferous annual with a short, thick taproot and an erect, smooth stem 6-16 inches tall. Its alternate, hairy leaves are finely cut and the flower head, a yellow center surrounded by white petals, resembles a small daisy.

Dog Fennel or Tall Dog Fennel (Eupatorium capillifolium (Lam.))

This is a 3-9 foot perennial or annual crowded with branches near the top. Its leaves are divided into several fine, thread-like portions and it produces numerous white flowers. These flowers are followed by extremely small seed. Tall Dog Fennel grows in old fields and unimproved land.

White Heath Aster (Aster pilosus Wild.)

This perennial, grows fibrous roots developed from a hard, thickened stem base. Its 1-3½ foot stems are erect, often branched and usually hairy. The lower leaves are oblong and usually hairy but the upper leaves are more smooth and more narrow. Its small, white or purplish flower heads appear singly on short branches. The seeds are pointed, small, light brown, and have a tuft of long, soft, white hairs at the top. It usually grows in unimproved pastures and along roadsides.

Figure III-17. Perennial Sow Thistle. (Iowa Department of Agriculture Photo 1965, E. P. Sylwester.)

Bitterweed or Yellow Dog Fennel (Helenium tenuifolium Nutt.)

This is an annual with a short branching taproot, erect stems, 4 inches to 2 feet tall, that branch in the upper portion. Its numerous, thread-like leaves are small and smooth. The flower heads have a yellow dome-like central portion. Its seeds are brown, wedge-shaped and hairy on the edges. Bitterweed grows in wastelands.

Common Ragweed (Ambrosia artemisiifolia L.)

This annual develops branching roots, 1 to 4½ foot erect, hairy, rough, branching stems. Its leaves are smooth and deeply cut into several portions. The plant produces two kinds of flower heads: pollon producing heads that are borne in slender clusters at the branch tips and are small; and seed-producing heads, few and inconspicuous when young, borne at the base of the leaves and in the upper branch forks.

Giant Ragweed or Horseweed (Ambrosia trifida L.)

This is a large, coarse, 6-12 foot annual. Its leaves are large with 3-5 lobes borne in pairs at the stem joints. It inhabits fields and wastelands.

Common Burdock (Arctium minus (Hill) Bernh.)

This is a coarse biennial with a long taproot and thick, 3-5 foot stem. The leaves are very large, oval or heart-shaped, dark green, dull, and wooly on the underside. The purple headed flowers mature into bristle-covered brown burs. The Common Burdock may be found around old buildings, in fence rows, and yards.

Wild Lettuce or Prickly Lettuce (Lactuca scariola L.)

This annual, has a fleshy taproot and is further distinguished by the milky juice contained in the plant. The erect 2-4½ foot stems may branch somewhat in their upper portions. The arrow-shaped, lower leaves have deep, spiny cuts in their edges with a prickly midvein on the underside. Upper leaves are longer and more spiny than the lower leaves, but are not so deeply cut on the edges. The plant bears numerous flower heads with yellow, petal-like flowers. The Wild lettuce is found in both cultivated fields and pastures or unkempt areas.

Figure III-18. Cocklebur. (Iowa Department of Agriculture Photo 1965, E. P. Sylwester.)

Cocklebur (Xanthium pensylvanicum Wallr.)

This is an annual with a deep, woody, stout taproot. The ½ to 3½ foot tall stems are erect, rough, and hairy. Branching in the upper portion, the stems are characterized by small, dark red spots. The leaves have a variable outline, often heart or egg shaped, occasionally the edges are cut to resemble a maple leaf. These leaves are rough and hairy with dark or yellow-green upper surfaces contrasting with a pale green lower surface. The Cocklebur produces two types of seed heads: small, quite round, pollen producing heads that are found in small clusters at the tips of the branches which produce the burs, and oval, ¾-1 inch long, hard seed producing heads that produce two seeds per bur and are found in crowded clusters on short branches. The Cocklebur grows in cultivated fields, unimproved pastures and along roadsides. The seedlings of this and related species of the Cocklebur are reported to be very poisonous to cattle if ingested.

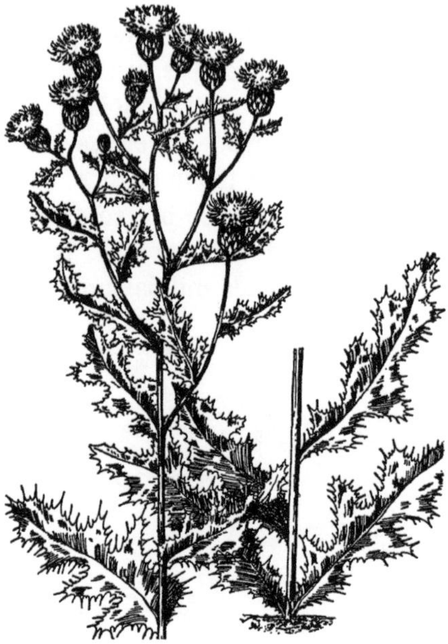

Figure III-19. Canada Thistle. (Iowa Department of Agriculture Photo 1965, E. P. Sylwester.)

Canada Thistle (Cirsium arvense (L.) Scop.)

This is a perennial with spreading roots that develop from short root-stocks. The erect, 1½-4 foot tall stems branch in their upper portions and are rather smooth or slightly hairy. The spear-shaped or oblong leaves have irregularly cut edges, bear sharp spines, and are hairy on their underside. The rose to purple flowers are borne on a branching cluster atop the plant. The oblong, smooth seeds are spread by the wind. The Canada Thistle can also reproduce through creeping root stems.

Bull Thistle (Cirsium vulgare (Savi) Tenore) or (Cirsium lanceolatum Scop.)

This biennial has a fleshy taproot and 2-6 foot, hairy stems. The bases of the leaves are very prickly as are the pinkish-purple, spiney flower heads. The Bull Thistle may be found in old fields and pastures.

White Snakeroot (Eupatorium rugosum Houtt.)

This ia a bunchy perennial, spreads by seeds and short rhizomes and commonly grows in wooded areas. Its 1-3 foot stems are smooth and branched near the top. The coarsely toothed, oval-shaped leaves have three main veins and taper gradually to a point. White flowers are borne in clusters near the branch tips. The small, black seeds are equipped with hairy tufts to enable them to be disbursed by the wind. The white Snakeroot is poisonous to cattle who injest it and to humans who consume milk from affected animals. The disease in cattle and horses is called "trembles" while the human form is called "milk sickness". It is believed that Abraham Lincoln's mother died of "milk sickness". This disease was one of the many problems coped with by Midwestern pioneers.

Figure III-20. White Snakeroot. (Iowa State University Photo, E.P. Sylwester.)

Dandelion or Lions-tooth (Taraxacum officinale Weber)

This perennial grows from a thick, long taproot and reproduces by seeds as well as new shoots from roots. The dandelion stem is short and contains a milky juice. Its deeply toothed leaves grow 3-10 inches long while its yellow flower heads have a diameter of 1-2 inches. Silky tufts of hair transport the elongated seeds from place to place via the summer breezes.

IV

Weed Control

The control of weeds by the use of chemicals may assume a variety of forms. Some crops may have the application of an herbicide made in conjunction with the planting of the crop. These "pre-emergence" applications are designed to destroy the seedling weeds, while not harming the emerging crop. The value of these as weed control agents depends on the existence of proper climactic conditions during the period of time between the application of the herbicide and the establishment of the "stand" of crop. Crops that are susceptible to post-emergence spraying of herbicides are especially benefited by the application of these pre-emergence chemicals. Some crops, such as corn and the grasses, are able to resist severe injury from the application of post-emergence herbicides and the application of these selective herbicides can be used to keep the field essentially weed-free during the early growing season.

Other herbicides are designed to kill most of the vegetation to which they are applied. These herbicides, often called "brush killers", are employed along the railroad right-of-way, irrigation ditches, industrial sites, and other areas where the lack of vegetation is desired, as in some war zones.

Non-chemical methods for the control of weeds include deep fall plowing to expose the roots of perennials and biennials to the weather, hoping for the destruction of the plant. The use of shallow, mechanical cultivation and the "rotary hoe" are important tools in the control of weeds in row crops. "Flame weeding" is also becoming popular in some areas and for some crops in which the method is economically and technically advantageous. Hand hoeing and the treatment of localized spots of noxious weeds with herbicides that are sufficiently powerful to injure the crop may also be used. Mechanical weed control suffers from the same limitation as does chemical weed control. If the weather is too wet, the field may not be sufficiently traversable to accomplish mechanical control, and the wrong weather conditions can reduce the effectiveness of chemical control techniques. The great "ounce of prevention" in the control of weeds is preventing the weeds from "going to seed" to infest the crops of the next year.

Figure IV-1. This small ash tree exhibits injury from exposure to 2,4-D. (Illinois Natural History Survey, Urbana, Illinois.)

Figure IV-2. Flame Cultivation of Grapes. (Humble Oil Refining Company.)

Figure IV-3. Corn Being Sprayed with Herbicide and Weeded with Mechanical Cultivator. (Gulf Oil Corporation.)

An outline of the herbicides that are appropriate for use with certain crops is presented below. Since the recommendations of the proper dosage change from one year to the next, precise details of the application of these chemicals should be obtained from the manufacturer or the nearest agricultural extension agent at the time of application. All dosages are presented in pounds of active ingredients per acre.

Corn

Corn that is grown on a mineral soil may be treated with one of the herbicides shown in Table IV-1.

TABLE IV-1

TIME OF APPLICATION	*AGENT*	*AMOUNT*	*TO CONTROL*
Pre-emergent or Pre-plant	Atrazine[1 2]	2 #/acre	annual broad-leafed weeds and annual grasses
Pre-emergent	Atrazine[1] PLUS Linuron (Lorox)	1 #/acre 1 #/acre	annual broad-leafed weeds and annual grasses
Pre-emergent	Atrazine[1] PLUS Propachlor (Ramrod)	1 #/acre 3 #/acre	annual broad-leafed weeds and annual grasses
Pre-emergent	Atrazine[1] PLUS Prometryne (Primaze)	1 #/acre 1 #/acre	annual broad-leafed weeds and annual grasses
Pre-emergent	Propachlor	4 #/acre	annual grasses
Post-emergent	2,4-D[3] (Amine)	½ #/acre	annual broad-leafed weeds
Emergent	Atrazine[1 4] PLUS Oil	2 #/acre 1 gallon/acre	annual broad-leafed weeds and annual grasses
Pre-plant	Atrazine[1] PLUS Butylate (Sutan)	1 #/acre 3 #/acre	annual broad-leafed weeds and annual grasses

[1]Commercially available as AAtrex
[2]The following year, field beans, oats, sugar beets, or vegetables should not be planted
[3]Apply before the corn is 6-8 inches high
[4]Apply before the weeds reach a height of 1½ inches

Corn that is grown on organic soils may be treated with the chemicals shown in Table IV-2 to control the weeds:

TABLE IV-2

TIME OF APPLICATION	*AGENT*	*AMOUNT*	*TO CONTROL*
Pre-emergent	CDAA (Randox)	4 #/acre	annual grasses
Pre-emergent	Propachlor (Ramrod)	5 #/acre	annual grasses
Post-emergent	2,4-D[1]	½ #/acre	annual broad-leafed weeds
Post-emergent[2]	Atrazine (AAtrex) PLUS Oil[3]	3 #/acre 1 gallon/acre	annual broad-leafed weeds and annual grasses

[1]May be applied until the corn is 6-8 inches high, when the corn is 8-18 inches tall, drop nozzles must be used. [2]
[2]Before the weeds are 1½ inches tall
[3]High-grade, not-phytotoxic crop oil sold specifically for this purpose, NOT left-over oil

The control of Quackgrass in the fields of corn may be accomplished by one of the following procedures:

1. The application of two pounds of Amitrole-T either in the fall or the spring followed by plowing within seven to ten days (if freezing weather occurs within seven to ten days the fall plowing is unnecessary). This treatment is followed by the application of two pounds of Atrazine per acre as a pre-emergence herbicide. The following year, vegetables, or sugar beets, etc. (see Corn, mineral soils, footnote #2) must not be planted in the treated fields.
2. The application of four pounds of Atrazine is made in the spring when the Quackgrass is 4-8 inches tall, followed by plowing ten days later. Alternatively, two pounds of Atrazine per acre may be used as a pre-plowing treatment and two pounds per acre may be applied as a pre-emergent herbicide to control the annual weeds at the same time.

If considerable amounts of Quackgrass are present, the application of the herbicide in the fall is recommended. The application of four pounds of Atrazine per acre in a single year requires that corn be grown on the field for two consecutive years (or the land be allowed to lie fallow).

Figure IV-4. Corn Row Containing Grass Killed by Spraying with Herbicide. (Gulf Oil Corporation.)

SMALL GRAINS

When seeded with Legumes, the broad leaf annual weeds may be controlled by the application of one of the following sprays when the grain is six to eight inches high: three-eighths pound of MCPA or three quarters of a pound of DNBP per acre. Sweet Clover is very sensitive to applications of MCPA, but usually the canopy of weeds and grain will prohibit injury to the seedlings of alfalfa.

Barley and Wheat that are not seeded to Legumes can have their broad-leaved weed population controlled by the application of one half pound of 2,4-D (amine) per acre when the grain becomes "fully tillered", but before it reaches the "boot stage" (this is usually when the grain is approximately 6-8 inches tall). The control of perennial weeds may be accomplished by the application of one half pound of 2,4-D per acre when the grain is fully tillered, but before it reaches the boot stage.

Figure IV-5. Wild Oats (inside white rings) are optimal size for treatment with Herbicide. (Gulf Oil Corporation.)

Broad-leaved annual weeds in Oats that are not seeded to Legumes may be controlled by the application of one quarter pound of 2,4-D (amine) per acre when the Oats are 6-8 inches tall and fully tilled. Alternatively, the broad-leaved annual weeds may be controlled by the application of three-eighths pound of MCPA per acre. The advantage of the application of MCPA in the place of the 2,4-D is the reduction in the injury of the Oats that is attained by the application of the MCPA.

LEGUMES AND PASTURES

Annual weeds may be controlled in Alfalfa, Clover and Trefoil seedlings, that are unaccompanied by small grain covers, by working three pounds of EPTC (Eptam) per acre into the soil immediately before planting the crop.

Broad-leaved annual weeds may be controlled when the aforementioned crops are in the two to three leaf stage with the application of three quarters of a pound of 2,4-DB per acre.

Yellow Rocket and broad-leaved winter annual weeds may be controlled in an established stand of Alfalfa by the application of one of the following: a spray containing one half pound per acre of MCPA that is applied in the late fall after a killing frost has put the legumes into the dormant stage, or one and one quarter pound of Simazine per acre that is applied in the late fall after a killing frost. The MCPA must not be used in the fall of the year in which the legumes were planted. The Simazine must be applied to Alfalfa that is at least one year old, since Alfalfa will sometimes sustain injury from the Simazine if the soil is deficient in organic matter (such as sandy soils).

Red Clover

Yellow Rocket and broad-leaved winter annuals that infest Red Clover may be controlled in the same year that the clover was seeded by the application of a spray in the late fall after a killing frost. A recommended formulation for the spray consists of one half pound of MCPA dispersed in water according to the directions of the manufacturer per acre of cropland to be treated.

Pasture

Biennial and perennial weeds that infest the pasture grass may be controlled by the application of one pound of 2,4-D (ester) per acre. The spray should be applied either in the spring or in the fall.

Perennial weeds in legume pastures may be controlled by the application of one pound of 2,4-D (ester) in the late fall after a killing frost. Heavy infestations of weeds may be treated in isolated areas with the 2,4-D (ester) during the growing season, but the legumes may also be killed.

Hoary Alyssum and broad-leaved annual weeds in either hay or pasture that is composed of grass or legumes, may be controlled by the application of a spray in the early spring when the Hoary Alyssum is in the two to four leaf stage. A suitable spray for this purpose consists of one pound of 2,4-DB (ester). The hay must not be harvested nor grazing allowed in the treated pasture for at least thirty days after the application of the spray.

Sugar Beets

Annual weeds can be controlled in fields of Sugar Beets by the application of one of the following herbicides at the time that the seed is planted: four pounds of Endothall plus eight pounds of TCA per acre, OR four pounds of Pyrazon (Pyramin) plus six pounds of TCA per acre. The Endothall-TCA combination sometimes yields an erratic control of Lambsquarters. The application of either spray may be made in a band above the seed to reduce the expense of the chemical control of the weeds in the rows.

Field Beans

Annual weeds can be controlled by either applying three pounds of EPTC (Eptam) per acre that is worked into the soil immediately after the application OR the use of a two pounds of Amiben plus two pounds of CDAA (Randox) per acre as a pre-emergent spray.

Some annual, broad-leaved weeds, with the exception of Mustard, Ragweed, and Smartweed, may be controlled by the application of three quarters of a pound of Trifluralin (Treflan) per acre. This spray will also control annual grasses. The proper application of the spray includes the double disking of the field twice, in different directions, immediately after the application of the chemical. The dosage that is given above must be reduced to one half pound per acre when the crop is grown either in loam or sandy soils that are low in organic material so that the risk of injury to the crop is reduced.

Soybeans

Among the pre-emergence sprays that are available, three pounds of Amiben per acre may be used to control annuals, with the dosage increased to four pounds per acre if the Amiben is applied on muck soils. For the control of annual broad-leaved weeds and grasses, a pre-emergent spray that contains either one and one half pounds of Linuron (Lorox) OR two pounds of Amiben plus one pound of Linuron per acre may be used. In the latter spray, the Linuron concentration should be reduced to three quarters of a pound per acre if the spray is used on sandy soils.

Before the crop is planted, three quarters of a pound of Trifluralin may be disked into the soil to control annual grasses and annual broad-leaved weeds. *Mustard, Ragweed,* and *Smartweed* will not be effectively controlled by this spray. The concentration of the Trifluralin should be reduced to one half pound per acre of sandy soil and increased to one pound per acre of clay loam.

At the time of planting, one gallon of NPA plus CIPC may be applied per acre for the control of annual weeds. This treatment is not especially effective when used on muck soils, but injury to the crop may occur on soils that contain a low amount of organic material.

Peanuts

The preparation of the seedbed includes the plowing of the plant debris at least three inches deep to control both the weeds and the Southern Blight disease. Two to two and one half pounds of Vernolate may be incorporated into the upper three inches of each acre of the soil either just before planting or up to two weeks prior to planting. This application will control most annual weeds and suppress *Nutsedge*, but it often will temporarily stunt the peanut and cause Leaf Fusion. To reduce the danger of these side-effects, good Peanut seed should be used.

One and one tenth to one and one half pounds of Benefin may be incorporated into the upper two to three inches of the soil any time from three weeks prior to planting until the time of planting to control many annual broad-leaved weeds and most grasses that originate from seeds.

Any of the following sprays may be used after the crop is planted but prior to the emergence of the Peanut plant to effect the control of a majority of the small-seeded annual weeds. Some injury to the Peanut plant is likely to be sustained if these sprays are used, especially if the weather is wet and cold at the time that the crop emerges. Most of the injury is temporary, but Peanut seed possessing good vitality should be used to minimize the effects of the spray on the Peanut plant. Sprays that may be used include: 2.7 pounds of Sesone, three pounds of 2,4-DEP, nine to twelve pounds of DNBP, OR two to four pounds of NPA per acre of cropland.

Potatoes

Before the seed is planted, four pounds of EPTC (Eptam) per acre may be incorporated into the soil to control annual weeds. The concentration of the EPTC may be increased to six pounds per acre if Nutgrass is of significant concern. Alternatively, one of the following chemicals may be applied after the weeds have emerged, but before the potatoes have emerged to control annual grasses and annual broad-leaved weeds: two pounds of Linuron (Lurox), OR three pounds of DNBP plus two and one half pounds of Dalapon, OR one pound of 2,4-D(ester) plus two and one half pounds of Dalapon per acre. The 2,4-D - Dalapon combination must not be used on fields that are being used to grow certified seed. For maximum effectiveness of the treatment, the fields should be leveled soon after the crop is planted.

VEGETABLES

In the fall or early winter, after the final cultivation of the year, the application of three and two-tenths pounds of Diuron per acre will control the weeds that infest the Artichokes.

Weeds in Asparagus may be controlled by the application of one of the following sprays: during the winter a dormant spray consisting of three and two-tenths pounds of Diuron per acre, OR the pre-emergent application of three pounds of Amiben per acre, OR a pre-harvest or post-harvest spray composed of either seven to fourteen pounds of Dalapon or one to three and two tenths pounds of Monuron per acre.

Either six pounds of Dalapon or three quarters of a pound of Trifluralin may be incorporated into each acre of soil four to five weeks before planting to control the weeds in Lima Beans. Alternatively, a pre-emergence spray that contains wither two pounds of Amiben or three to nine pounds of DNBP per acre may be used.

Four to five weeks before the Pole Beans are planted, six pounds of Dalapon may be incorporated into each acre of soil OR two to six pounds of CIPC per acre may be used as a pre-emergent herbicide. Three pounds of EPTC may be incorporated into the soil either as a pre-emergent or a post emergent treatment for the control of the weeds.

The weed control program for Snap Beans may consist of the pre-emergent application of either four pounds of CDAA or six to ten and one half pounds of DCPA per acre of the crop. Alternatively, four to five weeks before the planting in the spring, one of the following sprays may be used: six pounds of Dalapon, OR three to four pounds of EPTC, or three quarters of a pound of Trifluralin per acre of the Snap Beans.

One of the following pre-emergence herbicides may be applied per acre of Beets in a weed control program: six to nine pounds of TCA (sodium salt), OR six and one half pounds of Endothall, OR four pounds of Pyrazon.

A pre-plant or pre-transplant treatment for Broccoli is four to eight pounds of CDEC or three quarters to one pound of soil incorporated Trifluralin per acre. An effective pre-emergence and post-transplant application of either six pounds of FW-925 or ten and one half pounds of DCPA is also reported.

A pre-emergent or post-emergent application of six to eight pounds of CDEC or ten and one half pounds of DCPA per acre of Brussels Sprouts can be used in the weed control program. The application of three quarters of a pound of Trifluralin that is incorporated in each acre of the soil may be used as either a pre-plant or pre-transplant method for the control of the weeds.

A weed control program for Cabbage may include the pre-emergent and/or post-transplant application of either four to six pounds of CDAA, OR four to eight pounds of CDEC, OR ten and one half pounds of DCPA, OR six pounds of FW-925 per acre. Alternatively, a pre-plant or pre-transplant application of three quarters of one pound of Trifluralin may be incorporated into each acre of the soil.

Four pounds of CDEC per acre of Cantaloupe may be used as a pre-emergent weed control agent.

One to two pounds of Linuron per acre may be applied as a pre-or post-emergent agent for the control of the weeds in Carrots, or ten and one half pounds of DCPA may be applied at the time of either planting or transplanting.

A pre-emergent or post-transplant treatment of four to eight pounds of CDEC OR six to ten and one half pounds of DCPA, OR six pounds of FW-925 per acre provides a means of controlling weeds that infest the Cauliflower field.

Celery crops may be protected from the effects of uncontrolled infestations of weeds by the application of four to six pounds of CDEC per acre as a pre-emergent or post-transplant control measure. A post-transplant treatment may also be comprised of either four pounds of CDAA OR seven pounds of CIPC per acre of the crop.

A variety of chemical means of controlling weeds in Sweet Corn fields are available. Among the pre-emergent treatments are the application of three to four pounds of Atrazine, four pounds of CDAA, seven and one half to nine pounds of DNBP (alkanolamine salt), five pounds of Propachlor, or two to four pounds of Simazine per acre. Sensitive crops should not be planted following Corn that is treated with Atrazine. If the corn is treated with Propachlor, the forage should not be used as feed. The Simazine may be incorporated into the soil. Six pounds per acre of 2,4-DEP may be used immediately after the crop is planted. The herbicide 2,4-D may be used in either of two ways; by the application of one and one half pounds per acre as a pre-emergent herbicide on clay soils, OR one quarter to three quarters of a pound per acre before the corn becomes two feet tall. When the corn is eight inches tall, one and one half pounds of Cypromid per acre can be applied for the control of the weeds if the stalks and foliage are not used as a feed.

The application of two to six pounds of the sodium salt of NPA may be used to control the weeds in fields of Cucumbers. This herbicide may be used in either a pre-emergence or a post-transplant program.

One and one tenth pounds of Benefin may be incorporated into the soil two to three weeks before the Lettuce is planted. The directions provided by the manufacturer should be rigorously followed, with special note being given to the feeding of the crop to livestock and the use of crop rotation. Alternatively, either two to four pounds of CDEC may be applied per acre as a pre-emergent or pre-transplant control agent, OR three pounds of CIPC per acre can be used as a pre-emergent treatment.

Either one pound of Trifluralin per acre may be incorporated into the soil before the Okra is planted or five pounds of Diphenamide per acre may be used at the time of the planting for the control of the weeds.

For the control of weeds in fields of Onions, one of the following herbicides may be used as either a pre-or post-emergence plan: six pounds of CDAA, four to eight pounds of CIPC, ten and one half pounds of DCPA, OR sixteen to twenty pounds of KOCN. All of the above concentrations are given in pounds of the active ingredient per acre of the crop.

A number of chemical agents are available for use in the control of weed pests in fields of Peas. A pre-plant application of either four to six pounds of CIPC per acre, OR one and three tenths pounds of Diallate, OR four to eight pounds of IPC per acre can be used to control the weeds. Diallate may be incorporated into the soil, but pea plants that have been treated with Diallate should not be used for grazing. The IPC may also be used as a post-emergent herbicide. Either four pounds of CDAA, OR four to six pounds of CIPC, OR six to nine pounds of DNBP(alkanolamine salt) per acre of the crop may be used to form a pre-emergent control of weeds. If the pea stubble is not used for grazing, four tenths of a pound of Barban may be used per acre for the control of Wild Oats. After the Peas are two to six inches tall and have acquired four to six nodes, with certain reservations, either seven tenths of a pound of Dalapon OR one and one half pounds of 2,4-DB per acre may be applied. The use of the 2,4-DB precludes feeding the crop to cattle, while vines that are treated with Dalapon cannot be fed to livestock. The 2,4-DB should not be applied within thirty days of the harvest and the Dalapon must not be applied within twenty-five days of the

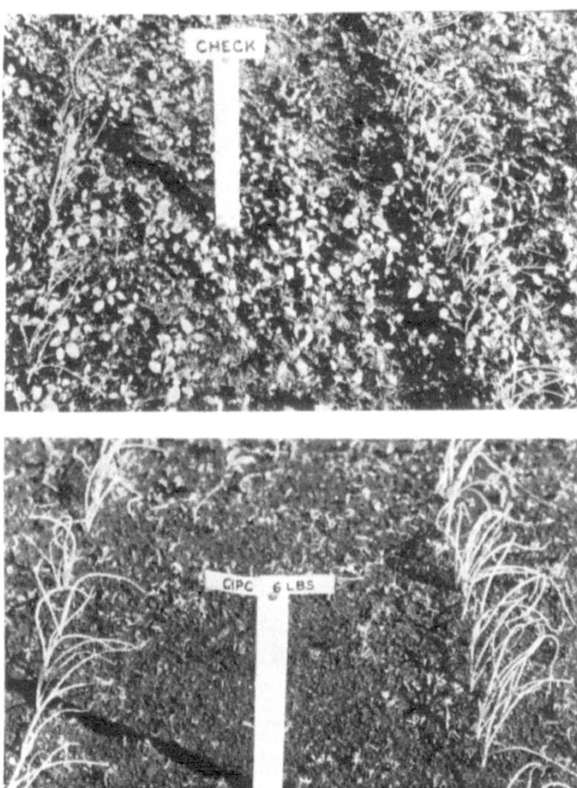

Figure IV-6. The lower photo shows onions treated with chloro IPC. The upper photo is the Control Group. (Pittsburgh Plate Glass Company.)

harvest of the Peas. One and one half to two pounds of DNBP (alkanolamine salt) per acre may be used as a post-emergent treatment, as may four to eight pounds of IPC per acre. Three quarters of a pound of MCPA (amine or sodium salt) per acre may be employed as a post-emergent treatment before the Peas have bloomed. During the early post-emergent period is the best time for the application of one quarter to three quarters of a pound of MCPB per acre.

The weeds in the fields of Southern Peas may be controlled by the application of one quarter to one pound of Trifluralin that is incorporated in each acre of soil before the crop is planted, OR the pre-emergent treatment of the field with six to ten and one half pounds of DCPA per acre.

During the last cultivation before the emergence of the Peppermint, either two and four tenths of Diuron per acre or eight tenths to one and six tenths pounds of Terbacil per acre may be applied as part of the weed control program.

Weeds in Pepper fields may be controlled by the application of six pounds of CIPC per acre as a pre-emergent control OR one pound of Trifluralin per acre as a pre-transplant control. Five pounds of Diphenamide may be applied immediately after the Peppers are seeded or within thirty days after the crop is transplanted. Three to four pounds of Amiben per acre may be used at the time of transplanting or six to ten and one half pounds of DCPA per acre can be used in the post-transplant control of the weeds.

The pre-emergent application of three to four pounds of Amiben per acre of soil can be used in the program to control the weeds in the fields of Pumpkins.

One to two pounds of CIPC per acre may be used in the pre-emergent control of weeds in crops of Spinach, while either two to four pounds of CDEC OR three to six pounds of IPC per acre can be used as either pre- or post-emergent control agents.

The pre-emergent application of three to four pounds of Amiben per acre of *Squash* can be used in the weed control program. Alternatively, clean, mechanical cultivation can be maintained during the early growth of the crop and, four to six weeks after the seeding of the field, ten and one half pounds of DCPA per acre may be applied for the control of the weeds during the middle and last portion of the growing season.

At the time that Sweet Potatoes are transplanted, either three pounds of Amiben OR four to six pounds of Diphenamide per acre may be used in the program to control the weeds. The use of four pounds of CDAA per acre as a post-transplant control measure is also recommended.

The control agents that are available for use in a program to control weeds in Tomato crops are shown in Table IV-3.

TABLE IV-3

APPLICATION	*AGENT*	*CONCENTRATION (pounds per acre)*
pre-transplant*	Trifluralin	one
pre-transplant*	Pebulate	four
pre-emergent	DCPA	ten-and-one-half
pre-emergent	CDAA	three
pre-emergent	CDEC	four to six
pre-emergent[1]	Diphenamid	four to six
post-transplant[2]	CDAA	six
post-transplant	CDEC	four to six
post-transplant[3]	Diphenamid	four to six
post-transplant	Pebulate	four
post-emergent	CIPC[4]	four
post-emergent	Trifluralin	one

*Incorporated into the soil
[1] Apply immediately after seeding
[2] Within forty-eight hours of the transplanting
[3] Within thirty days after transplanting
[4] In the form of the granules.

FRUIT TREES

A program for the control of weeds in the Apple can be formulated from the herbicides shown in Table IV-4.

TABLE IV-4

APPLICATION	*AGENT*	*CONCENTRATION (pounds per acre)*
Poison Ivy in full leaf	AMS	54 #/100 gal. water
spots of grass[1]	Dalapon	11 pounds
general weed control[2]	Diphenamid	6 pounds
general weed control[3]	Dichlobenil	4-6 pounds
general weed control[4]	Diuron	3.2 pounds
general control of small weeds[5]	DMBP (butyl-phenol)	1.9 pounds
pre-emergent[6]	Simazine	2-4 pounds
general weed control[7]	Terbacil	1-3.2 pounds
patches of young weeds[8]	2,4-D[9]	2 pounds

[1] The trees must be at least four years of age
[2] After clean cultivation and not within three months of the harvest
[3] Early in the spring or after clean cultivation
[4] Applied in the winter or the spring
[5] Applied as a coarse spray to the ground cover
[6] Applied in the spring before the weeds emerge
[7] Trees must be at least three years old, applied as a single application in the spring
[8] Sprayed directly on the patches
[9] Acid or oil-soluble amine

The weed control program for Apricot trees may consist of either the application of forty to one hundred gallons of an aromatic oil emulsion per acre or the use of Dalapon (See Apple). The aromatic oil emulsion is sprayed directly on patches of the young weeds.

Either Dichlobenil (see Apple) or Simazine (see Apple) may be used in a program to control the weeds in Cherry orchards. The Simazine spray should be used in the fall or the spring when the crop is dormant and only if the trees are at least one year old.

The control of weeds in Olive groves may be accomplished by the application of an emulsion of aromatic oil (see Apricot). Alternatively, one and six tenths pounds of Diuron per acre may be applied in a split application, used during the fall or the spring, OR two to four pounds of Simazine per acre may be sprayed in a single application in the late fall or mid-winter.

Weeds in orchards of Peach trees may be controlled by the application of four to six pounds of Dichlobenil per acre in the early spring OR four pounds of Simazine that are applied per acre to the soil around established trees before the weeds emerge. The Simazine may be applied anytime from after the harvest until early spring. Alternatively, one of the following agents for the control of weeds may be applied: aromatic oil emulsions (see Apricots), Dalapon (see Apple), Diphenamid (see Apple), DNBP-butyl-phenol (see Apple, or Terbacil (see Apple).

The weeds that are a pest in Pear groves may be controlled by the application of four pounds of Simazine per acre that is applied as a directed spray around well established trees. Other control agents that may be used include: AMS (see Apple), aromatic oil emulsion (see Apricot), Dalapon (see Apple), Dichlobenil (see Peach), Diuron (see Apple and Olive), DNBP (see Apple), or 2,4-D–the acid or oil-soluble amine (see Apple).

Four to six pounds of Simazine per acre of Plums may be used to control annual grasses and broad-leafed weeds, if applied as a pre-emergent treatment. The application of Simazine granules between November 15 and January 1 can be used to control perennial weeds. Alternatively, one of the following agents may be employed in the weed control program: Aromatic oil emulsion (see Apricot), Dalapon (see Apple), Dichlobenil (see Apple), or DNBP (see Apple).

NUT TREES

The weed pests around Almond trees can be effectively controlled by the application of an emulsion of an aromatic oil (see Apricot).

The application of the emulsion of an aromatic oil (see Apricot) can be used in the control of weeds aroung English Walnut trees. Four pounds of Diuron per acre may be applied in either the late fall or the early spring if the trees are well established to control the weeds. For less well-established trees, the dosage should be divided between the fall and the spring.

SUB–TROPICAL NUT AND FRUIT TREES

The use of the aromatic oil emulsion (see Apricot) is used in the control of weeds in groves of Avocado trees. Six pounds of Dichlobenil per acre can be applied as a directed spray in the early spring after the clean cultivation of the area. A third control agent is the application of one and six tenths pounds of Monuron per acre under trees that are well-established. The Monuron may be applied in the spring or fall, but only in certain selected areas of California. Local governmental authorities should be consulted before Monuron is used in the control program. Finally, Simazine (see Apple) may be used after the final preparation of the grove.

A large variety of chemical agents can be used in selecting the agent that is to be used in the control program in Citrus groves. Generally the application of an aromatic oil emulsion (see Apricot) is satisfactory in the control of many weeds. Eight pounds of Bromacil per acre of Grapefruit or Orange is used in some weed control programs. The Bromacil should be used only in two year old, established commercial groves in Texas. This chemical, especially, should be used as directed, in a single spot spray during the growing season, but only for two years. All precautions concerning crop rotation and reuse of the chemical that are supplied by the manufacturer should be followed. The Bromacil is also used at the concentration of six and four tenths pounds per acre as a single, spot directed spray between the winter and early summer in two year old, established groves in California and Arizona only. Again the above warnings and cautions should be heeded.

Three pounds of Dalapon per acre of Grapefruit, Limes, Oranges or Tangerines may be used as a spot treatment to control grass around these trees.

For the control of weeds aroung nursery and non-bearing stock, one year after transplanting, four to six pounds of Dichlobenil may be either sprayed on or incorporated in the soil.

In selected areas of California and Arizona, the spring and fall application of one and six tenths pounds of Diuron per acre can be used as a soil treatment in groves that have been established for at least one year.

Two pounds of Simazine per acre may be applied on established plantings of Lemons in California and Arizona during the rainy season. Orange groves in Florida and Texas are treated with nine and six tenth pounds of Simazine per acre.

The application of a single, directed spray of four to eight pounds of Terbacil per acre may be made to two year old, established groves of Oranges and Grapefruit in Texas. The herbicide should be applied in the spring before the weeds have emerged. Orange and Grapefruit trees that reside in Arizona or California should receive only three and two tenths pounds of Terbacil in the control of their weeds.

Six pounds of Dichlobenil may be used per acre of Mango grove if the spray is applied in a directed manner in the early spring or after the clean cultivation of the grove.

SMALL FRUIT PLANTINGS

Table IV-5 shows the chemical agents that may be used for the control of weeds in plantings of small fruits. The dosages* are presented in pounds of active ingredients per acre. Many sources recommend the application of the herbicide in the form of a low, directed spray so that the fruit plants will not be injured. If the application by such a spray is not practical, the use of granules is recommended.

TABLE IV-5

FRUIT	*AGENT*	*DOSAGE*	*NOTES*
Blackberries	CIPC	2-6	soil treatment when berries are dormant
Raspberries	Diuron	1-2	soil treatment on year old, dormant plantings
	Diuron	2-4	directed spray in late winter or divided spring and late fall
	Simazine	2-4	when berries are dormant
Blueberries	Diuron	1-2	soil treatment when year old berries are dormant
Cranberries	CIPC	8	post-harvest on dormant berries
	Dalapon	8.8	directed spot spray on grasses, post-harvest
	Dichlobenil	2.6	post-harvest or pre-bloom (granules)
	Ferrous sulfate	800	between spring and mid-summer
	Simazine	2-4	soil treatment when berries are dormant (Mass. only)
Grapes	CIPC	10-14	directed spray early in the spring to dormant crop
	Delapon	15	spot spray on grasses during vegetative growth of Grape
	Diuron	1-2	soil treatment on three year established, dormant crop
	Diuron	4.8	around well established plants
	Diuron	1.6	split application: early spring and late fall
	Diuron	9.6	directed, spot spray once in four years to established plantings in heavy soil
	DNBP (alkanolamine salt)	1.25	soil under trellis
	Simazine	2-4	soil treatment on three year established crops between late fall and early spring

FRUIT	*AGENT*	*DOSAGE*	*NOTES*
Strawberries	Chloroxurion	4	fall after beds are renovated; spring at least 60 days before harvest
	CIPC	1-3	to dormant plants to control Chickweed
	DCPA	9	soil treatment after transplanting
	Diphenamid	6	2-6 weeks after transplanting; when rows become established; not more than one year prior to harvest and at least 60 days post-harvest
	DNBP (amine salts)	3-4.5	early spring to dormant plants: post-harvest; northeastern states only
	EPTC	4	30 days after planting new beds; after harvest on established beds following clean cultivation
	Sesone	3-4	soil treatment post-harvest or post-transplant. Must not be used when plants are blooming or bearing
	2,4-D	1.5	post-harvest, spray young weeds
	2,4-DEP	4	post-harvest, established beds; pre-plant or post-transplant in fields not to be harvested until following year

Figure IV-7a & b. Before and after treatment for Aquatic Weeds. (Penn Salt Chemicals Corporation.)

FIELD CROPS

Sorghum

A rotary hoe may be used between the time the Sorghum is two inches tall and when the stalks start to stiffen for control of the shallow-rooted, seedling weeds.

One to two pounds of Atrazine may be used after the crop is planted but before the Sorghum has emerged to control a majority of the small-seeded annual weeds. The Sorghum should be planted at least one inch deep to minimize the injury to the crop. Atrazine should not be used on Sorghum in soils that have a coarse texture or a low concentration of organic matter.

After the Sorghum has emerged, two to three pounds of Atrazine per acre may be broadcast or applied as a directed basal spray. This application will effect the control of most annuals that are less than one inch tall at the time of application.

Most annual broad-leaved weeds can be controlled by a one quarter to one half pound application of 2,4-D per acre. Although this chemical may be applied as a foliage spray when the crop is four to twelve inches tall, the injury to the Sorghum should be reduced if a basally directed spray is employed when the crop reaches six to twelve inches in height. At best, 2,4-D is likely to be responsible for some injury to the crop of Sorghum.

After the Sorghum is a foot tall, the tops of a majority of the small weeds may be killed by the basal direction of a flame. The day prior to flame weeding, a trial-run should be made on a few rows so that the equipment can receive the proper adjustment and the results of the test run can be evaluated the following morning before the entire field is covered.

Cotton

Any time from several weeks prior to planting until immediately before planting, the time is appropriate for the incorporation of one of the two following herbicides into the top one to two inches of the soil for the control of small-seeded, broad-leaved, annual weeds and most of the grasses that originate from seeds: eight to ten and one half pounds of DCPA, or one half to one pound of Nitralin per acre. One to one and one half pounds of EPTC per acre may be placed on either side of the row at a depth of one inch in bands that are six inches wide and centered seven inches from the drill row. This application will control Nutsedge, Johnson Grass (from seed), and a majority of the annual weeds. This treatment is to be used in combination with a pre-planting application of trifluralin, OR a pre-emergent application of Diuron, and to be followed by post-emergent treatments of the field with naptha, DSMA or some other post-emergent herbicide.

After the crop is planted, but before it has emerged from the soil, one of the following agents may be employed to control a majority of the small-seeded annual weeds: four to ten pounds of CIPC per acre either as a spray or granules that are incorporated into the top one-half inch of the soil, OR eight pounds of DCPA which is especially preferred per acre of fine soil, OR four tenths to one and six tenths pounds of Diuron per acre, OR one and six tenths pounds of monuron per acre which is preferred for very coarse solids, OR three quarters to two and one half pounds of norea, OR two to two and four tenths pounds of Prometryne.

Occasionally, Naphtha is recommended as a spray for cotton from the time the crop is two inches tall until before the cracks appear in the bark, but this material can cause extensive damage to Cotton that is either diseased or injured.

In the irrigated Cotton Belt (only) of the western United States, three quarters of a pound of Trifluralin is applied between the rows and is immediately worked into the upper one to two inches of the soil. This agent is applied after the Cotton has emerged, but before the first irrigation of the crop. The Trifluralin controls Johnson Grass (from seed) and a majority of the small-seeded annuals that exist between the rows.

One of the following treatments may be applied once or twice after the smallest cotton is at least three inches tall, but prior to the appearance of the first blossoms. This spray is applied as a basally directed spray and ordinarily one pint of a surfactant per twenty-five gallons of spray is used to improve the wettability of the spray. These treatments include: 1) three pounds of DSMA OR two pounds of MSMA, 2) two pounds of MSMA OR three pounds of DSMA *plus* eight tenths of a pound of norea*, or 3) two pounds of MSMA or three pounds of DSMA *plus* eight tenths of a pound of Fluometuron*. All concentrations were given in pounds of active ingredient per acre and the noted(*) combinations are more effective over a range of weather conditions than either herbicide used alone. Alternatively, one to two applications of the following sprays may be made between the time when the smallest cotton plants are at least six inches tall and the first blossoms open as the basal spray containing one pint of surfactant per twenty-five gallons of spray: 1) either two pounds of MSMA OR three pounds of DSMA *plus* two tenths of a pound of Diuron or 2) two tenths to four tenths of a pound of Diuron per acre. Again, the combination of two herbicides provides a control that is less dependent upon the weather conditions than either herbicide alone. The Diuron may yield poor control of the weeds during conditions of drought.

Flame weeding that is directed at the base of the cotton plant can be used any time after the cotton is ten to twelve inches tall until the cotton is too large for the equipment to pass down the rows. Flame weeding between the rows may be initiated without regard to the size of the cotton provided the flames are carefully adjusted. This method for the control of the weeds is most effective if it is used when the weeds are small. The most obvious danger in flame weeding is the igniting of dry plant material.

After the completion of mechanical cultivation due to the size of the plants, three tenths to three quarters of a pound of Diuron per acre may be applied as a basal spray. This spray may be applied when the cotton is one foot tall and re-applied when the plant reaches fourteen to twenty-five inches of height. One pint of surfactant is commonly added to each twenty-five gallons of spray to improve the wettability of the spray.

When the cotton is fifteen to twenty-four inches tall, a single application of one of the following may be applied as a basal spray (with the addition of one pint of surfactant per twenty-five gallons of spray if emerged weeds exist): eight pounds of CIPC, one half to one and one half pounds of Diuron, one half to one and one half pounds of Linuron, one half to one and two tenths pounds of monuron, OR one pound of Prometryne. The above concentrations of active ingredient per acre are used to control any weed that germinates after the application in addition to the control of any weeds already emerged that are less than three inches tall.

Spot spraying of Johnson Grass may be conducted with either slightly less than three quarters of a pound of Dalapon per five gallons of water that is used to wet all of the grass foliage, OR one pound of TCA per five gallons of diesel fuel that is used to wet the lower four to six inches of the clumps of Johnson grass or used at one half strength to thoroughly spray the Johnson Grass that is less than one foot tall.

Tobacco

To effect the control of annual weeds in the plant bed, one pound of commercial urea plus one half pound of calcium cyanamide should be incorporated into the upper three inches of each square yard of soil. Alternatively, one pound of methyl bromide per one hundred square feet of pulverized soil should be added under a gas-proof cover. Since the methyl bromide is very toxic, it should always be applied by a trained operator. Three pounds of Pebulate per acre may be incorporated into the top two to three inches of the soil to control several annual weeds and suppress the growth of Nutsedge. The directions of the manufacturer concerning the use of Pebulate should be followed very closely.

An application of four to six pounds of Diphenamid per acre that is made immediately after the seeded tobacco is planted or after transplanting and before the emergence of the weeds, will control most annual grasses as well as some of the small-seeded broad-leaved weeds. The incorporation of one and two tenths to one and a half pounds of Benefin per acre into the top two to three inches of soil after the beds are formed but before the Tobacco is set will control many annual grasses and some broad-leaved weeds.

LAWNS AND TURF

Two to three days before the new lawn is seeded, the application of one pound of methyl bromide per one hundred square feet of lawn can be conducted by a trained operator who strictly adheres to the directions provided by the manufacturer. The soil temperature should be above 65°F and the fumigation should be conducted under a gas-tight cover since the methyl bromide is very toxic. Before, during, and after the new lawn is seeded, it may be treated with one quarter of a pound of Siduron per one thousand square feet to control Crabgrass or other annual grasses. The Siduron should be applied only to lawns that are composed of Bluegrass or Fescue. If rain does not fall within three days of the application of the Siduron, the lawn should be watered.

In established lawns, one and one half pounds of Benefin per acre may be applied prior to the emergence of the weeds to control Crabgrass if the lawn is not re-seeded for at least ninety days. Alternatively, one quarter of a pound of DCPA per one thousand square feet of lawn may be applied in early April.

One quarter of a pound of Dalapon per gallon of water that is applied per one thousand square feet of lawn will generally kill the lawn grasses, and may be used for the following two purposes: 1) to control infestations of Bermuda Grass, the spray is applied in July and repeated in three to four weeks, with watering of the area between the applications; the lawn may be re-seeded in three to six weeks depending upon the temperature during this period; 2) infestations of Orchard Grass, Quack Grass, Nimbleweed, or Timothy may be spot-treated with the spray and the area can be re-seeded within three to six weeks.

Orchard Grass, Quack Grass, Nimbleweed, and Timothy may be spot-treated with a coarse spray of full-strength petroleum naptha at the rate of one gallon per one thousand square feet.

Two applications of one half to one pound of SMDC per one hundred square feet of lawn in early August will control Bermuda Grass provided the spray is re-applied in four to six weeks and the directions supplied by the manufacturer are carefully observed. The lawn may be re-seeded in three to four weeks after the last application of the spray.

The following weeds, when they are small, can be controlled by the application of two pounds of Endothall dispersed in fifty to one hundred gallons of water per acre of lawn: Burclover, Henbit, Knotwee, and Pennywort. Children and pets must be kept away from the area that is treated until the lawn is watered.

In the early spring, Chickweed, Clover, Henbit, Knotweed, and Red Sorrel, may be controlled by the application of one half pound of 2,4-D plus one half pound of Silvex low volatile esters per acre of lawn.

When the weeds are actively growing, any one of the following treatments may be made to the lawn:

One and one half pounds of Silves per acre of lawn for the control of Burclover, Chickweed, Ground Ivy, Henbit, Knotweed, Pennywort, Violet, White Clover, and Wood Sorrel; the chemical can be applied as a spray or the weeds may be spot-treated with a sponge attached to a stick or similar implement.

One quarter of a pound of Dicamba per acre for the control of Chickweed, Clover, Henbit, Knotweed, Pursone, and Red Sorrel. The recommended dosage of this herbicide should not be exceeded and, as with all other herbicides, it should not be allowed to drift onto desired trees or shrubs.

One pound of 2,4-D (amine salts or low volatile esters, either in the liquid or granular form) may be used as a spray or spot treatment for the control of Buckhorn Plantain, Curly Dock, and Dandelion; the lawn should not be mowed for one to two days following the use of the 2,4-D.

Knotweed, Wild Garlic, and Wild Onion may be controlled by several successive treatments with one to two pounds of 2,4-D (amine salts or low-volatile esters) per acre; following the directions of the manufacturer. This agent should be applied in the late fall and again in the late winter or early spring.

LAND THAT IS NOT USED FOR CROPS

The control of perennial sedges on drainage ditchbanks may be accomplished by the use of one of the following agents according to the directions that are supplied by the manufacturer: Amitrole, Amitrole T, Dalapon, or 2,4-D. Care should be exercised in the application of these materials to avoid the contamination of either irrigation water or potable water.

Unwanted woody shrubs and small trees may be chemically controlled by the use of one of the following: 2,4-D, OR 2,4,5-T, OR AMS, OR Fenuron TCA, OR picloram, OR Silves, OR 2,3,6-TBA. The continuous adherence to the directions and recommendations that are supplied by the manufacturer is necessary to avoid poisoning either humans or animals, harming desirable species of plants, or causing secondary and long-lived damage through the alteration of the ecology or the watershed. In addition, the adherence to the directions provided by the manufacturer is necessary to the achievement of the best results from the product.

V

Insects

ORDER THYSANURA

The *Thysanura* Order of wingless insects, the Bristletails or Silverfish, contains a few species that are significant household pests. These species rely on bookbindings, paper, wallpaper, other starch-sized articles and, in general, vegetable matter which contains proportionally large amounts of carbohydrates. These swift-running, carrot-shaped, nocturnal pests do not have a metamorphosis. The young (nymphs) grow to adults without a gross change in either appearance or form; only the size of the insect is altered through growth. The adult insect is one-third to one-half inch long with a ten or eleven segmented abdomen. The body is soft and covered with hair or scales, giving the pest a shiny appearance. The anterior end of the insect has mouth parts of the chewing type and two long, many-jointed antennae while the posterior end is distinguished by two or three long, bristled, tail-like appendages.

Members of this order are reproduced by eggs. The nymphs resemble the silvery, greenish-gray, or brownish (occasionally faintly spotted) color of the adult. These pests are found generally throughout the United States and Canada in North America.

The **Silverfish**, or **slicker**, (*Lepisma saccharina* Linne) is silvery or greenish-gray and less than one-half inch long. This common pest prefers temperatures less than 80°F. and is commonly found in damp areas, next to the soil (*e.g.* basement rooms, porches).

The **Firebrat** (*Thermobia domestica* Packard) is mottled with patches of whitish and blackish scales. This insect won't lay its eggs at ordinary room temperatures, preferring 98° to 102°F. and 70 to 80 percent relative humidity. The Firebrat will hatch in one or many weeks, depending upon the temperature. Under favorable conditions this pest will breed rapidly, laying oval, whitish eggs that produce white young who are the minature of the adult. The Firebrat is often called an "asbestos bug" since it will feed on the sizing in asbestos insulation that is found around steam pipes. The Firebrat further finds the conditions in furnace rooms and around heating pipes ideal for sustenance and reproduction.

ORDER COLLEMBOLA

Frequently dark colored, the wingless **Springtail** and **Snow-flea** insects of this order have chewing mouth parts and antennae with few segments. The young insects resemble the form of the adult and develop without metamorphosis. These tiny insects, rarely one-fifth of an inch long, have six abdominal segments, the first bearing the ventral tube. The fourth abdominal segment usually contains a pair of appendages that form a "springing organ" held by the short appendages found on the third abdominal segment. The springing organ is folded up under the body until pressed against the surface on which the insect is resting; when it extends, it throws the insect into the air with a peculiar hopping or leaping motion. Most species are found under decayed leaves, under stones, in crevices of bark, on herbage in damp areas, and, in general, in the vicinity of decaying matter. Representatives of this order occasionally become pests around maplesap buckets, in mushroom beds, and on seedlings in greenhouses. These insects may cause injury to tender plants, such as mushrooms, by feeding on them.

ORDER ORTHOPTERA

Representatives of this order **(Grasshoppers, Crickets, Cockroaches)** are distinguished by two pair of wings. The venated fore-wings are rather thickened and cover the rear set of wings when the insect is at rest. At rest, the hind-wings are folded in plaits resembling a closed fan. A few members of this order have vestigial wings or no wings at all. The nymphs pass through five *instars* in their gradual metamorphasis. The adult insects are generally large with long legs. Many species of the order *Orthoptera* make music, noise, or sounds (depending upon your personal taste) by rubbing the rough parts of the body together. Usually the male representatives of the species are involved in this music-making activity. In temperate latitudes, most species in this order spend the winter months in an egg stage.

Family Locustidae

The **Grasshoppers** and **Locusts** comprise the Family *Locustidae*, formerly called *Acrididae*. These dark colored and variously mottled insects have prominent heads, that are deeper than they are wide, with large eyes. The antennae are shorter than the body. The ovipositor thrusts the eggs into the soil or soft wood. Masses of twenty to over one hundred of such eggs are covered with a sticky, foamy material that solidifies into a protective coating. The grasshoppers and locusts usually produce only one generation per year, which is more than sufficient to supply enough representatives of the species for anyone who considers them a pest.

The **Spur-throated Grasshoppers** of the Subfamily *Cyrtacanthacridinae* are considered among the most harmful of the grasshopper clans. Occasionally rapidly reproducing and thriving, large numbers of these grasshoppers are capable of traveling great distances, causing tremendous damage along their route. These insects only migrate, it is reported, when their numbers reach some high level and cease the migration when their numbers are reduced. Between 1874 and 1877 hordes of these grasshoppers appeared on the plains east of the Rocky Mountains in the United States and migrated to the Mississippi Valley and Texas, destroying crops along the route. It is no small coincidence that one of the plagues with which Egypt was purportedly cursed in Biblical times, was the plague of the Locusts.

From 1862 to 1877, and for a few years after 1930, serious crop losses west of the Mississippi River were cause by the **Rocky Mountain Locust**. This pest is a form of the **Lesser Migratory Grasshopper,** *Melanoplus mexicanuae* (Sauss.). This strong flying pest is approximately one inch long and lays its eggs mainly on cultivated fields.

The **Locusts** are members of the Family *Locustidae,* formerly called *Acrididae.* The female of many species of this family works her ovipositor one to two inches into the soil (often sod), depositing eggs that are bound together with a fluid. The fluid later hardens the clusters together within a protective pod. Twelve or more pods of eggs may be produced in several weeks by an enterprising female during the late summer or fall. The eggs hatch the following spring and the young work their way out of the ground. Young locusts will molt several times before becoming adult.

Family Tettigoniidae

Katydids and similar insects comprise the Family *Tettigoniidae*. Members of this family formerly called *Locustidae*, are not generally considered as pests.

The **Long-horned Grasshoppers** have gracefully formed wings with a delicate color. The very long and slender antennae add an ornamental appearance to the commonly green body of the insect. The long-horned grasshopper is similar in form to the true grasshopper and is often nocturnal. The grasshopper's ovipositor is sword shaped and it lays one generation of eggs per year, singly or in rows on twigs, leaves, stems, or the soil.

Family Gryllidae

Crickets have long tapering antennae and, when compared to other members of the Order Orthoptera, small wings, if wings are present at all. The front wings are bent down over the sides of the body. Many species of these nocturnal insects are "songsters". The needle- or spear-like ovipositor is used to lay the eggs in the ground or in vegetation. Most crickets pass the winter in the egg stage of development.

The black or brown **Field Cricket** combines a relatively long ovipositor with a generally indiscriminate appetite to endear it to pest classifiers. The field cricket dines on vegetable matter, animal matter, (human) food, clothing, and linens.

Slender, one-half to three-quarters of an inch long **Tree Crickets** are often found on trees and shrubs, singing beautifully. The rows of punctures that these insects make in twigs for the purpose of deposition of eggs entitles them to the classification of pests.

Mole Crickets have well developed front legs for burrowing and cutting roots. These pests are larger and stouter than Field Crickets and are covered with fine, velvety brown hairs. Mole Crickets feed on plant roots, other insects, and earthworms to enable them to burrow in the soil for their nests that are usually near a source of water. The Mole Cricket, *Scapteriscus vicinus* Scudder, attacks the roots of crops, especially tobacco, in the southeastern United States and damages seedbeds in this region.

Family Phasmidae

Members of the *Family Phasmidae* (**Walking Sticks** or **Devil's Darning Needle**) have long, stick-like bodies that resemble twigs of the surrounding environment. Only one species, *Diapheromera femorata* Say, of this slow moving family is widespread, although fifteen to twenty species of Walking Sticks are found in the United States, mostly in the Southern areas. These brown or green insects usually have long and slender legs and antennae, but are either wingless or have only a vestige of a true wing structure. Feeding on foliage, these insects may strip the foliage from acres of forests if conditions allow them to become sufficiently abundant. The Walking Sticks are rather casual about the future generations since the mother will drop her eggs from wherever she happens to be at the moment during the autumn. It is reported that this phase of the reproductive process has been audibly compared to falling rain, if sufficient Walking Sticks are similarly occupied. Those eggs which survive this cavalier treatment may hatch the following spring, or the spring after that. Most members of this family produce a sound reported to be musical by those considered competent to judge such matters.

Family Blattidae

The generally dark colored (black, brown, or green) **Roaches, Waterbugs,** or **Cockroaches** belong to the Family *Blattidae*. These broad and flattened insects have their head bent under the body so that the eyes look down and the mouth opens backwards. They possess long, slender, many segmented antennae and long, spiny legs, well adapted for running. The cockroach's wings are usually well developed in the adult insect, with the rear pair folded once when in the resting position. However, some species have either no wings at all or else short wings. Found primarily in warm, humid climates, these pests are active at night and retreat to small cracks of other dark regions during the day. This nocturnal habit may conceal the presence of these pests until they have become abundant. Many species of roaches live in the houses built by man while other species prefer the pleasures of the outdoor life. In northern climates, the only outdoor preferring species are known as **Wood Roaches** and live under and in decaying wood or similar habitats.

Roaches achieve the designation "pests" from their disposition to consume foodstuffs freely, as well as chewing on leather goods or anything containing sufficient starch or paste to spark their appetite. These unwelcome visitors further achieve their classification as a pest by the odor they leave behind wherever they have been.

The roaches' eggs are layed in packets, having been carried in a case that projects from the body of the female for a period of days or even weeks. When hatched, the young will molt several times on their path to adulthood, being active and feeding freely all the while.

Family Mantidae

Praying Mantis, Mule Killer, or **Soothsayer** is a member of the Family *Mantidae* (Mantids). These large insects are distinguished by their long, narrow body and a broad head that freely moves from the thorax. The forepart of the body (prothorax) has two large legs, used for the grasping of the prey. These legs are held in a characteristic position that causes the insect to appear as if it were praying, when in fact it is preying. The Praying Mantis has four other legs which are used in walking. Well developed wings are also used for locomotion, with the rear pair of wings folded under the fore-wings on the back of the abdomen when the wings are not in use. The Praying Mantis is represented by fifteen to twenty species in the United States and is a very valuable insect friend, since it carnivorously dines on flies and other insects generally recognized as pests. Found mainly in warmer regions, some species that are found in the tropics exhibit remarkable development. While some achieve a very large size, others in this beneficial climate resemble local leaves, both in color and form. In the fall, the Praying Mantis lays its eggs in frothy capsules attached to plant stems. These capsules dry quickly to form a protective environment for the eggs and may be more than one inch long.

ORDER DERMAPTERA

Members of this order are commonly called **Earwigs** due to the mistaken belief that they are in the habit of crawling into the ears of sleeping persons.

Most abundant in warm climates, both winged and wingless species of this order exist. If wings are present, they are always tough and the front pair is shorter than the hind pair. At rest, the hind pair are first plaited, then folded twice across their length before being brought under the forewings. To aid this complicated wing-folding procedure, forcep-shaped, leathery *cerci* placed at the tip of the abdomen are accasionally used. Earwigs stay under leaves or in crevices during the day, emerging at night to feed upon plants and decaying vegetable matter.

The **European Earwig,** *Forficula aurecularia* Linn., became established in the United States at Portland, Oregon in or about 1909. In the autumn, the female European Earwig lays white eggs in masses on the ground. The adult remains with the eggs to brood over them and care for the nymphs until their first molting. Wheras the nymphs feed on green plant shoots, the adults eat flower stamens, the bases of petals, fruit, and vegetables. However, the damage that results from the appetite and taste of the earwig is generally not too serious. The chief annoyance that these pests cause in the home results from their crawling over things at night and hiding, or resting, under chair cushions and similar homey acts about the house during the daylight hours.

ORDER EPHEMEROPTERA

The **May Flies, Lake Flies**, or **Shad Flies** belong to this order and are of no great economic importance as pests. These insects have a delicate appearance and may live for one to three years beneath the surface of water in the nymph stage, but last from only a few hours to a few days as adults before perishing. The nymphs subsist on both live and dead aquatic vegetation while breathing through tracheal gills. These nymphs are an important source of nourishment for the surrounding fish population. The metamorphosis from newly formed nymph to adult is quite gradual and may consist of up to twenty molts. The adult fly has four net-like, triangular wings that are folded upright over the back when the insect is at rest, like those of a butterfly. The hind pair of wings are smaller than the front pair. These flies are also recognized by very short antennae and two or three long, slender, many-jointed tails.

Attracted by the lights of towns and cities that are built near lakes and streams, the adult flies quickly become a nuisance as their bodies accumulate on the streets. This nuisance value, rather than any economic destruction, earns this insect the title of pest.

ORDER ODONATA

The large conspicuous insects, such as the **Devils's Darning Needle**, **Dragonfly** and **Snake Feeder**, belong to the Order *Odonata*. These insects have slender bodies, but large heads and compound eyes. The antennae are short and inconspicuous, a contrast to the relative dimensions of the rest of the insect. These insects are carnivorous, often grabbing other insects in flight. Both the chewing type mouth parts and the four wings of these insects aid in the capture and disposal of the prey.

Suborder Zygoptera

The suborder *Zygoptera* consists of the slender bodied **Damselflies**. These flies have bright colored bodies, but are not vigorous in flight in spite of their two pair of equal-sized wings.

Suborder Anisoptera

The (true) **Dragonflies** belong to the suborder *Anisoptera* and often have a shorter and heftier body than the Damselflies, but are similarly brilliantly colored. The Dragonflies are accomplished in flight, catching other insects (flies and mosquitoes are considered a delicacy) in flight as well as mating in flight, an act of virtuosity if not one of deportment. The dragonflies fly in sunny weather, laying their eggs either in water or attached to water plants. The eggs will either hatch in a few days, or wait until the spring if they are laid in the fall. When hatched, the young nymphs remain at the bottom of the water and feed on insects and small, aquatic animals. When grown, the nymphs emerge from the water, molt for a final time, develop wings, and thus become adults.

The Dragonflies are mentioned here in spite of their *not* being pests. These beneficial "bugs" feed on harmful insects but neither bite nor infect man nor beast. These insects should be welcomed, rather than considered for control and destruction.

ORDER PLECOPTERA

The members of this order, commonly known as **Stone Flies**, customarily dwell near water, preferable swiftly moving streams. The adult Stonefly usually has four membranous wings, chewing mouth parts (often vestigial), and long *cerci*. The nymphs often dwell under stones in the water and frequently become fodder for indigenous fish.

ORDER ISOPTERA

Member of the Order *Isoptera*, the **Termites** or **White Ants**, are distinguished by their colonial life, which is similar to the organization of ant communities. A colony consists of four different types of termites: (1) dark-bodied and long winged kings and queens, (2) lightly pigmented and short winged males and females, (3) pale colored and wingless workers, and (4) wingless soldiers. All types have a thin skin and soft body, an abdomen that is wide where it joins with the thorax, and a pair of small *cerci* at the end of the abdomen. The Kings and Queens are capable of producing all castes in the colony and are the only caste with this ability. Their four dark-colored wings are long, narrow, and somewhat opaque with indistinct veins. The similar sized pairs of wings are laid flat over the back of the bearer when the insect is not in sight. A King and Queen pair are necessary for the existence of a colony. The short winged males and females may reproduce their own kind as well as member of the lower castes, but since they are incapable of producing Kings and Queens, they are incapable of initiating a new colony. The soldiers' duties are self-evident, and the workers occupy their time by building nests, caring for eggs, supplying food, and feeding the young nymphs. The common food is primarily wood, frequently dead hardwood. The protozoa in the digestive tract of the termite digest the cellulosic wood and the termite subsists on the products of this primary digestion. Since the wingless castes have an aversion to open air, they usually are confined to their nest or the soil or wood in which they are working. Living predominately in a dark, sheltered environment, the lower castes are thin-skinned, pale, and blind.

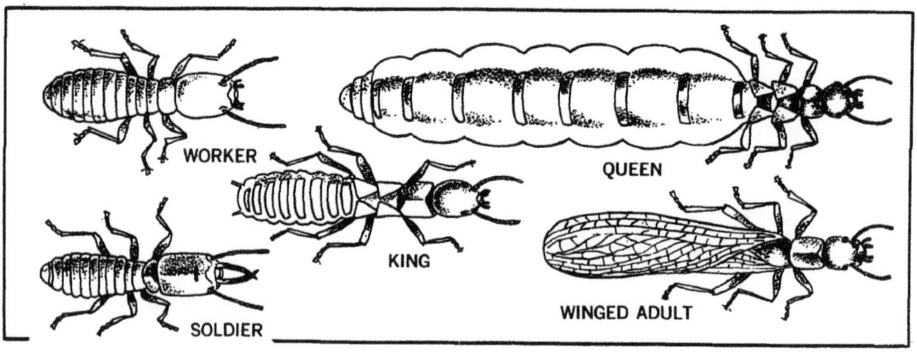

Figure V-1. Termites. (Shell Chemical Company.)

ORDER CORRODENTIA

The Order *Corrodentia*, or Order *Psocoptera*, consists of tiny, soft bodied, inconspicuous insects which eat refuse, molds, and other tasty dishes. All species in this order have long, slender antennae and chewing mouth parts. the winged species usually have four wings. Some species are capable of making a tapping noise, similar to the quiet ticking of a watch, and may be heard in old, infested buildings. The wingless **Booklice** often nibble on paste and paper sizing and are a member of this order.

ORDER MALLPHAGA

The **Chewing Lice**, or **Bird Lice**, that comprise the Order *Mallphaga* feed by biting pieces of the hair, feathers and/or scales from the skin of their host. These lice can be recognized by their large, flattened head, wingless body, and hard exoskeleton. A louse's eggs are attached separately to the hairs or feathers of the host. The newly hatched representatives of this order resemble the adults and molt during maturation, a process completed within a few weeks. Being totally dependent on the host for survival, the demise of the host is soon followed by the death of the lice that he had been supporting. Lice are very selective in their associations, each species of louse infesting either one species of host only, or several closely related species of hosts.

Many species of lice are common to domestic fowl, some preferring the environment of the head of the bird, while others opting for a habitat on the body proper. In either case, the fowl become irritated at the presence of the lice. Severe infestations of lice may disturb hens sufficiently to reduce egg production.

ORDER THYSANOPTERA

The Order *Thysanoptera* consists of over three thousand known species, commonly known as **Thrips.** Many of these species are capable of causing serious damage and some Thrips are reported to be carriers of diseases of plants.

The Thrips are small, slender insects whose mouth parts are used for both rasping and sucking. These insects' four narrow, long wings have few veins and are often fringed with long hairs. A Thrips feed on plant juices by puncturing and/or rasping the surface of the plant to expose the juices prior to sucking them. The white marks or streads on the plant leaves that result from the dried, empty plant cells are an indication of the visitations of Thrips, as are the small black dots of excrement on the stems of the host plant. Thrips may attack blossoms (often blighting them and preventing the setting of the fruit), stems, and leaves (particularly the underside) of a variety of plants. By injuring the stems and leaves of grain and most particularly cereals, the Thrips inhibit the growth of the top of the plant.

Lacking the desire, as well as the means, for major migrations, Thrips generally hide in flowers, crevices of leaf joints, and, for a few species, under the loose bark of trees. This lack of exhibitionism makes the Thrips difficult to reach with insecticides and thus to control in general.

The life cycle of the Thrips begins somewhat differently, depending on the suborder to which the Thrips belongs. Females of the suborder *Terebrantia* possess an ovipositor which is used to saw slits in the epidermis of a plant preparatory to the placement of an egg. Since the females of the suborder *Tubulifera* lack an ovipositor, the eggs are laid on the surface of the food substrate. Finally, some species of Thrips are capable of autogeneration of the new life form *parthenogenetically.* The nymph, somewhat resembling the adult, molts two to four times before leaving its initial food source for a well-protected area. In the enclave, the nymph molts into a prepupal stage in which wing stubs appear and other basic bodily changes are begun. The next molt begins the pupal stage in which the insect ceases feeding temporarily and becomes quiet while continuing to more closely resemble the adult. On completion of the pupal development, a single molt remains for the development of the adult.

In addition to the farmer, the natural enemies oof the Thrips include driving rain, lady beetles, and other insects that fee upon the Thrips.

Four species of Thrips have made themselves sufficiently obnoxious to warrant specific discussion: the **Flower Thrips**, *Frankliniella tritici* (Fitch), the **Onion** and **Tobacco Thrips**, *Thrips tabaci* Lindeman, the **Citrus Thrips**, *Scirtothrips citri* (Moutlon), and the **Pear Thrips** *Taeniothrips inconsequens* (Uzel).

The **Flower Thrips** dine on a variety of plants, including apple, strawberries, wheat, alfalfa, and roses. As a result of this wide range of digestible materials, the Flower Thrips are almost commonly known as **Strawberry Thrips** and **Wheat Thrips**. The attack of Thrips on strawberry blossom can result in the stunting fruit (buttons) to be formed. Attacked leaves often curl and become malformed, frequently turning brown and dying. These adult Thrips are yellowish and, in favorable climates such as the southern United States, the life cycle requires about twelve days and several generations can be produced in each season.

The light yellow, young **Onion Thrips** become brown as they age. Both the nymphs and the (up to one twenty-fifth of an inch) adults pass the northern winters in protected areas, such as in discarded onions that were left in the field or in clover or grass near the ground. In the spring, the insects attack young, growing plants such as peas, beans, clover, and alfalfa, but especially onions. In more temperate climates, such as the southern United States, these Thrips feed during the winter as well as during the summer months. A severe infestation of Onion Thrips will result in little growth of the affected plant, and little growth of the onion bulb if the onion is attacked. Infested onions are characterized by whitened leaves, a result of the removal of the plant liquids by the Thrips, followed by paling and bending downward of the leaves.

The light amber colored **Citrus Thrips** grow to one fiftieth or one twentieth of an inch long. The young of this active, running species appear in March when they gather on the leaves and fruit of the Host plant. Remaining on this initial home until midsummer, the Thrips then migrate to more tender plants, which they leave in August and September to return to the citrus trees to lay their eggs. Whereas the eggs will hatch the following spring, the adults may perish in the cold winter weather. Passing the pupal and prepupal stages in crevices in the oils or among the rubbish at the base of the host tree, the Thrips remain quiet until reaching the adult stage.

The very dark brown **Pear Thrips,** *Taeniothrips inconsequens* (Uzel), emerges from the ground in the spring to enter pear or cherry fruit buds when they open slightly. Working their way into the partly opened bud, the Thrips eat the delicate parts of the bud. The adult lays its eggs on one of the young leaves, young fruit, or stems of its host. Hatching in approximately eight days, the nymphs feed on the leaves and young fruit, forming a type of scab on the affected fruit. The Thrips remain on the tree two to three weeks after which they fall to the ground and burrow five to seven inches into the soil. Remaining two to six months in the soil, the Thrips enter a final stage before becoming an adult. This final molt occurs in late fall or winter, but the adult remains in the ground until early spring at which time the cycle is repeated.

ORDER HOMOPTERA

Insects of the Order *Homoptera* exhibit very diverse characteristics although all members of this order feed on plants. Most of the members of this order have sucking mouth parts equipped with four piercing stylets as well as a beak that rises from the back of the head. The adults of a few species have either vestigial mouth parts or lack them entirely. The winged members of this order usually have four wings. Generally members of the Order *Homoptera* have well developed compound eyes.

Suborder Auchenorhynchi

The large or medium sized insects of the Suborder *Auchenorhynchi* are active and free moving insects possessing small antennae that are tipped with a bristle.

Family Cicadidae

The **Locusts** comprise the Family *Cicadidae* and are large insects, generally growing to one to two inches with stout bodies. The long, generally transparent wings usually have prominent veins, may be variously colored, and have a spread of up to six inches. The egg-laying adult female locust is responsible for the principal damage done by these pests. The adult female makes slits in twigs with her ovipositor, the receptacle for her eggs. The emerging nymphs drop to the ground to seek the same from roots of the plant, their source of nourishment. The adult male may be recognized, in part, by the shrill sounds that he produces with a complicated group of sound organs that are located beneath the abdomen.

Figure V-2. Periodical Cicada-Adults. (Illinois Natural History Survey.)

The **Periodical Cicada**, or **Seventeen Year Locust**, *Magicicada septendecim* (Linn.), has a life cycle of seventeen years in northern climates and thirteen years in southern climates. Stout, black bodied, one-and-one-half inch long adults appear in May or June, particularly in the vicinity of wooded areas. These adults are characterized by red eyes, reddish legs, and orange veins in their wings. The females lay eggs in twigs. These will hatch in six to seven weeks. Upon emerging, the nymphs descend to the ground in order to burrow to the roots of the plant. These nymphs develop in the ground until the spring of their seventeenth year when they crawl up to just beneath the surface of the soil. The larvae wait beneath the surface of the soil until time for the final molt. Crawling out of the ground at night, the larvae molts for the final time and is ready for flight by the morning. This adult stage lasts approximately one month, a short life for such a long adolescence.

Various species of the Family *Cicadidae* are known as the **Dog-Day Cicadas**. These large, blackish insects who often have greenish markings are generally found during July and August.

Family Cercopidae

The usually brown or gray, about one-half inch long **Froghoppers**, or **Spittle Bugs**, are hopping insects that feed on herbaceous plants, shrubs, grasses, and some pine trees. The common name, Spittle Bug, is derived from the manner in which the nymph insures itself of a moist habitat. The numph produces a fluid, and injects air into it before surrounding itself in the resulting froth. *Philaenus leucophthalmus* (Linn.) is a meadow species that seriously stunts clovers while *Aphrophora saratogensis* (Fitch) and *Aphrophora parallela* (Say) are pests to pine trees.

Figure V-3. Periodical Cicada has produced the holes on the twig while depositing her eggs. (Illinois Natural History Survey.)
Figure V-4. Eggs of Periodical Cicada within the tissue of a twig. (Illinois Natural History Survey.)

Family Membracidae

The small, less than one-half inch long, jumping insects of the Family *Membracidae* are commonly known as **Treehoppers**. If observed closely, the peculiarly large and unusually developed *pronotum* gives these insects a grotesque appearance. Otherwise, these insects somewhat resemble buds on the tree if observed in a casual manner. Generally the damage to plant life caused by the Treehoppers occurs when the insect lays its eggs. For instance, the **Buffalo Treehopper**, *Ceresa bubalus* (Fabricius), generally makes slits in the bark and the twigs of apple and other trees, causing some injury to the tree.

Family Cicadellidae or Family Jassidae

The **Leafhoppers** of the Family *Cicadellidae*, or *Family Jassidae*, are small, active insects with an appearance similar to that of the Froghoppers. The Leafhoppers may be serious pests since their feeding may interfere with the normal growth of many plants and they may be vectors of organisms that are capable of causing diseases in plants.

Family Fulgoridae

The **Planthoppers** of the Family *Fulgoridae* feed on a variety of plant types, but generally only a few plants of any one type. These pests are of minor economic importance and no further detail is necessary.

Suborder Sternorhynchi

Family Psyllidae or Family Chermidae

The **Jumping Plant Lice** are small insects, generally reaching two to five millimeters. The winged adults have strong legs for jumping. Feeding on the juices of their plant hosts, the Jumping Plant Lice may resemble tiny Cicadas.

The **Tomato**, or **Potato**, **Psyllid**, *Paratrioza cockerelli* (Sulc), is the carrier of the virus that causes **Psyllid Yellows** in Tomatoes, Potatoes, Peppers, and Eggplants.

The **Cottony Alder Psyllid**, *Psyllia floccosa* (Patch), is considered a pest based upon the destructive abilities of the nymphs. These nymphs commonly feed on the twigs of the alder during the early summer, producing a wax that causes groups of them to resemble cotton.

The adult **Pear Psylla**, *Psylla pyricola* Forster, has a black body with reddish markings. These adults overwinter in sheltered areas of a tree, emerging in the spring to lay their eggs on the twigs of the tree. The oval, broad, and flat nymphs that hatch from these eggs suck the sap from the leaves and stems of the tree. This insect is particularly a pest of pear trees and was introduced into the United States from Europe.

Family Aphididae

The small, soft-bodied, often pear-shaped insects of the Family *Aphididae* are commonly known as **Aphids** or **Plant Lice**. Although some species are brown, pinkish, black, or yellowish, these generally green pests suck the sap of a great variety of plants. The honeydew that the Aphid excretes is an attractive media for the growth of fungus. In addition, the sap that is removed weakens the host plant, and the Aphid can be a carrier of various plant diseases. Aphids enjoy an international distribution, being found throughout the world and few plants have not been chosen as their host.

The life history of an Aphid is a trifle complicated. Usually the Aphid survives the the winter in the form of a small, black, oval, fertilized egg. In the spring, female nymphs hatch from such eggs to mature into wingless adults. These "stem mothers" give birth to wingless female nymphs *ovoviviparously* (already hatched from the egg) without the benefit of mating. Many nymphs are produced in this manner during the ensuing few weeks, resulting in a number of generations being produced within a span of a few weeks. One of these generations will contain some winged females, "spring migrants", that fly to a new type of plant, generally an annual, that is used as a summer host. The spring migrants, in turn, initiate further generations until a generation is produced in the late summer containing winged females again, the "fall migrants", as well as winged males. The fall migrants give birth to wingless forms of female nymphs, often back on the perennial plant from which the species began the cycle in the spring. These wingless nymphs must mate if they are to produce further generations of the Aphid. After successful mating, the female nymphs lay a few large fertilized eggs in a sheltered spot on the plant. Later they fatally collapse to end their function in the life cycle, while providing for the continuance of the species during the next year.

Subfamily Eriosomatinae. The **Woolly** and **Gall-Making Aphids** of the Subfamily *Eriosomatinae* have abundantly productive wax glands as well as an ovipositor and related system that produces a single egg. In the sexual forms of these Aphids, no mouth parts are present. Most species of this subfamily overwinter on either a tree or a shrub, while living on a herbaceous plant during the summer months. The common name of this subfamily is derived from the galls produced by many species on their winter host.

The **Woolly Apple Aphid,** *Eriosoma laigerum* (Hausmann), is distinguished by the characteristic woolly wax on its purplish body. Overwintering on the elm, the spring migrants pass to the apple and related trees. In the fall while some individual Aphids migrate to the elm for sexual reproduction and a winter home, others migrate to the roots of the apple tree and form galls in which several generations may spend the winter or even several years. Another member of this subfamily worthy of slight note is the *Pemphigus populi-transversus* Riley which forms galls on the stems of the leaves of the poplar tree.

Subfamily Aphidinae. This subfamily of **Typical Aphids** has less productive wax glands when compared to the Woolly or Gall-Making Aphid. The sexual female has functional mouth parts, and the ovipositing female is able to produce more than one egg.

The **Corn Root Aphid,** *Anuraphis maidi-radicis* (Forbes), overwinters in the egg stage in the nest of field ants, generally of the genus *Lasius*. The ants transport the young Aphids in the spring to the roots of weeds such as the Smartweed. The Aphids feed on the weeds until corn roots are ready for the menu. The ants tunnel to the roots of the corn to deposit these greenish-blue Aphids to their new lining room for the remainder of the season. In the autumn, the ants gather the eggs of the Aphid to return them to the next for the winter. The ants perform these shepherding duties in exchange for the excreted honeydew that is provided by the Aphids.

The **Rosy Apple Aphid,** *Anuraphis roseus* Baker, is greenish with a purplish tinge. Overwintering on apple and related trees, this Aphid moves to the narrow-leaved plantain for its alternate, summer host.

Migrating in early summer to grasses, including grains such as oats and wheat, the **Apple Grain Aphid,** *Rhopalosiphum prunifoliae* (Fitch), spends the winter on its primary host, the apple tree.

The **Apple Aphid,** *Aphis pomi* DeGeer, is a green, common pest of apples, and is loyal to its host extending its tender mercies to no other plant.

The **Cotton**, or **Mellon, Aphid,** *Aphis gossypii* Glover, is a very serious pest of cotton plants. In addition to sucking the sap from the plant and thus weakening it, this pest excretes a honeydew that is used as a nutrient for a sooty mold. The mold, in turn, interferes with the photosynthesis of the plant in addition to staining the cotton lint. Severely infested melon plants are characterized by downward curling leaves, wilting, and occasionally death of the host. This Aphid will also feed on citrus fruits, okra; asparagus, eggplant, strawberry, spinach, beets, and beans. Various weeds are also used by this Aphid as hosts, including shepherds' purse, pigweed, peppergrass, and dock.

The long-legged, pale green **Pea Aphid,** *Macrosiphum pisi* (Kaltenbach) is a serious pest of peas as well as alfalfa and other legumes. Overwintering as an egg on perennial legumes, this Aphid migrates to annuals during the summer.

Figure V-5. Left: Cloudy-winged whiteflies. Right: Sooty mold fungus. (James Brogdon, Florida Cooperative Extension Service.)

Family Aleyrodida or Family Aleurodidae

The minute **Whiteflies** of the Family *Aleyrodida*, also known as the Family *Aleurodidae*, resemble moths. The wings of the adult Whitefly are covered with a white dust or powdery substance. These insects are abundant in subtropical and tropical regions throughout the world.

The sooty fungus that grows on the excreted honeydew establishes the **Citrus Blackfly,** *Aleurocanthus woglumi* (Ashby), as a pest of citrus trees where it is commonly found in the West Indies and Mexico.

The **Common Citrus Whitefly,** *Dialeurodes citri* (Ashmead), is a common pest of the citrus trees in Florida. The feeding practices of this pest reduces the growth of the tree as well as the yield of the fruit. This Whitefly also produces a honeydew on which a sooty fungus becomes established.

Family Coccidae

Scale Insects and **Mealybugs** are the common names of members of the Family *Coccidae*. Whereas the females of this family are without either legs or wings, the males have one pair of wings but lack mouthparts. In most species the first instar nymphs are commonly referred to as *crawlers* as a result of their general mobility, having both legs and antennae. During the first molt, the legs and antennae are often discarded after which activity is curtailed, needless to say. At this point in the life cycle, the insect secretes a waxy, scale-like coating. The females of the species remain within this covering to produce either their young or their eggs. During the final instar of the male, commonly called a pupa, the wings are developed for later use.

The **Cottony Cushion Scale,** *Icerya purchasi* Maskell, is a serious pest of citrus trees in the western United States. Having arrived in California in 1868, this immigrant attacked orange trees with such vigor that hundreds of thousands of trees were killed and the citrus industry of that state was threatened with extinction. An entomologist, sent to Australia, returned with a natural enemy of this scale, the **Lady Beetle**, or Vedalia *Rodolia cardinalis* Muls. From the basis of one hundred-forty specimens of this natural enemy of the scale, a sufficient population was developed within one and one half years to check the epidemic of the Cottony Cushion Scale. The female Cottony Cushion Scale is a large scale insect possessing a brown, red, or yellowish body. At the rear of this body is a large cottony mass, fluted on the top, in which her four hundred to one thousands eggs are laid.

Subfamily Pseudococcinae or *Eriococcinae.* Members of the Subfamily *Pseudococcinae*, also known as the Subfamily *Eriococcinae*, are commonly, if not particularly aestheticly, known as **Mealybugs**. The segmented, oval female covers herself with a waxy secretion that has the appearance of meal from which the common name of the insect was derived. In most species the females deposit their eggs in a large cotton-like mass that is secreted by their bodies. Although the adult males have two wings, the females are wingless. Round Mealybugs suck sap or juices of the leaves or the fruit of the host plant.

A list of the examples of some members of this family includes *Pseudococcus citri* (Risso) and *Pseudococcus fragilis* Brain, that are pests of greenhouse plants as well as citrus trees and the **Grape Mealybug**, *Pseudococcus maritimus* (Ehrhorn).

The **Black Scale,** *Saisetia oleae* Bernard, a tropical species, is a pest of olive, apricot, and general types of citrus trees, as well as other plants living in the warm parts of the United States. The black, nearly hemispherical, one-eighth to one-quarter inch long female is characterized by the "H" design of the ridges that are present on her back. The young of the species attack the leaves and twigs of the host, remove sap from the host, and leave a honeydew that supports a sooty fungus that interferes with the growth of the host as well as spoiling the fruit.

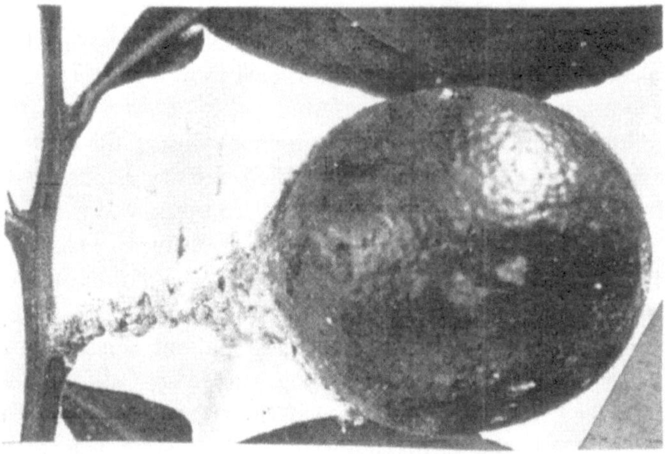

Figure V-6. Mealybugs on fruit. (Dr. James E. Brogdon.)

The **Cottony Maple Scale,** *Pulvinaria Vitis* (Linn.), lays its eggs in a cottony mass that protrudes from the end of the scale.

Subfamily Diaspidinae. Members of the Subfamily *Diaspidinae* are commonly known as **Armored Scales** because the insect lives under a scale that is generally not attached to it. The soft-bodied, small, legless females lack eyes and live under a scale covering that is usually unattached to the insect. The winged, antennaed, and legged males have a scale covering that is usually smaller and longer than that of the female. In the first instar stage, the young insects, referred to as *crawlers*, are both active and able to exist for a few days without food. These insects are able to travel some distance from the location of their birth as a consequence of these two attributes. On deciding upon a site for a future residence, the crawler inserts its mouthpiece into the host that has been selected and becomes inactive as it matures. These insects then suck the sap from their tree or shrub host, earning the description as pest.

The **San Jose Scale,** *Aspidiotus perniciosus* Comstock, first was found in the United States (circa 1870) in California. Since that time, this pest has spread throughout most of the United States. The San Jose Scale is a serious pest in fruit orchards although it also attacks currants, ornamental cherries and plums, dogwoods, roses, lindens, poplars, and lilacs. The grayish-brown adult female is approximately the size of the head of a pin, resembling a flattened nipple. The male has a similar appearance, but its shape is more oval than the round shape of the female. Although this insect may overwinter in nearly any stage of its development, not all overwintering members of the species survive the rigors of the cold. Surviving insects resume feeding in the spring after which the males emerge in a short while to mate with the females. In approximately four weeks, the young are born to repeat the life cycle of the species. A female may produce four hundred yellow colored young during her short lifetime. These young crawl until they decide to settle and feed. Then they begin the production of their scale covering while proceeding through the necessary number of molts on their static journey to adulthood. Whereas northern insects produce three generations per year, those located in the more hospitable southern regions may produce four more generations per year.

Figure V-7. Cottony Maple Scale. (Illinois Natural History Survey.)

The common name of the **Oyster-Shell Scale,** *Lepidosaphes ulmi* (Linn.), is derived form the shape and color of the brownish, adult, one-eighth inch long female. Whereas the female feeds on any part of a tree or shrub that is covered with bark, the male scales may even infest the leaves of the plant. A list of plants that commonly are subject to the attack of the Oyster-Shell Scale includes the pear, apple, and ash as well as other similar trees. The females of this species lay eggs that overwinter under the shell of the mother. These eggs hatch in May or June to begin the new generation of scales.

The **Pine Needle Scale,** *Phenacaspis pinifoliae* (Fitch) primarily attacks pine trees, although it will occasionally reduce itself to accepting the hospitality of spruces, hemlocks, and cedars. An abundant population of this pest may cause infected branches to appear as if they were sprinkled with white paint. The male is white; the female is off-white and about three times as long as the male. The purplish crawlers of this species usually appear in both the middle of May and early in September in the northern states of the United States.

Figure V-8. San Jose Scale. (Illinois Natural History Survey.)

Figure V-9. Oystershell Scale. (Illinois Natural History Survey.)

The citrus groves in California and Texas have sustained severe damage from the attack of the **California Red Scale,** *Aonidiella aurantii* (Maskell). This red bodied insect resides under a transparent circular scale and has a life history similar to that of the San Jose Scale.

The **Purple Scale,** *Lepidosahes beckii* (Newman), is an important pest in the citrus groves in Florida and other gulf states as well as in areas of southern, coastal California. This scale also attacks avocado, fig, eucalyptus, and olive as well as other plants. The results of an infestation of this scale include yellowed leaves, stunted and badly flavored fruit, and a potential entrance for fungal infections. The life cycle of the Purple Scale begins and ends, for two separate generations, when the female lays two eggs under her scale before she dies. After the eggs hatch, the very pale crawlers roam about the tree, seeking a semi-shaded portion on which to settle and feed. As the scale grows, she becomes thicker and purplish-brown or reddish, molting twice at three to four week intervals. The male, however, matures in approximately two months during which time four molts are accomplished. When mature, the male, using his new wings, flies to seek his mate and initiates the procreation of the new generation.

ORDER HEMIPTERA

Members of this order are known, familiarly, as **"true bugs"** and generally have flat bodies that are used, among other things, as a resting position for the wings that most species possess. These bugs have piercing and sucking mouth parts to gain the sustenance necessary for their existence. The life cycle of the members of the Order *Hemiptera* is an incomplete metamorphosis during which the nymphs, resembling small adults, hatch from eggs and reach adulthood through three successive molts.

Suborder Cryptocerata

Members of the Suborder *Cryptocerate* are commonly known as **Short Horned Bugs** and **Aquatic Predaceous Bugs.**

Family Corixidae

The **Water Boatmen** are oval, rather flat, and frequently gray bugs that have back legs which resemble oars. These insects are frequently found in ponds and along the edge of quiet waters.

Family Nepidae

The **Water Scorpions** have peculiar front legs that are adapted for capturing aquatic animals. These slow moving bugs are outfitted with a breathing tube for use in their aquatic environment.

Family Belostomidae

The **Giant Water Bugs** that comprise the Family *Belostomidae* are brown, flat, and oval and grow to two inches or more in length. When they leave the water and fly, these bugs may be attracted to electric lights, or any other type of bright illumination. These insects often will feed upon other insects and very small animals.

Suborder Gymnocerta

The Suborder *Gymnocerta* consists of the insects that are commonly called the **Long Horned Bugs.**

Family Reduviidae

Several species of the Family *Reduviidae*, also known as the **Assassin Bugs**, are very undesirable. The **China Bed Bug,** *Triatoma protracta* (Uhler), the **Masked Hunter,** *Reduvius personatus* (Linne), and the **Mexican Bed Bug,**

Triatoma sanguisuga (Leconte), are all capable of attacking man by biting him. Besides inflicting pain upon the human, these bugs may be vectors of serious human diseases.

Family Cimicidae

The **Bed Bugs** are oval, quite flat bugs that are able to crawl into small crevices for the purpose of hiding during the day as a result of their configuration. These pests are active during darkness and inactive during the daylight hours. The common Bed Bug that bites humans belongs to the species *Cimex lectularius* Linn. Bed Bugs may live several months in the adult form, and, therefore, are not a brief annoyance.

Family Pentatomidae

Approximately five thousand species comprise the Family *Pentatomidae*, commonly known as the **Stink Bug**. The common name of this family is derived from the disagreeable odor that accompanies the presence of this pest. Many of these Stink Bugs have a shield shape.

The **Harlequin Terrapin**, or **Callicoback Bug**, *Murgantia histrionica* (Hanh), is gradually migrating northward from its formerly southern confinement, feeding on cruciferous plants as it moves its locations of infestation. The one-half inch long, dark blue or black adult has bright orange or red markings. Laid on the underside of leaves, the eggs hatch in three to eleven days, releasing the nymphs. These nymphs spend thirty to sixty days sucking sap from their plant host in the process of reaching the adult stage.

A second species of the bothersome Stink Bug, the **Green Stink Bug**, *Acrosternum hilare* (Say), is about five-eights of an inch long and feeds on legumes and cotton, sucking the sap from the cotton buds and balls of the latter.

Family Miridae

The **Leaf** or **Plant Bugs** that comprise the Family *Mirridae* include several species that are worthy of mention as pests.

The **Tarnished Plant Bug**, *Lygus oblineatus* (Say), attacks many garden plants and fruit trees, stunting their growth. This pest has also been accused of carrying the bacteria that causes the *Fire Blight* of pears. Overwintering as an adult, in the spring, the female places her eggs in the leaves, stems, or blossoms of the selected host plant. The nymphs hatch and begin a month of feeding on the sap of the plant host in the process of maturation to the adult stage. The one-quarter inch long, oval, brownish adults are marked with yellow and dark brown or black. These adults proceed to attack the buds of such fruit trees as pear, peach, and apple.

The **Dark Apple Redbug**, *Heterocordylus malinus* Reuter, has a piercing, sucking mouth that feeds on the sap of the leaf or the juice of the fruit of the host. Infested apples become lumpy and deformed with hardened areas in the location of insertion of the beak of the insect.

Family Phymatidae

The **Ambush Bugs** of the Family *Phymatidae* are usually no more than one-half inch long, but may prey upon insects as large as a bumble bee. The Ambush Bugs love to hide on goldenrod while waiting for their dinner to pass by.

Family Lygaeidae

The most notorious pest of the Family *Lygaeidae* is the **Cinch Bug**, *Blissus leucopterus* (Say). Reported to be a native of Tropical America, the Cinch Bug is a pest in corn and sorghum as well as many cereals and several types of grass. The dark gray or black-bodied adult may have white, black spotted, full sized wings (such as the varieties found in the Mississippi Valley, Atlantic states, and some Midwestern States of the United States), or a short-winged variety, found in the latter two areas. The short-winged variety overwinters on grassland, migrating on foot in the spring to the appropriate host crop. Able to fly, the long-winged variety prefers the aerial method of locomotion in moving to the about-to-be-infested area. Three to four weeks are spent depositing its eggs on the base or root area of the host plant. Within one to two weeks, the eggs hatch to release nymphs that suck the sap of the host during their development period. If the nymphs are interrupted during this period, such as by the harvest of their host, they will migrate, on foot so to speak, to seek millet, corn, or other new grass-type crops to devastate. On attaining the adult stage, the insect will begin laying eggs to begin the repetition of the life cycle.

Family Pyrrhocoridae

The **Cotton Stainer**, *Dysdercus suturellus* Herrich-Schaeffer, is the most significant member of the Family *Pyrrhocoridae*, gaining its classification by feeding on partially developed cotton bolls. The Cotton Stainer breeds on the Spanish cockleburr or the hibiscus that grow along the edges of the cotton fields.

Family Coreidae

The **Squash Bug**, *Anasa Tristis* (DeGeer), dines on pumpkin, melon, and cucumber in addition to the squash from which its name was derived. Overwintering under refuse, the three-quarter inch long, dark brown adult is marked with light brown or gray. The adult emerges in the spring, feeding during the period when it lays its brown eggs, on the underside of the leaves of the host. Five molts are required for the production of the one generation of adults per year.

Family Tingidae

The small **Lace Bugs** of the Family *Tingidae*, usually one eighth of an inch long, have sculptured markings on the upper surface of their bodies. These markings give the Lace Bugs the appearance of being constructed of lace. This pretty facade is evident only when the bug is in the adult stage; the nymphs are rather spiney. Although the nymphs of the Lace Bug are often black, the adults are usually light colored. The adult Lace Bugs often feed upon the leaves of shrubs and trees.

ORDER ANOPLURA

The small, one twenty-fifth to one quarter of an inch long, **Sucking Lice** of the Order *Anoplura* are wingless parasites of mammals. These lice feed by sucking the blood of the host from the exterior of the mammal. The head of the louse is narrower than the thorax, and contains an interesting mouth structure. The mouth contains a flexible proboscus, which, when extended, turns "inside out" and exposes hook-like structures. These "hooks" attach to the skin of the host. After attachment, lances, lodged in the head of the louse, penetrate the skin of host in the area encircled by the proboscus. The sucking lice have either vestigial eyes or none at all. The legs are attached to the thorax and are tipped with claws that are used to grasp the host.

Family Haematopinidae

The Family *Haematopinidae* consists of **Sucking Lice** that are pests of domestic animals. The contrcl of these pests is important since the lice that infest an animal can become so numerous that the physical condition of the host is weakened. In effecting any control, it is important to consider that the control agent may be either absorbed through the skin of the animal or licked off by him. Therefore, the toxicity of the agent to the animal must be carefully considered.

Infesting beef cattle and mature dairy cattle, the **Short-Nosed Cattle Louse,** *Haematopinus eurysternus* (Nitzsch), is the largest species of cattle louse. These dark-gray lice have a short, rounded head. Light or moderate infestations of these lice may be found in patches on the head, neck, or inner side of the leg of the host.

The **Long-Nosed Cattle Louse,** *Linognathus vituli* (Linne), is a bluish, long beaked pest, often found on young stock and dairy cattle. Both this louse and the **Hog Louse,** *Haematopinus adventicius* Neum., which is confined to swine, succumb to the same sort of pest control. The Hog Louse resembles the Short-Nosed Cattle Louse, but is much larger, often reaching one-quarter inch long. This brownish-gray louse has black markings on its edges and frequents either the area of the head of the hog or its legs.

Family Pediculidae

The Family *Pediculidae* consists of three species of Lice that are found on the human body: the **Head Louse**, the **Body Louse**, and the **Crab Louse.**

The **Head Louse,** *Pediculus humanus humanus* Linne, is found amongst the hairs of the human head. The grayish, one-eighth inch long, adult requires one to four weeks to develop from the egg and will live for approximately another thirty days. The adult attaches its eggs to the hairs of the host. All stages of development of the Head Louse subsist by sucking the blood of the host.

The **Body Louse,** *Pediculus humanus corporis* DeGeer, is very similar to the Head Louse, except that the former often lays its eggs on clothing. Since neither louse can exist long from its host, especially at cool temperatures, both types of lice require either direct personal contact or the use of the clothing, bedding, personal items, etc., of an infected host shortly after his use to maintain their existence. Therefore, infestations of these lice are easily controlled where ordinary conditions of cleanliness prevail. Conversely, infestations of these pests can be found in such places as military camps, where World War I "doughboys" called them "cooties". These lice have been reported as carriers of typhus fever and relapsing fever, negative desirability further indicating the necessity for their control.

The **Crab Louse,** *Phthirus pubis* (Linne), is shorter, smaller, and broader than the previous two lice of the order *Anoplura.* Whereas the forelegs of the Crab Louse are slender, the hind pairs are stout and each hind leg contains a powerful claw that is utilized in gripping the hair of the host. This louse, as would be expected from the foregoing, is found on the hairy parts of the human body, excepting the head. The common name of this louse is derived from the sideways, or crab-like, movement that is used to traverse from hair to hair.

ORDER COLEOPTERA

The Order *Coleoptera,* whose members are commonly known as **Beetles**, is the largest order of insects since these Beetles account for over forty percent of the known species of insects. The members of this family have diverse habits and habitats and have been studied extensively.

A very characteristic type of wing is common to members of this order. The *elytra* or front wings, of the mature insect, usually thickened by *chitin*, are used as shields for the hind wings and the back of the abdomen when the insect is at rest. The generally delicate hind wings are used for propulsion during flying.

Both the mature beetle and its larvae possess mouth parts of the chewing type. The maturation of the larvae takes members of this order through a complete metamorphosis. The larvae of this order are commonly known as grubs and/or borers. Most species of the beetles produce one generation per year, although species that produce several generations per year, and those that require more than a single year to complete a life cycle are also present. Generally beetles overwinter as adults, pupae, or partially grown larvae. Very few overwinter in the egg stage.

Suborder Adelphaga

The Suborder *Adelphaga* consists basically of the **Terrestrial Predaceous Beetles** and the **Aquatic Predaceous Beetles**, for the purpose of this discussion. In each of these two groups of beetles, two families are worthy of note.

Family Cicindelidae

The **Tiger Beetles** of the Family *Cicindelidae*, **Terrestrial Predaceous Beetles**, are active and usually either irridescent or metallic in color. Often found on sandy, sunny beaches, these beetles feed on small insects and may inflict painful bites on the person who disturbs them.

Family Carabidae

This family of **Terrestrial Beetles** contains hundreds of species of **Ground Beetles**, most of whom are shiny, dark, and rather flattened in shape. These beetles are generally found under logs, leaves, rocks, or debris and will run quickly, rather than fly, if disturbed. Usually hiding during the daylight hours, these insects will emerge during the night to feed on other insects. Whereas some species dine on snails, others, of the genus *Calosoma,* feed on caterpillars. As an example of the beneficial side of the life of this order, *Calosoma sycophanta* (Linn.) is a bright green beetle that was brought to the United States from Europe to help control the **Gypsy Moth.**

Family Dytiscidae

The **Predaceous Diving Beetles** are members of the **Aquatic Predaceous Beetle** group and thus are aquatic, living on ponds or slow running streams. The usually greenish, brownish, or rather black adults vary in the length of their oval bodies. The smooth, very hard bodies are equipped with hairy, flat hind legs that are used to provide aquatic locomotion. These beetles are able to store air in a special chamber under their elytra enabling them to remain submerged for a period of time. Both the larvae and the adults of this family feed upon many different small aquatic forms of life.

Family Gyrinidae

The oval, black beetles of the Family *Gyrinidae*, commonly known as **Whirligig Beetles**, are rapid swimmers and may be seen flitting rapidly over the surface of still water. Insects that fall into the water are the primary foodstuffs of the adult Whirligig Beetles, while the larvae subsist on small aquatic forms of life. The Whirligig Beetles deposit their eggs on the underside of the leaves of aquatic plants and pupate either on these plants or in cells of mud at the shore.

Suborder Polyphaga

Family Staphylinidae

The **Rove Beetles** of the Family *Staphylinidae* generally live in or around decaying animal or vegetable matter, such as dung and carrion. These beetles often prey upon other insects that may be found in such an environment.

Family Silphidae

A majority of the members of the Family *Silphidae*, commonly known as **Carrion** or **Burying Beetles**, are rather large, growing to a length of one-half to one-and-one-half inches. The two basic types of insects that comprise this family are the *Necrophorus* and the *Silpha*. The large, stout *Necrophorus* has a rather cylindrical body and red-marked elytra that are often shorter than its abdomen. The *Silpha* has a broad, flattish body, usually less than one inch in length. These insects' elytra are generally black. Many of both the adult and larval members of the Family *Silphidae* feed on the corpses of animals.

Suborder Clavicornia

Family Hydrophilidae

The generally oval, black members of the Family *Hydrophilidae*, commonly known as the **Water Scavenger Beetles**, swim on the surface of quiescent water. Upon submerging, these beetles take their air with them as a silvery film on the under side of the body. Although the adult beetles are scavengers, the larvae feed on a variety of aquatic animals.

Family Cucujidae

The very flat beetles of the Family *Cucujidae*, the **Flat Bark Beetles**, are either brownish, reddish, or even yellow in color. The flat larvae have a grub-like appearance. Found under the bark of freshly cut logs from trees such as the ash, birch, elm, maple, or poplar, these beetles prey upon the mites and other small insects that are found under the bark.

Family Coccinellidae

Members of the Family *Coccinellidae*, widely known by their common name of either **Lady Beetles, Ladybird Beetles**, or **Ladybugs**, are small, convex, oval insects who are often brightly colored. These insects generally overwinter as adults. With the exception of two species of *Epilachna*, these insects are generally beneficial since both the beetles and their larvae prey upon other insects, particularly Aphids and Scales.

The **Ladybird Beetle**, *Rodolia cardinalis* (Mulsant), was imported from Australia to the United States to aid in the control of the **Cottony Cushion Scale**, a pest of the California citrus trees.

The **Mexican Bean Beetle**, *Epilachna varivestis* (Mulsant) existed for many years in the semi-arid regions of the United States and Mexico. Since arriving in Alabama circa 1920, this beetle has become a significant pest to bean farmers in many states because of its proclivity for ingesting the underside of leaves of plants until the leaves appear skeletonized. This yellowish pest may be recognized by the eight spots that can be found on each elytron. Generally the infestation of the Mexican Bean Beetle is limited to areas east of the Mississippi River within the United States. Overwintering in sheltered places, the adult bean beetle deposits its yellow-orange eggs in large clusters on the undersides of the leaves. The emerging orange, fuzzy grubs begin their existence by dining on the leaves of the plant. Whereas one generation of beetles per year is produced in the northern states, three to four generations per year are not uncommon in warmer climates. Near the end of the summer, the last generation of beetles leaves the bean fields to seek shelter for the winter, preferably in wooded areas.

The pale yellow-orange **Squash Beetle,** *Epilachna borealis* Fabricius, has twelve spots arranged in two rows on the elytra in addition to the large black dot placed near the tip of the elytra. Although the oval, yellow larvae are spiny, they have few other interesting characteristics.

Family Dermestidae

The Family *Dermestidae*, commonly known as **Skin Beetles**, has some destructive members who eat fur, feathers, hides, cloth products, food items, and carrion. The larvae of the species are frequently the destructive culprits in the life history of the pest. The small, convex, oval, adult *dermestids* may be either dark and dull or decorated with a colorful pattern over their either hairy or scale-covered bodies. The larvae of this family are often long-haired and brown.

The **Larder Beetle,** *Dermestes lardarius* Linn., is also known as the **Pantry Pest** and attacks food of a fatty nature, such as ham and cheese. The dark brown or black adult is slightly more than one quarter of an inch long and has a tannish or yellowish band across the elytra near their base. The brown larvae possess many long black hairs.

An attractive, small, black and white beetle with irregular red markings on its back is the **Buffalo Carpet Beetle** or **Buffalo Moth,** *Anthrenus scrophulariae* (Linn.). The adult Buffalo Carpet Beetle feeds on the pollen of flowers, especially flowers of the Spiraea. The brownish, shiny larvae are covered with tufts of hair and attack animal products, such as wool, leather, silk, and feathers.

The grayish black **Black Carpet Beetle,** *Attagenus piceus* (Olivier) is similar in habit and activities to those of the Buffalo Carpet Beetle. The larvae of the Black Carpet Beetle, however, are golden or reddish brown with a tuft of hair at the tip of the body.

Suborder Serricornia

Members of the Suborder *Serricornia* are commonly known as **Saw-Horned Beetles**. The only family of this suborder worthy of further mention is the Family *Cleridae.*

Family Cleridae

The **Checkered Beetles** of the Family *Cleridae* are covered with a dense, pubescence and are often marked with bright colors. Most of these beetles are predaceous both in the larval and adult stages. Many Checkered Beetles prey upon the Bark Beetles or other wood-boring insects and thus perform a beneficial service to both plant and man. A few of these beetles are pollen eaters, while others prey upon the nests of bees and wasps, a habit that is not utterly without redeeming social value.

ORDER COLIOPTERA

The **Dung Beetles** of the Order *Colioptera* are represented by a number of families and sub-families. Generally these scavengers are beneficial to mankind, but no particular species is of great economic significance.

Family Scarabaeidae

The **June Bugs**, or **May Beetles**, are common names for various species of the genus *Phyllophaga* (or *Lachnosterna*) in the Family *Scarabaeidae*. Adult June Bugs are stoutish, one half to one inch long beetles. These brown and lustrous insects are attracted to lights in the early summer. The adults are night feeders, preying upon flowers and leaves. The larvae are common white grubs that feed on the plant roots beneath the soil. These grubs are characterized by their stout hind end, brownish head, and their usual posture of being curled into a rather circular configuration. The life cycle of this genus is lengthy, usually requiring three years for its completion.

The **Rose Chafer**, *Macrodactylus subspinosus* (Fabricius), is a long-legged, buff-colored beetle. This insect thrives on the foliage and flowers of roses, grapes, and a host of other plants. This insect is a particular pest in areas that have sandy soils.

Figure V-10. Japanese Beetle. (Illinois Natural History Survey.)

The **Japanese Beetle**, *Popillia japonica* Newman, is a noteworthy pest belonging to the Subfamily *Rutelinae*. Nursery stock imported to the United States from Japan, circa 1916, was reported to have carried this beetle to the United States. This beetle is now found throughout the United States in areas east of the Mississippi River and in Ontario, Canada. The Japanese Beetle attacks shrubs, lawns, fruit trees, ornamental trees, garden plants, and, needless to say, is a serious pest. The pretty adult beetle is approximately one half inch long. The head and front part of its body are green, the elytra are rather brown with green tinted fringe areas, and two white spots are visible at the rear end of the beetle. This beetle overwinters in the soil in the larval stage.

The **Grape Pelidnota,** *Pelidnota punctata* (Linn.), is a yellowish beetle, more than one inch long, and has three black dots on each elytron.

Family Cerambycidae

The **Roundheaded Apple Tree Borer,** *Saperda candida* Fabricius, is a white beetle that has three large brown bands on its back. The larvae of this beetle bore into trees, especially apple trees, as might be deduced from its name.

Suborder Lamellicornia

The members of the Suborder *Lamellicornia* are often referred to as the **Leaf-Horned Beetles**. Seven families in this suborder are worthy of mention for our purposes.

Family Lucanidae

The **Stag Beetles**, also known as the **Pinchbugs**, received their second common name from large mandibles that are possessed by the male members of the family. The title "Stag Beetles" is derived from these branched mandibles that resemble the antlers of a stag. These beetles, sometimes attaining a length of one-and-one-half inches, usually inhabit woodlands. The larvae of this family are often found in decaying wood in the same area as the adults.

Family Scarabaeidae

The **Lamellicorn Beetles**, also known as the **Scarab Beetles**, consist of a great many species of diverse size, habits, and appearance. Generally these beetles are ovalish, convex, and rather stout. The Scarab Beetles are divided into the **Dung Beetles** that feed upon decomposing substances, and the **Plant Feeders** whose adults eat flowers or leaves and whose larvae feed on roots or decaying wood.

Family Lampyridae

Evenings in early summer are oftened brightened by the long, soft-bodied **Lightning Beetles** or **Fireflies** who fly about blinking their "lights". Many members of this family have a segment near their tail that is capable of emitting light. These luminous segments are usually a greenish-yellow during those times when they are not glowing. The flashing of these insects is thought to be a part of the mating procedure of the insects. Some species in this family are distinguished by wingless females who crawl on the ground and are commonly known as "glowworms".

Family Cantharidae

The long, soft-bodied **Soldier Beetles** resemble Fireflies, but the Soldier Beetle doesn't glow with light. These insects are often brown, black, or yellowish and are found on flowers. Members of this family have some beneficial side to their character in the larvae who feed on other insects.

Family Buprestidae

The hard bodied adult **Flatheaded Borers**, also known as the **Metallic Wood Borers**, have a metallic appearance to their green, copper, black, or blue coloring. The white larvae have a yellowish head. Boring into the trunks of trees or the berry cones, the larvae cause a swelling at the point of entrance and residence. Some species of this family are also known as **Gall Makers** or **Leaf Miners** after the activities of the larvae.

The **Flatheaded Apple Tree Borer**, *Chrysobothris femorata* (Olivier) seeks fruit trees, shrubs, and hardwood trees, looking for those in a weakened condition. This serious enemy of fruit trees is widespread throughout the United States and southern Canada. Soon after the period when the apple tree is in blossom, the dark brown, one half inch long beetles appear. Their bronze-tinted, broad body is marked with gray bands and spots. The underside of the insect is a metallic blue-green and bronze in color. The adults lay their eggs in the bark of the tree that is selected as the host. In two to three weeks, the hatching larvae bore into the inner bark of the tree, where they grow vigorously. Weakened trees are selected by the beetle for its larvae, since an outpouring of sap from a healthy tree will either kill the larvae or drive them to outer portions of the bark. Upon attaining a length of one inch, the larvae find suitable sapwood in which to begin feeding. In the autumn, the larvae enter the pupal stage and remain in this stage in the tree during the winter. The maintenance of healthy trees is necessary to attempt to control an infestation of Flat Headed Borers.

The genus *Agrilus* contains many species whose larvae attack raspberries or other shrubs. The adult beetles of this genus are narrow, long, and generally of dark metallic coloring with some species possessing light colored markings. *Agrilus ruficollis* (Fabricius) causes galls in the blackberry and raspberry plants. Twig girdling is the manner in which *Agrilus arcuatus* (Say) earns its distinction as a pest.

One notable genus, *Brachys,* has members that feign death in order to escape some forms of destruction. If disturbed, these beetles draw their legs up and drop to the ground as if they had perished.

Family Elateridae

Many species of this family of **Click Beetles** are economically significant pests since their larvae, called **Wireworms**, feed on the roots of various plants, including grasses, grains, and corn. Most Wireworms are slender, one half to one-and-one-half inches long, and have a brown or yellow coloring. These often shiny larvae are covered with a hard cuticle and are capable of causing considerable damage to the crops they infest. These larvae may live in the ground from two to six years before the life cycles of the species is completed. While some larvae feed upon the aforementioned plant roots, others feed upon the planted seeds, decaying wood, or even fungae. The **Wheat Wireworm**, *Agriotes mancus* (Say), is the most notable pest of this family of pests and is especially significant to wheat farmers. The adult Click Beetle is a long, slender insect whose sides are nearly parallel. Often of a dark hue, the spine of these beetles is specially constructed to enable the beetle to right itself if it lands on its back. Upon landing on the back, the spine snaps itself and thrusts the beetle into the air with a clicking sound, hoping to land in a more desirable position. Members of this family commonly pupate in the late summer and the frailness of the pupa is exploited in some manners of control.

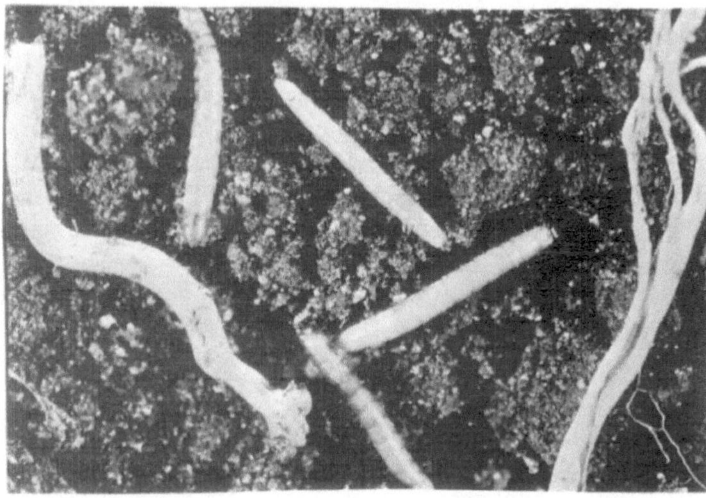

Figure V-11. Wire worms. (Shell Chemical Company.)

Suborder Heteromera

Members of the Suborder *Heteromera* consist of beetles with different-jointed *tarsi*, or feet.

Family Tenebrionidae

Generally the **Darkling Beetles** are either brown or black and some species have red markings on their elytra. Several species that eat bracket fungi are dark brown and scruffy, that is, they have a texture similar to that of bits of bark. Most members of this family subsist on plant substances. Two examples of members of this family that are pests include members of the genus *Tribolium* and the genus *Tenebrio*. The oblong, brown beetles of the genus *Tribolium* inhabit flour, cereal, corn meal, and dried fruits. Both the larvae and the adult forms of the beetle of this genus are a pest in these foodstuffs. The black and dark brown beetles of the genus *Tenebrio* devour stored grain. The larvae of this genus, commonly called **Mealworms**, resemble the Wireworms.

Family Meloidae

The **Blister Beetles** of the Family *Meloidae* obtained their common name from their body fluids that are capable of causing blisters if they come in contact with the skin of humans. The long gray or black adults are rather soft and some members may be decorated with gray or yellow stripes on the margins of the wings. The tip of the abdomen of these beetles projects beyond the wing cones. The **Striped Blister Beetle,** *Epicauta vittata* (Fabricius), used to be commonly known as the **Old-Fashioned Potato Beetle** since it was the principal pest of potatoes prior to the spread of the Colorado Potato Beetle.

Members of this family overwinter as a pseudo-pupa in the soil. The true pupal stage is passed in the spring followed by the emergence of the adult beetles in July. These active beetles commonly dine together in swarm-like congregations on garden plants, flowers, and even weeds. In the summer, the females deposit their eggs in the soil. In less than a month, the larvae hatch and search in the soil for grasshopper eggs. The larvae of this family undergo four rather unusual molts in their hypermetamorphosis, resulting in their attaining a pseudo-pupal stage for the overwintering.

Subfamily Criocerinae. The Asparagus is the victim of several species that are included within this subfamily. New shoots of the plant are damaged by the black headed and gray larvae of *Crioceris asparagi* (Linn.). The adults of this species are distinguished by greenish blue elytra that are splotched with pale yellow. The **Spotted Asparagus Beetle,** *Crioceris duodecimpunctata* (Linn.) has six black spots on each elytra. This brownish red pest only feeds in the berries of the plant and is relatively harmless. Potatoes and other crops are attacked by the orangish *Lema trilineata* (Olivier) that has three black bands on the elytra.

Family Chrysomelinae

The **Potato Beetle**, also known as the **Colorado Potato Beetle** and more formally as *Leptinotarsa decimlineata* (Say) is native to North America and specifically to the region of the state of Colorado. Prior to the invasion of the area by European settlers, this species was content to feed on various indigenous species of Nightshade (*Solonum*), especially the varieties known as Buffalo Burr. With the arrival of settlers and their potato plants, the beetles abandoned the weeds for the potato patch. Traveling eastward via potato patches, these beetles have become widespread pests. The yellow adult beetle has black stripes on the elytra. Overwintering as an adult, the yellow-orange eggs are laid on the underside of a leaf in the spring. The brick-red, black spotted, humpbacked larvae hatch in approximately a week to begin chewing the leaves of their host. When the larvae have attained their full growth, they pupate in the soil and emerge as adults within five to ten days. One to three generations may be formed in any year depending upon the climate and the weather.

Cucumbers and related plants, such as squashes and other cucurbits, often are severely damaged by the **Striped Cucumber Beetle,** *Diabrotica vittata* (Fabricius), and the **Spotted Cucumber** Beetle, *Diabrotica decimpunctata* (Fabricius) or *Diabrotica undecimpunctata howardi* Barb. Whereas these beetles feed on the plants, the larvae eat the roots and underground stems. To add insult to injury, these beetles are vectors of **Cucurbit Wilt**. The Spotted Cucumber Beetle is also known as the **Southern Corn Rootworm**. Besides feeding on cucurbits, this pest can be a serious problem to growers of corn. The greenish-yellow, one quarter inch long adult has black spots on its elytra. After overwintering in the adult form, this insect lays its eggs. The hatched grubs feed upon the roots of the corn as well as drilling into the stems of the crop. This beetle may also attack crops of grain and alfalfa.

Family Mylabridae or Family Bruchidae

The **Bean** and **Pea Weevils** are stout, short beetles that have rather short elytra and are usually dull colored. The **Pea Weevil**, *Mylabris pisorum* (Linn.) or *Bruchus pisorum* (Linn.), is a dark brownish beetle whose color is mottled with white and black. The one-fifth inch long adult overwinters to emerge in the spring and lay its eggs on the young pea pods. Upon hatching, the larvae bore into the young peas and feed. When the larvae are full grown, they pupate and become adults. Whereas in the northern climates the adult overwinters in the pea pods, in the south the weevil leaves the peas and overwinters in some protected area.

The small, mottled gray-brown **Bean Weevil**, *Acanthoscelides obtectus* (Say) lays its eggs on the bean pods. The life cycle of this weevil resembles that of the Pea Weevil, except that the Bean Weevil will also attack dried beans in addition to the beans that grow in the field.

Other destructive weevils of this family include the **Broad Bean Weevil**, *Bruchus rufimanus* Boheman, the **Cowpea Weevil**, *Callosobruchus maculatus* (Fabricius), and the **Southern Cowpea Weevil**, *Callosobruchus chinensis* (Linne).

Family Cyladidae

The **Sweet Potato Weevil**, *Cylas formicarius elegantulus* Summers, is also known as the **Sweet Potato Root Borer**. Resembling ants, these beetles have a brown-red coloration on the front of their heads and blue-black elytra.

Superfamily Rhynchophora

The Superfamily *Rhynchophora* consists of several families of beetles commonly known as the **Snout Beetles**.

Family Curculinonidae

The **Snout Beetles** of this family are more specifically known as the **Typical Weevils** and are represented by two subfamilies that are worthy of some mention.

Subfamily Curculioninae. The mottled, dark colored **Plum Curculio**, *Conotrachelus nenuphar* (Herbst), has shiny lumps on its elytron. This insect can be a severe pest of fruit such as the peach, plum, cherry, apple, and nectarine. The female eats a small hole in the fruit in which her egg is deposited. This hole is then expanded until it is crescent-shaped. The hatched larvae feed in the fruit for approximately two weeks. At this point, the fruit usually falls to the ground and the larvae emerge and burrow into the ground to pupate. The emerging adult overwinters in some protected place to renew the species in the spring

The **Cotton Boll Weevil,** *Anthonomus grandis* Boheman, is one quarter of an inch long and dark gray to black upon maturation. Overwintering in the adult form, the Weevil lays her eggs in holes that she eats in the blossom buds (squares) of the cotton or in its bolls. The white larvae have brown heads and feed on the plant while ruining the cotton in the boll. The larvae may even cause the boll to drop from the plant if their infestation is sufficiently severe. Pupation occurs in the boll and several generations may be produced each year. Native to the American continents, this pest entered the United States (circa 1890) from Mexico by way of Texas.

The three sixteenths of an inch long **Alfalfa Weevil,** *Hypera postica* (Gyllenhal), is a dark colored beetle. Overwintering as an adult, this pest lays its eggs in the stems of the alfalfa in the spring. The larvae feed in the stems and on new leaves, causing great damage to the plant in the process. In spite of the production of a single generation each year, this weevil may become a serious pest.

Subfamily Calendrinae or Subfamily Rhynchophorinae. The **Rice Weevil,** *Sitophilus granaria* (Linn.), is a pest of stored grain throughout the southern two thirds of the United States. The dark brown, one eighth of an inch long adult has four reddish spots on the elytra. Following the pattern of the other weevils, the female chews a hole in the kernels of grain and deposits the eggs in these holes. The hatched larvae feed upon the grain, causing further damage to the product.

The shiny, brown-red **Granary Weevil,** *Sitophilus granaria* (Linn.) is similar to the Rice Weevil in its life cycle and proclivity to attack stored grain. This pest ranges farther north than its cousin, causing damage to stored grain in the northern United States and into Canada.

Family Scolytidae

The **Bark** or **Engraver Beetles** include the **Smaller European Elm Bark Beetle,** *Scolytus multistriatus* (Marsham), and the **Native Elm Bark Beetle,** *Hylurgopinus rufipes* (Eichhoff). The Smaller European Elm Bark Beetle is one eighth of an inch long and a dark blackish-brown in color. The adults emerge in the early summer to feed on the bark of healthy elm twigs. These two types of Bark Beetles may be carriers of the **Dutch Elm Disease,** a fungus, which is one of the paramount reasons that they are considered pests in the United States.

ORDER STREPSIPTERA

The **Twisted-Wing Parasites** of the Order *Strepsiptera* consist of comparatively few members. These minute insects are parasites of other insects. Frequently the females are without either legs or wings, while the males are usually winged. As a result of her lack of motor apparatus, the female never leaves the host. After mating, she is instrumental in the production of the large numbers of larvae that feed upon and molt with the body of the host. The life history of members of this order is very complex and will not be explained here. These parasites prey upon various members of the orders *Hemiptera, Hymenoptera, Orthoptera*, and

Homoptera. The most frequently parasitized insects include the bees, leaf-hoppers, and wasps.

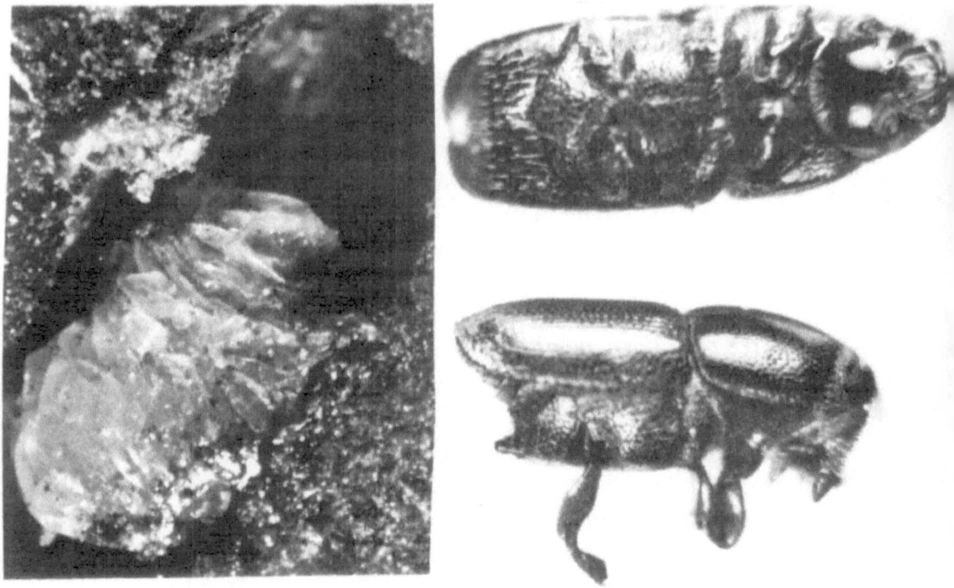

Figure V-12a. European Elm Bark Beetle, pupa hibernating in the Bark and also gallery hole. **12b.** Smaller European Elm Bark Beetle-Adults. (Illinois Natural History Survey.)

ORDER NEUROPTERA

Many members of this order are beneficial to mankind for several reasons. Some prey upon ants, others upon aphids, and still other members of this order dispose of the nests of some spiders. Other members of the Order *Neuroptera* contribute to the sustenance of mankind by serving as a food for fish. These insects are soft-bodied and have four wings.

ORDER MECAPTERA

The small and moderate-sized insects of the Order *Mecaptera* are of no particular economic significance. The common name for members of this order, **Scorpion Flies**, is derived from the abdomen tips of the male members of some species which are shaped so as to resemble the tail of a scorpion.

ORDER TRICHOPTERA

The adult **Caddisflies**, or **Caddice Flies** have four wings that are often covered with hairs. These insects are rather weak fliers and vaguely resemble moths. The mouth parts of these pests are not greatly developed. The larvae of this order, commonly known as **Caddice Worms**, construct a characteristic case that is carried for use as a habitat and, frequently, a pupation case. These larvae are aquatic and generally feed upon small aquatic vegetation.

ORDER LEPIDOPTERA

The Order *Lepidoptera* consists of the sub-orders of moths and butterflies and is one of the largest orders in the classification system. Members of this order are generally large, with four membraneous wings that are often covered by scales. The often brilliant colored adults possess mouth parts designed for sucking. The larvae, which have chewing mouth parts, undergo a complete metamorphosis before reaching the adult stage.

Suborder Heterocera

The Moths are usually night fliers whose wings assume various positions during the periods of rest. These insects are characterized by an often stout abdomen and antennae that lack the common enlarged "clubs" at their tips. A brief discussion of the most common and/or important families of this suborder follows.

Family Tineidae

The Family *Tineidae*, containing over one hundred and twenty five species in North America, consists of small moths with a wingspan of less than one inch. The most bothersome members of this family include the **Webbing Clothes Moth,** *Tineola bisselliella* (Hummel), the relatively rare (in the United States) **Tapestry Moth,** *Trichophaga tapetzella* (Linn.), and the **Case-Bearing Clothes Moth,** *Tinea pellionella* (Linn.).

The **Webbing Clothes Moth**, or **Naked Clothes Moth**, is a night flying pest of woolen cloth. The spotless, (no reference is meant to its personal habits) pale cream, buff, or straw colored adult has a one half inch wingspan and lays its eggs on either wool cloth, fur, mohair, feathers, or felt. The white larvae hatch in three to ten days and feed until the time of pupation when a silken cocoon is spun.

Family Psychidae

The members of the Family *Psychidae*, the **Bagworms**, construct silken bags around their bodies, thus giving rise to their common name. The wingless females of this family usually lay their eggs inside this bag. The most common species of Bagworm in the United States is *Thyridopteryx ephemeraeformis* (Haworth). A heavy infestation of these Bagworms may cause damage to the red cedar and **arbor vitae**, the favorite hosts of this species.

Figure V-13. Bagworm. (Illinois Natural History Survey.)

Family Gelechiidae

The Family *Gelechiidae* is best known for the **Angoumois Grain Moth,** *Sitotroga cerealella* (Olivier) whose larvae are capable of destroying wheat, corn, barley, and other grains. The common name for this moth is derived from the province, Angoumois, in France where the pest was first reported. The buff or pale brown colored, slightly black speckled adult has characteristically fringed light gray wings that spread one half to two thirds of an inch. The adult moth lays eggs on or near the grain to begin the production of approximately two generations per year in the cold climates, and up to six generations per year in warmer areas. The emerging caterpillars attack the kernels of grain to satisfy their voracious appetites. Within approximately three weeks, the larvae pupates, flying from the infested grain kernel a week later. The first generations of the year may be found in the small grain heads, while later generations may develop in storage areas if the climate is sufficiently warm.

The **Pink Bollworm,** *Pectinophora gossypiella* (Saunders), is a second unwelcome representative of the Family *Gelechiidae*. Believed to be a native of India, members of this species reached Mexico (circa 1911) purportedly in seeds that were transported from Egypt. Reaching Texas in 1917, this bollworm soon spread to other cotton growing regions of the United States. The brown adults, having a three quarters of an inch wingspan, hide during the daylight hours. These pests lay their greenish-white eggs chiefly on the cotton bolls. The larvae bore into the bolls and feed on the seeds, tunneling through the boll and soiling the lint. This attack can cause the growth of the boll to cease, the boll to rot, or fail to open properly. A heavy infestation can cause losses as high as fifty percent in the yield of the crop. During the cold months of the year, the larvae exist in cocoons in the cotton seeds, plant refuse, cracks in the soil, or bolls on the stalks. Some larvae hatch after feeding only for three to four weeks, while others may remain in the cocoons for as long as two years. The adults emerge from the cocoon from March to the middle of the summer, but usually in May. When conditions are "favorable", four to six generations may develop in any one year.

Family Aegeriidae

The Family *Aegeriidae*, the **Clear Wing Moths**, consists of brilliantly colored adults, resembling wasps in body structure, and whitish larvae. The moths fly, rapidly, during the warmest hours of the day, propelled by wings distinguished by the absence of scales on one or both pairs. Two species of this family deserve particular attention: the **Squash Borer**, *Melittia cucurbitae* (Harris), and the **Peach Tree Borer**, *Sanninoidea exitiosa* (Say).

The **Squash Vine Borer** is native to America, and attacks pumpkins, squashes, and, occasionally, cucumbers, melons, and gourds if the pumpkins and squashes are not available. Infestations of this pest are characterized by a sudden wilting of the leaves combined by wet and slimy larvae burrows. Heavy infestations, resulting in a loss of crops in excess of twenty five percent, are not unknown. Dark, metallic green forewings and transparant hind wings, both having a spread of one to one-and-one-half inches, and an abdomen ringed with black, red, and orange characterize the Squash Borer. As further identification, the hind legs are fringed with hairs, causing the legs to appear larger than they really are. The adults begin the reproductive cycle of their species by appearing from April, in the south, to July, in New England, and laying reddish-brown, oval, flattened eggs on leaf stalks, stems, and other parts of the plant, especially near the base of the stem. After one to two weeks, the eggs hatch to reveal tiny larvae that bore into the plant, making holes to the surface in an apparently random manner in order to expel their excrement. This excrement that is found in the vicinity of the base of the plant may be the first sign of the presence of a potential infestation. Reaching full growth in four to six weeks, the larvae burrow a few inches into the nearby ground where they wrap themselves in a dark, silken three quarter inch long cocoon to undergo pupation. This cocoon provides an over-the-winter home for many of the species, except for a few, particularly in the southern climates, which produce a second generation during the same season.

Another native of North America, the **Peach Tree Borer**, *Sanninoidea exitiosa* (Say), attacks peach trees, and occasionally plum, nectarine, cultivated and wild cherry, and apricot trees. The Peach Tree Borer has a blue body. The male is dark and ringed with several narrow yellow bands, and with a single orange band across the middle of the female insect. Whereas the male has transparent wings, bordered with blue, the female's forewings are blue. The moths may be observed flying about peach orchards during midday from May to October on their one inch wingspreads. About ten days after, the eggs are laid singly or a few at a time on the trunks of the trees near the ground level, the larvae hatch and bore into the sapwood at the base of the tree. These larvae feed until they are about one third grown, and overwinter in a quiescent state to resume feeding the next spring. Upon reaching full growth, the larvae prepare a cocoon of silk, covered with dirt and gum, after burrowing either into the surface of the trunk of the tree or into the soil. The adult moth hatches after three to four weeks of the pupal stage. The appearance of gum on the bark near the base of the tree is an indication of an infestation of the Peach Tree Borer.

Family Tortricidae

The large Family *Tortricidae* is formed from the small moths that are often destructive to shade, fruit, and other trees. The caterpillars of many species either roll the leaves of their host together or tie them to form either individual or colony-type nests. When disturbed, the larvae are reported to "wriggle" out of such a nest to descend on a silken thread. Pupation of members of this family generally takes place within a lightly constructed cocoon.

The **Spruce Budworm**, *Choristoneura fumiferana* (Clem.) or *Archips fumiferana* (Clemens), is a particularly destructive pest that causes significant damage to spruce and balsam-fir forests. The reddish-brown or gray moths have golden brown markings on the forewings and a white spot on the center edge of the forewing. Propelled on a one inch wing span, the Spruce Budworm lays pale green eggs on the foliage of its host. The young larvae, hatched from the eggs within eight to twelve days, spin a covering under the edges of the bark for protection during the overwintering period. Emerging in the spring, the larvae feed for three to five weeks by boring into the needles and opening buds of the tree. During the feeding process, the larvae ties the tips of several twigs together to form its characteristic nest. The dark brown, one inch long, full grown larvae can be recognized by their covering of yellowish tubercles and the yellowish stripes on either side of its body. At this point in its development, the larvae spins a loose cocoon. Pupating inside this cocoon for seven to ten days, the adult moth emerges during July and August to complete its one generation per year life cycle.

Another member of the Family *Tortricidae* worthy of mention is the **Red-Banded Leaf Roller**, *Argyrotaenia velutinana* (Walker). It is more or less indiscriminant in its feeding habits, but can become "pesky" in apple orchards.

The larvae of this species skeletonize the leaves of its host from the underside while folding and webbing the leaves together to form its nest. If sufficiently abundant, the larvae will eat irregular, shallow channels in the fruit.

Family Olethreutidae formerly *Eucosmidae*

Many of the members of this family bore into various parts of shrubs, trees and other forms of plant life. These insects are generally small, having wingspans in the order of three quarters of an inch, and often are serious pests.

The **Codling Moth,** *Carpocapsa pomonella* (Linn.) is a particularly severe pest, especially in apples, but is capable of infesting pears and other fruit trees including the English Walnut. A worm found near the core of a "wormy" apple is generally the larvae of this moth. The adult is characterized by brown or rosy forewings that exhibit brown and gray markings crossing the front of the wings and a brownish spot near the tips of the forewings. Flying only at night, the adult will lay between fifty and seventy-five translucent, flat, small eggs singly on the leaves. Within five to seven days, the eggs hatch and a light colored, dark headed caterpillar emerges. After feeding on the leaves of the tree, this caterpillar crawls into the blossom end of the fruit, tunnelling into the core to continue feeding until it attains its full growth of about three-quarters of an inch long, in approximately one month. This process usually causes the fruit to fall from the tree prematurely. Upon reaching full growth, the then pinkish, caterpillar burrows through the side of the fruit to crawl down the tree in search of a piece of loose bark. If this generation is to overwinter, this new, underbark home serves as the location for its silk lined cavity until the spring pupation. The spring pupation occurs one to two weeks after the petals fall. Otherwise, the pupae hatch into adult form to continue the life cycle. A second generation may then be created to attack the fruit, inflicting further damage. This generation may be present in the fruit at the time of harvest, in which case the overwintering will take place in the side of the fruit storage containers.

Physically resembling the Codling Moth, the **Oriental Fruit Moth,** *Grapholitha molesta* (Busck) is a pest of stone and pome fruits, especially peaches. Presently found throughout the United States, this pest is believed to have been introduced in shipments of flowering trees from Japan and was discovered near Washington, D.C. in 1913. This pest injures young tree shoots and ripening fruits; the latter by tunneling through the flesh of the fruit and around the stone. The full grown larvae overwinter in a cocoon on the bark of the tree or in the rubbish around the tree. In the spring, the larvae pupates and the adult emerges during the time when the peach trees are in bloom. When about five days old, the adult lays eggs singly on the underside of the peach tree leaf. The tiny larvae emerge to tunnel inside the young peach shoots. This tunneling results in the demise of the shoots. After feeding for eight to sixteen days, the larvae enter the pupation stage, resulting in a new adult generation a week later, which feeds upon the fruit of the host. In this manner, four to seven generations may appear during a single year. The parasites of the Oriental Fruit Moth have some value in controlling this pest.

Family Pyralididae

The **European Corn Borer,** *Pyrausta nubilalis* (Huber), was first noticed in the United States in eastern Massachusetts in 1917. Believed to have come to the United States in broom-corn imports between 1909 and 1914, this Corn Borer has now spread throughout large portions of the central and eastern United States. The yellowish-brown, dark marked adults appear in the late spring or early summer borne on wings with a one inch spread. After the eggs are laid in clusters, the larvae hatch and bore into the stems of the corn to feed. The varieties of European Corn Borer that infest the eastern seaboard proceed to produce two generations per year, the adult moth emerging in June or July to begin the procreative cycle again during the same year. The Corn Borers found in and west of New York State produce only one generation per year, and the nearly full-grown larvae hibernate in the stalk during the winter to emerge as a moth in May and June. As a result of its smaller stalks, sweet corn may be more severely damaged by this pest than is field corn.

The **Garden Webworm,** *Loxostege similalis* (Guenee) generally feeds on grasses, Alfalfa, and occasionally on grain and cotton crops. Especially a pest of alfalfa, this Webworm can also damage beans, peas, beets, and, as a mixed blessing, many weeds by spinning webs around the leaves. The yellow-green, black dotted larvae feed inside these webs. The pale brown moths have markings of gray.

The last two members of the Family *Pyralididae* are also significant pests, but are somewhat limited in their designation as a pest since they have a reasonably limited range of diet. The short and gray **Mediterranean Flour Moth,** *Ephestia Kühniella* (Zeller), spins webs in flour, making loose masses and messes for the miller. The **Indian Meal Moth,** *Plodia interpunctella* (Hübner), is a gray moth having two thirds of the forewings colored dark brown. This moth is reported to cause significant destruction of dried fruits and shelled nuts.

Living in beehives, the **Wax Moths,** *Galleria mellonella* (Linn.), feed on wax and spoil the honey. Stored combs are the main recipient of the attention of this pest since strong colonies of bees are able to prevent the entrance of the Wax Moth into the hive. The brownish front wings of this pest have a spread of approximately one inch.

A pest of greenhouses, and especially of soft leafed plants, the **Celery**, or **Greenhouse, Leaf Tier,** *Phlyctaenia rubigalis* (Guenee) may also infest outdoor plants, such as celery, spinach, beets, and beans after a mild winter. The slender, pale green caterpillar draws the leaves of its host plant together while feeding on the under surface of the leaves. The brownish moth flies at night borne on wings having a three quarters of an inch spread. The forewings are distinguished by two dark, irregular lines on their surface.

Family Sphingidae

The strong flying **Hawk Moths** of the Family *Sphingidae* generally are large moths that fly at dusk, hovering over the flowers to feed on their nectar through the extension of its long tongue. This insects' cigar shaped, stocky body is accompanied by large pointed antennae and often brightly colored wings. The larvae of the Hawk Moth are usually large and green.

The related species of **Tomato Hornworm,** *Protoparce quinquemaculata* (Haworth), and the **Tobacco Hornworm,** *Protoparce sexta* (Johannsen), have wingspreads of four to five inches. The forewings of the Tomato Hornworm are ash-gray and its abdomen has a row of five pair of yellow spots on them. The Tobacco Hornworm has six pair of yellow spots and brownish-gray forewings but is otherwise similar to its aforementioned cousin. Possessing similar life cycles, both species overwinter in the ground as pupae. Hatching in the spring, the moths lay their eggs singly on the leaves of their respective plant hosts. After hatching, the caterpillars feed for three to four weeks, attaining a full grown length of three to four inches. The greenish larvae of the Tomato Hornworm have greenish-white markings that extend down each side of their body while the posterior end of the beast has a green and black horn. The Tobacco Hornworm larvae have similar coloration, with the exception of less complex side markings and a reddish horn at the posterior of the body.

Family Geometridae

The **Inchworms** and **Measuring Worms** of the Family *Geometridae* have larvae that are distinguished by three pair of legs in the front, while only two pair in the rear. Given this arrangement, the larvae must move by bringing their hind end to the vicinity of their front, reaching the front end forward in search of a site on which to place its feet. Resembling twigs, the caterpillars are not readily visible and feed on the leaves of shrubs and trees. The various sized moths are generally delicate and fly either at dusk or during the night. Not really pests, members of this family are generally more of a "bother" or "nuisance" than anything else.

The eggs of the **Spring Cankerworm,** *Paleacrita vernata* (Peck), hatch during the time when leaves are opening in the spring. The brown-green or brown caterpillars drop down from their perch on a single thread if they are disturbed. When they are full grown, these caterpillars enter the ground to pupate. The adults emerge in the spring. At this time the wingless females crawl up the side of the tree to lay their eggs, frequently under loose bark. The male Spring Cankerworm has the same general appearance, that is one-and-one-quarter of an inch wingspread and light gray coloring, as the male **Fall Cankerworm,** *Alsophila pometaria* (Harris). The life cycles and details of existence of the Fall Cankerworm is similar to that of the Spring Cankerworm with the exception that while the latter adult moth "hatches" from its cocoon in the spring, the former hatches in the fall.

Family Lymantriidae

Although the Family *Lymantriidae* is commonly known as the **Tussock Moths**, it includes such well known pests as the **Brown-Tail Moth** and the **Gypsy Moth** in addition to the **White-Marked Tussock Moth**.

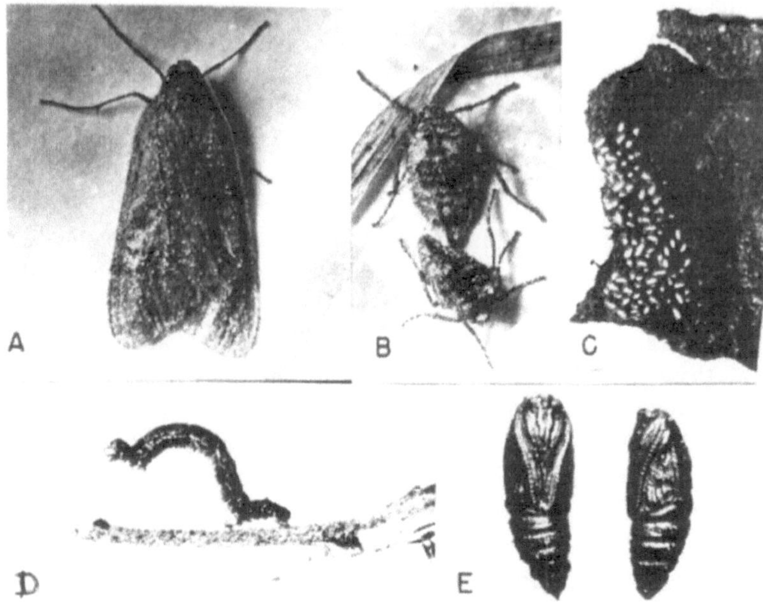

Figure V-14. Spring Cankerworm. **A.** Adult male, **B.** Wingless adult females, **C.** Eggs, **D.** Larva, **E.** Pupae (Illinois Natural History Survey.)

A pest of shade trees, the **White-Marked Tussock Moth,** *Hemerocampa leucostigma* (Smith and Abbott), occasionally attacks fruit trees. The wingless, stocky female is light gray while the gray male has wavy markings on wings that spread one inch. Overwintering in the egg stage, these pests hatch in the spring to feed upon foliage until they have reached their full growth of about one-and-one-half inches. The bright red headed caterpillar has grayish sides with a pair of yellow bands. Two red glands on the top of the body, four large tufts of black hair one half inch long projecting from the back behind the head, and a single tuft of hair near the tail distinguish this very colorful caterpillar. Pupation is accomplished on a twig, a piece of bark, or other suitable location within a thin, gray cocoon composed of silk and body hair. An emerging female moth crawls to the surface of the cocoon to deposit her eggs before covering them with a white froth that hardens to a white crust.

A white moth presenting a tuft of golden brown hair at the tip of the tail and on the abdomen and a one inch wingspread is a description of the **Brown-Tail Moth,** *Nygmia phaeorrhoea* (Donovan). The eggs that are laid by the female, generally on leaves, are covered with hairs from her abdomen. These eggs hatch in two to three weeks. The small, emergent caterpillars eat the green leaf parts, leaving a skeleton of the leaf intact. Upon arrival of early fall, the caterpillars migrate to the tips of the twigs where a three to four inch, tent-like cocoon is constructed. After the leaves fall, these cocoons become a conspicuous winter home for the caterpillars. In the spring, the caterpillars emerge to begin feeding on the delicate new leaves as they appear. The one-and-one-half inch long, mature caterpillar is nearly covered with fine, reddish-brown hairs. The body is also distinguished by two bright red *tubercles* that are placed in the middle of the back in the proximity of the tail. The small, brittle, larval hairs contain an irritant that can produce a painful rash to humans, particularly when copious members of them are shed into the air during the molt.

Figure V-15. White-marked tussock moth. (Illinois Natural History Survey.)

A pest of European origin, the **Gypsy Moth,** *Porthetria dispar* (Linne), immigrated to the Boston area of the United States (circa 1870). This pest has spread throughout New England and eastern New York State. Whereas the brown, yellow marked adult male moth flies during the day borne on wings of a one-and-one-half inch spread, the two inch wingspread of the heavy bodied female is purely ornamental, since the female does not fly very well. The wings of the female are marked with dark, narrow, wavy lines while its body is tipped with tan hair. The female lays her eggs on tree trunks, limbs, etc., in oval clusters and covered with her abdominal hair. The caterpillars hatch in the spring, after overwintering in the egg stage. Upon emerging, the caterpillars begin to eat the leaves of trees and shrubs, such as apple, birch, alder, willow, and oak. By July, the brown, hairy caterpillar may have reached its full growth of two inches and will be recognized by its back that is marked by five pair of blue spots and six pair of red spots. Crawling to the underside of a tree branch, or similarly suitable site, the caterpillar spins a few threads about its body for the purpose of pupation. After one to three weeks, the adult emerges to continue the life cycle of the Gypsy Moth.

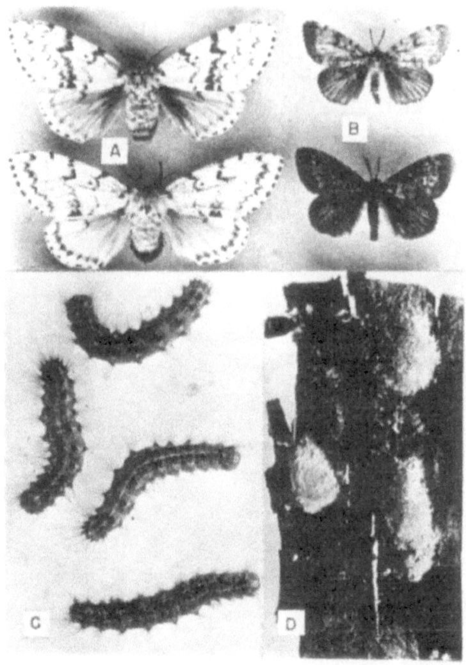

Figure V-16. Gypsy moth: A. Female Adults, B. Male Adults, C. Larvae, D. Pupae and hair covered egg clusters.

Family Notodontidae

The stout, hairy bodied, medium sized moths of the Family *Notodontidae* are known commonly as the ***Prominents***. Few species of Prominents are considered important pests. The larvae of the genus ***Datana*** feed on fruit, shade, and forest trees. Feeding in large groups of one to two hundred per branch, these insects may cause damage to their host trees. When disturbed, the two inch long caterpillar tips both ends of its body to assume the shape of an "U".

Family Noctuidae

The **Owlet**, or **Noctuid, Moths** of the Family *Noctuidae* comprise the largest family of moths that are found in the United States. These dull colored, medium sized, thick bodied, strong winged moths usually fly at night and are attracted to lights. The caterpillars of this family are often serious pests.

The **Corn Earworm, Tomato Fruitworm**, or **Cotton Bollworm**, *Heliothis armigera* (Hubner), is widespread throughout the world. In the northern United States, this pest attacks corn and tomatoes, while its southern cousins attack tobacco seed pods and cotton bolls as well. The pale reddish-brown to olive forewings of the moth have greenish tinged edges and darker markings while the aft wings are lighter with dark vein markings and a dark band across the tips of the wings. Both sets of wings have a one-and-one-half inch span. Having passed the winter in the pupal stage, the adults lay eggs on corn silks, other portions of the corn plant, or in similar locations on other hosts. Hatching in a week, the larvae find a tender portion of the plant to bore into for the initiation of the eating phase. In corn, the larvae attack the buds, developing leaves, tassels, or developing ears, while the Corn Earworm will bore into the fruit or eat the tips of the tomato plants or blossoms, the bud leaves or seed pods of tobacco, and the tender cotton bolls.

The **Cotton Leafworm,** *Alabama argillacea* (Hubner), is a native of Central and South America which flies to the southern United States yearly. The light reddish-brown moths have a one-inch wingspread and lay their eggs singly on the leaves of the cotton plant. The green-yellow caterpillars emerge to begin feeding. As the caterpillars grow, some develop a yellow line centered in a black stripe along the middle of the back as well as four black dots on the back. When full grown, the larvae pupates inside of a leaf that it has tied together. The emergent moths fly as far north as Canada in September and October. The spiney tongue of the moth can rupture the outside of soft-skinned fruits providing double destruction of crops by the same generation.

The **Armyworm,** *Pseudaletia unipuncta* Haw. or *Cirphis unipuncta* Haworth, appears in South America, but is widespread in the United States. The brownish-gray adult moth has a small white spot near the center of each wing, and darkish edges on the hind wings. Supported on a wingspan of one-and-one-half inches, these strong, night-flying moths are attracted to light. As a partially grown caterpillar, the Armyworm overwinters in the soil. Feeding on grasses and small grains, the caterpillar pupates in the ground to hatch into the adult moth in early summer. This adult lays its eggs on the grasses or other plants. The eggs hatch in eight to ten days to reveal the larvae that grow to maturity in three to four weeks. The one-and-one-half inch long, full grown, greenish larvae have three black stripes interspersed with two whitish stripes that extend the length of the back. Since innumerable larvae may hatch in an area during "favorable" conditions, the local food supply may be quickly exhausted. The swarms of larvae "marching" off to find additional food have given the Armyworm its common name. When the larvae have reached their maturity, they pupate in the ground to reappear as moths in the fall. These fall moths lay the eggs for the overwintering generations.

The **Fall Armyworm**, *Laphygma frugiperda* Abbott and Smith, is a general pest in the southern United States, but migrates into the northern states to cause some damage. The caterpillars are pests of grasses, grains, and garden plants, most especially the field crops of corn, rice, winter wheat, and sorghum. With a wingspread of one-and-one-half inches, the mottled gray front wings of the adult moth often have a white spot near their tip while the white hind wings have dark margins. The eggs, that are laid on grass blades, hatch within a few days into caterpillars that mature in two to three weeks. Pupating two to three inches into the soil for ten days, the adults emerge and many fly north to lay the eggs for ensuing generations. Since the Fall Armyworm cannot survive the frost in the northern areas, these areas are attacked only in the fall by the last generation of the year, while in southern areas, such as the Gulf states, as many as six generations per year may tax the patience of the most stoical of farmers. The life cycle, "marching" habits, and control of the Fall Armyworm are similar to those of the Armyworm.

Other members of the Family *Noctuidae* include the **Cabbage Looper**, *Trichoplusia ni* (Hubner) or *Autographa brassicae* Riley, and the **Celery Looper**, *Anagrapha falcifera* (Kirby).

The larvae of many species of the Family *Noctuidae* are commonly referred to as "**cutworms**" since they often feed on the stems or roots of a plant and thus sever the plant from its supporting structures at the surface of the ground. These gray, medium-sized moths have blackish or mottled brown forewings and white, unmasked hind wings that spread one to two inches. The Cutworms overwinter either as partly grown caterpillars or as pupae, emerging in the spring to hide in the soil during the day and to feed during the night. These caterpillars are generally hairless and dark colored with pale markings. Reaching one to one and one half inches when grown, the caterpillars often curl in a ball when disturbed.

Family Arctiidae

The members of the Family *Arctiidae* are commonly known as **Tiger Moths** or **Footman Moths.** Some of the lesser (in terms of classification as economically significant pests) members of this family include the **Salt March Caterpillar,** *Estigmene acraea* (Drury), the **Banded Woolybear,** *Isia isabella* (Abbott and Smith). and the **Striped Footman Moth,** *Hypoprepia miniata* (Kirby).

The **Hickory Tussock Moth,** *Halisidota caryae* (Harris), feeds upon hickory, apple, and walnut trees. The one-and-one-quarter inch long larvae are covered with black and whitish hairs as well as having a band of black tufts of hairs down their back. Overwintering as pupae, the moths hatch in late summer. These yellow moths have brown dots and streaks on their narrowly pointed forewings while the off-white hind wings are rather transparent.

The larvae of the **Fall Webworm,** *Hyphantrai cunea* (Drury), are ugly pests of fruit and shade trees. The moths hatch in the late spring from a thin cocoon on the surface of the ground. The pupae have previously overwintered in this cocoon. These moths have pure white wings, occassionally marked with black spots, that have a spread of one inch. The eggs, which hatch in approximately ten days, are laid in masses on the underside of the leaves. The one inch long, pale yellow or green larvae have a black head, and a slightly hairy body that is marked by a dark stripe down the back with two yellow stripes on either side. Feeding together, the larvae spin a thin, white web around a few leaves, after which the leaves are devoured on the inside of this chamber. When these few leaves have been eaten, the web is enlarged to include more leaves. The Fall Webworm can be distinguished from the Eastern Tent Caterpillar since the latter becomes noticeable in the branches of the trees early in the summer and forms its web at the crotch of the branches while the former becomes noticeable in the late summer and feeds inside the web that is placed at the tips of the branches.

Figure V-17. Fall Webworm. (Illinois Natural History Survey.)

Family Lasiocampidae

The larvae of the **Eastern Tent Caterpillar,** *Malacosoma americanum* (Fabricius), commonly infest wild apple and cherry trees, and occasionally many shade trees. Considerable damage to these trees may be caused by an infestation of this pest. The reddish or yellowish-brown, stout bodied, night flying adult has two light colored vertical lines on its forewings. Whereas the wingspread of the male spans one-and-one-quarter inches, the female's wingspan extends an additional one half inch. In the early summer, the adult lays many eggs in wide bands that circle a twig. The bands are covered with a sticky, brown substance that soon hardens into a protective covering. The larvae remain in the egg stage within the protective covering during the winter. In the spring, the larvae emerge from the egg shell and creep to a fork of the tree to construct their communal web. Emerging from the web to feed during the day, the caterpillars spin a thread that trails them down their limb. As the caterpillars mature, the web is continually enlarged. After six weeks of feeding, this two-inch long, hairy pest is black with a white or yellowish stripe that runs the length of the back, decorated by rows of alternating blue and black spots flanking it on either side. At maturity, the caterpillar leaves the communal web seeking a sheltered area in which to spin a white cocoon. The moth hatches from this pupation cocoon in approximately three weeks.

Figure V-18. Persimmon Tree invaded by Tent Caterpillars. (Illinois Natural History Survey.)

A serious defoliator of many trees, the **Forest Tent Caterpillar**, *Malacosoma disstria* Hubner, feeds on maple, oak, poplar, and other trees. The habits, appearance, and control of the Forest Tent Caterpillar are similar to those of the Eastern Tent Caterpillar with several exceptions. The moths of the former are pale brown with dark markings on the forewings. The Forest Tent Caterpillar is distinguished by a row of oval white spots extending down its back, as well as pale blue sides with wavy yellow markings. The Forest Tent Caterpillar, in contrast to its name, does not construct the communal tent home, but rather is characterized by swarming over an infested tree in larger groups than does its cousin.

Family Saturniidae

The **Giant Silkworms** of the Family *Saturniidae* include the **Polyphemus Moth**, *Telea polyphemus* (Cramer), and the **Promethia Moth**, *Callosamia promethea* (Drury). The bluish-green Polyphemus Caterpillar is distinguished by the two rows of tubercles, resembling tiny red, yellow, and blue horns, that are found on its back. The brown wings of the moth have black edges. The forewings are marked with a yellow edged, clear spot ringed with a black band. The hind wings have an enlarged mark. The Promethea Moth has a similar appearance in both the adult and caterpillar stages to the Polyphemus Moth. The light blue-green Promethia caterpillar has black, red, and yellow tubercles on its back rather than the red, yellow, and blue tubercles of its cousin. The dark brown male adult moth has lighter wing edges and forewings that are distinguished by a black "eyespot" with a blue margin on one edge. In contrast the lighter brown female has triangular white spots in the middle of each wing as well as a whitish band near the edge of the wing. The Promethia moth caterpillar feeds on a variety of trees, but favors the ash, sassafras, and wild cherry trees. After completion of its feeding cycle, the full grown caterpillar chooses a leaf that is firmly attached to a twig. Wrapping itself inside this leaf, the caterpillar spins a cocoon in which to spend the winter while swinging from the tree.

Family Bombycidae

The **True Silkworms** of the Family *Bombycidae* are represented by the **Silkworm**, *Bombyx mori* (Linn.). This cream colored moth has pale markings on the forewings that spread one-and-three-quarters to two inches. Feeding on mulberry leaves, the larvae spin a silk cocoon composed of a single thread as long as one thousand feet. Having been domesticated in the Orient for centuries, the silken cocoons can be laboriously unraveled by hand after the larvae have been killed, usually by hot water. These silkworms are generally a laboratory curiosity in the United States, and generally are not considered as a pest.

Family Papilionidae

The **Black Swallowtail**, **Parsleyworm**, or **Celeryworm**, *Papilio ajax* Linne, is a pest of parsley, celery, dill, and related plants. The eggs are laid on the intended victim and hatch to reveal black caterpillars in approximately ten days. Feeding for up to several weeks, the mature, two inch long larvae becomes green with yellow spots and black bands that mark its back. If the larvae is disturbed, it will project a fleshy horn, the *osmetaria*, located immediately behind the head that emits a disagreeable odor. Completing its feeding, the caterpillar forms its *chrysalid* to last during the ten to fourteen day pupal stage. The black emergent adult possesses a three to four inch wingspan marked by two rows of yellow spots arranged along the outer edges of their surface. The hind wings are further distinguished by a blue coloration between the two rows of yellow spots. The Black Swallowtail overwinters as a pupa in the northern United States and as an adult butterfly in the southern areas.

The adult **Orange Dog**, *Papilio cresphontes* Cramer, has deep brown wings that are marked with yellow. The larvae of this member of the Family *Papilionidae* are occasionally harmful to the orange and other citrus trees in the southern United States.

Family Pieridae

A native of Europe, the **Imported Cabbageworm**, or **Cabbage Butterfly**, *Pieris rapae* (Linn.), is believed to have reached Quebec (circa 1860) and is now widespread throughout the continental United States. The white adult has a gray tip on the front wing. The male has a black dot in the middle of the front wing in addition to a black smudge on the front edge of the hind wing. The female is characterized by a black spot on each front wing. Supported on a two inch wingspan, the adult lays its eggs in the spring on the leaves of cabbage and other Crucifers. Within seven days, a pale green caterpillar emerges to feed for two to three weeks. Pupating for one to two weeks, the adult butterfly emerges to commence the next generation. Two to six generations per year may be developed, dependent upon the climate and location. The Cabbage Worms will overwinter as a three-quarter inch long *chrysalis*. The larvae are characterized as pests since they initially feed on the under side of the cabbage leaves, often stripping the leaf to leave only the veins. At this point the larvae will attack the heads of the cabbage, chewing into them. If sufficiently abundant, these pests are capable of causing considerable damage to cabbage crops.

The **Alfalfa Caterpillar** or **Orange Sulfur**, *Colias philodice eurytheme* Boisduval, changes from its initial brown color after hatching to a dark green color with a white stripe down each side as it matures. Favoring clover, alfalfa, and various other legumes, this pest produces from two to six generations per year, depending upon the climate. The two-inch wingspread of the adult is yellow orange with dark outer edges. A black dot in the middle of the forewings and two adjacent red orange spots on the hind wings further aid in the identification of this pest.

Family Danaidae

Members of the Family *Danaidae* are commonly known as **Milkweed Butterflies**. The familiar **Monarch Butterfly**, *Danaus plexippus* (Linn.) may be seen flitting through the air on a summer's day in generally all parts of the United States. The yellow-green caterpillars have black bands on their bodies and yield the beautiful adult. The orange-brown wings of the adult are marked with characteristic black lines as well as white spots near the outer edge of the wing.

ORDER HYMENOPTERA

The Order *Hymenoptera* is comprised of hundreds of known species. Adults of this order have four membraneous wings, the front pair being larger than the rear pair and both pair being laced with few veins. While most members of the order have mouth parts that are designed for chewing, the mouth parts of some species are also capable of lapping. Generally females of the Order *Hymenoptera* possess either an ovipositor, or an ovipositor that is present in the form of a stinger. Members of this order undergo complete metamorphosis. Many species of this order are beneficial to mankind, particularly the bees and also some insects that are parasites of harmful insects. On the other hand, the order is represented by ants and wasps as well as many pests of fruits, berries, and other food crops.

Suborder Symphyta

Most members of the Suborder *Symphyta*, also known as either the Suborder *Chalastogastra* or the Suborder *Sessiliventres*, feed on leaves. The ovipositors of these insects are adapted to either sawing or boring rather than stinging. The major pests of the Suborder *Symphyta* belong to the Family *Cephidae* **(Stem Sawflies)** or the Family *Tenthredinidae* **(Typical Sawflies)**.

Family Tenthredinidae

The **Pear Slug,** *Caliroa cerasi* (Linné), is a native of Europe that has become widespread in the United States. This insect is a pest of cherries and plums, as well as the pears from whence its common name was derived. The black bodied adult, approximately one fifth of an inch long, saws slits in leaves during the spring when the leaves are developing. The eggs placed in these slits hatch into larvae which cover themselves with a brown, glossy substance. These larvae skeletonize the tissue of the leaf in the process of their feeding through the time required for four molts. After the fourth molt, the now yellow-orange larvae cease feeding to enter the ground for the purpose of pupation.

The **Imported Currantworm,** *Nematus ribesii* (Scopoli), feeds on the leaves of currants and gooseberries. The small, whitish immature larvae become green with a yellow head and tail as they reach their full growth of three quarters of an inch long. The one-third-of-an-inch long adult fly has either a reddish yellow or pale colored stout body with black spots.

Other minor pests of the Family *Tenthredinidae* include the **Rose Slug** whose small larvae are dark colored, the **Birch Leaf Miner,** *Fenusa pusilla* (Lepeltier), a troublesome insect in the northeastern United States, and the **Elm Leaf Miner,** *Fenusa ulmi* Sundevall.

Suborder Apocrita

Many species that comprise the Suborder *Apocrita*, also known as *Clistogastra*, are parasitic to other insects. The females of this suborder are distinguished by their piercing ovipositor.

Family Ichneumonoidae

The Family *Ichneumonoidae* consists of a number of species, many of which resemble wasps except that members of this family do not sting. Two parasites worthy of brief notice belong to this family and are commonly known as the **Ichneumon Flies.** *Itoplectis conquisitor* (Say) is a parasite of the caterpillars of many harmful insects, such as the **Cotton Leafworm** and the **Eastern Tent Caterpillar.**

Family Braconidae

The **Braconid Flies** of this family are distinguished by three members that are particularly valuable parasites. The *Apanteles congregatus* (Say) is a parasite of the **Sphingid Caterpillars** while the *Apanteles glomeratus* Linne was brought to the United States from England to parasitize the **Imported Cabbageworm.** The **Strawberry Leaf Roller** is parasitized by the **Braconid Fly,** *Macrocentrus ancylivorus* Rohwer, that also attacks the larvae of many other species in the Order *Lepidoptera.*

Family Chalcididae

Several species of this family of **Chalcid Flies,** a broad group of insects, are of sufficient economic interest to mention.

Pteromalus puparum (L.), thought to be a native of Europe, is a parasite of the pupae of the Imported Cabbageworm. By parasitizing the eggs of pests such as the **Angoumois Grain Moth** and the **Codling Moth,** *Trichogramma minutum* (Riley) can be of significant economic interest.

The **Wheat Strawworm,** *Harmolita grandis* (Riley) is a tiny black insect. Resembling an ant, this pest is of significant economic importance west of the Mississippi River in crops of wheat. Overwintering in the pupal stage in the stubble of the wheat, a small wingless form of the pest emerges in the spring to lay its eggs in the young plants of the wheat. The larvae soon hatch to begin feeding on the stem, reaching the crown of the stem at the point of their full growth. The attack on the crown of the wheat destroys the head and prevents the formation of the grain. Passing through a brief pupal period, the larvae emerge as winged adults and may fly to a new field. The adults lay more eggs on the tender joints near the heads of the wheat. The resulting larvae feed on the heads of the wheat, reducing the potential yield of the grain.

The **Wheat Jointworm,** *Harmolita tritici* (Fitch) is similar to the Wheat Strawworm, except that the Jointworm produces only one generation per year, but is a greater pest east of the Mississippi River in the United States. This insect produces gall-like swellings on the stems of the wheat, an activity that earns the Jointworm the designation as a pest.

The larvae of the **Clover Seed Chalcid,** *Bruchophagus gibbus* (Boheman), feed within the seed of clover crops such as alfalfa. This pest may cause considerable damage in those areas in which the clover crop is used for the production of seed.

Family Scoliidae

Several species of the genus *Tiphia* are beneficial to man since they are parasites of the larvae of several harmful beetles. Currently several members of this family are being evaluated as a means of control of the Japanese Beetle.

Family Vespidae

The Family *Vespidae* includes the **Typical Wasps**. The large black and white, **Bald-Faced Hornet**, *Vespula maculata* (Linn.), may be found throughout North America. Noted for its fierce sting, this Wasp builds a paper covered nest. A similar species, *Vespula maculifrons* (duBuysson), is commonly known as the **Yellow Jacket**. This small yellow and black Wasp nests on or in the ground. The three quarters of an inch long, dull black **Paper Wasp**, *Polistes fuscatus pallipes* Lepeletier, has yellow markings and is common in the eastern United States.

Family Formicidae

Social insects that are found throughout the world, commonly known as **Ants**, form the Family *Formicidae*. The colonies that are formed by these insects are composed of female queens, males, and workers. Although the males and females often have wings during a period of their life cycle, the workers are wingless. The queen of the colony produces eggs that are cared for by the workers who also tend both the larvae and feed the queen. At certain times in the life of a colony, the winged young males and females will fly and perform aerial mating. After the females get their feet back on the ground, they will remove their wings, prepare a nest, and commence the production of the eggs that will staff the new colony.

The shiny yellow-red **Southern Fire Ant**, *Solenopsis Xyloni* McCook (or MacCook), nests in mounds and is commonly found in the Gulf states of the southern United States. These pests have a severe sting, capable of injuring young poultry and birds.

The **Argentine Ant**, *Iridomyrmex humilis* Mayr, is a native of South America that has become fairly widespread throughout New Orleans during the late nineteenth century. This small brown Ant goes "everywhere" in search of food and is a significant household pest.

The small insects of the genus *Lasius* often cultivate the Corn Root Aphid or other Aphids to obtain their honeydew.

A number of Ants are of sociological interest due to their peculiar habit of enslaving other species. An example of such slave-masters is the **Shining Amazon Ant**, *Polyergus lucidus* Mayr.

A tiny, red insect, the **Pharaoh Ant**, *Monomorium pharaonis* (Linne), is particularly fond of both sweets and protein foodstuffs. These pests may be seen marching in lines from their nest to a source of food in a house.

The **Black Carpenter Ant**, *Camponotus herculeanus pennsylvanicus* (DeGeer), is the largest of the common Ants; these workers are often one quarter to one half of an inch long. Nesting in wood, this pest may damage older buildings or utility poles. One means of post-mortem identification of the Black Carpenter Ant is through the odor of formic acid that may be sensed after this pest is crushed.

Superfamily Apoidea

The Superfamily Apoidea consists of many types of bees in addition to the well known **Bumble Bee** and the **Honey Bee**. Some of the species of bees do not build hive-type dwellings that are associated with bees in general. The role of the bee in the cross-pollination of the blossoms of so many plants that are of either aesthetic or economic importance to man has established the bee as a significant friend of man, not as a pest. The **Bumble Bee** (*Family Bombidae* or *Family Bremidae*) and the **Honey Bee** (*Family Apidae*) are social insects that live in colonies consisting of queens, workers, and male drones.

Family Andrenidae, Family Megachilidae

The Family *Andrenidae* includes the **Mining Bees** while the **Leaf-Cutter Bees** belong to the Family *Megachilidae.*

Family Apidae

The **Honey**, or **Hive**, **Bee**, *Apis mellifera* Linn., is represented by two races or strains in the United States today. The German Bee, distinguished by its black abdomen, is known for its temper—or rather the lack thereof. This strain was reasonably vulnerable to insect pests and proved not to be a particularly industrious gatherer of honey. Many of the colonies of the German Bee became "wild" and, as a consequence, a second race, the Italian Bee, was introduced into the United States for the purpose of honey gathering.

A colony of Honey Bees often consists of thousands of bees, each developed to perform one of three sets of tasks. The single queen occupies her time by the laying of eggs after she is fertilized by the drones. Although the queen may live for several years, her consorts have a limited life span since they are executed by the workers after they have fertilized the queen. With the exception of the generation of workers that exists during the winter, the workers generally live one to two months. These workers gather food, build the honeycomb, care for the young, and perform any other necessary manual labor in their spare time. During certain times of the year, a queen and some workers leave a full hive in a "swarm" to find a location for a new colony. At the new location, the queen lays eggs singly in the bottom of a special brood cell. In approximately three days, white larvae hatch and are fed by the worker bees. After five days of care, the brood cells are closed while the larvae spin silk cell lining, pupating for ten days until they become adults.

ORDER DIPTERA

The Order *Diptera* consists of a multitude of what are commonly known as **Flies**. The classification of many of the members of this order as pests is a very common, and usually correct, generalization. Over a dozen families that represent this order will be briefly discussed to provide the reader with a brief introduction to the order, similar to the brief excursions into a few representatives of the preceeding orders.

Family Culcidae

The members of the Family *Culicidae* contain several of the most common, and most dangerous, pests in the Order *Diptera*. This family is commonly known as the **Mosquitoes** and includes the **Northern House Mosquito,** *Culex pipiens* Linn., the **Common Malaria Mosquito,** *Anopheles quadrimaculatus* Say, the **Yellow Fever Mosquito,** *Aedes aegypti* (Linn.), and the **Salt-Marsh Mosquito,** *Aedes sollicitans* Walker. Members of the Family *Culicidae* are small insects; the females of some species feeding on blood, while the males, with weaker mouth parts, feed on plant juices and nectar. Eggs of this family are laid on or near water and hatch to yield legless larvae that continue to live in the water, breathing air through a respiratory tube. After eating tiny water animals or algae, the larvae, known as Wrigglers in most of the species, molt to become pupae, often called *Tumblers*, for a few days before proceding to the adult stage.

The **Northern House Mosquito,** *Culex pipians* Linn., is prevalent in the northeastern United States where it distinguishes itself as a "bothersome" pest to humans during the evening hours. The adult rests its unspotted wings with its body parallel to the object chosen for its, seemingly, infrequent resting area. Overwintering in sheltered areas, the adult lays its eggs on the surface of water in clusters. Hatching in a few days, the larval stage will last between one and two weeks during which time the larval respiratory tube projects through the surface of the water a great deal of the time.

The **Common Malaria Mosquito,** *Anopheles quadrimaculatus* Say, is distinguished by the angle that its body is held in relation to the substrate that is used for resting. Overwintering as an adult, the mosquito lays its eggs on the surface of water during the larval stage while breathing through the respiratory tube. Both dawn and dusk are the prime times for the attack of these pests. After an *Anopheles* has bitten a human infected with malaria, it becomes, after a time, capable of transmitting the disease. Although the southeastern United States is fairly free of infestations of this Malarial Mosquito through the continued use of DDT, other areas of the world remain infested and some strains of the pest that are resistant to DDT are becoming more common with each year.

A small mosquito possessing white bands on its legs and silvery lines on the top of the thorax is the **Yellow Fever Mosquito,** *Aedes aegypti* (Linn.). Living near or in houses, the adults feed in the late afternoon or early morning. The dread of Yellow Fever until the end of the nineteenth century is well known, aided by the notoriety achieved by contributing to the failure of the French in their attempt to build a Panama canal.

Family Cedidomyiidae

The tiny flies of the Family *Cedidomyiidae*, the **Gall Midges**, have long antennae and many cause plant galls. Some species overwinter as larvae inside the gall. The larvae of most species are small and bright colored.

The **Hessian Fly,** *Phytophaga destructor* (Say), attacks wheat, barley, rye, but rarely attacks wild grasses. A native of Europe, the Hessian Fly was believed to have been introduced to the United States in the straw bedding of the Hessian Troops during the American Revolution. Having spread throughout the North American wheat growing regions, the Hessian Fly is a serious pest. Soon after the emergence of the fall wheat, the female deposits her reddish eggs, frequently on the upper surface of the young leaf. Small, pinkish maggots hatch in a few days, working their way between the stem and leaf until they are above the joint of the plant. Sucking the sap of the plant for nourishment, the maggots become lighter in color while the wheat fails to develop new shoots in the vicinity of the banquet. Infested wheat turns a dark greenish-blue, become coarse and erect. Later an infested field will be disfigured by yellow and drying plants, when the uninfested plants are green and vigorous. After nearly a month of feeding, the larvae begin pupation by forming a flaxseed stage (so-called because of the appearance during the pupal stage) to overwinter in the young plant. Hatching in the spring, the adults lay their eggs on the leaves of the growing wheat to continue the life cycle. The pupal stage is again reached during the harvest time, when many Hessian Flies remain behind in the wheat stubble. Delaying fall planting until the summer hatch of flies is dead is a recommended control measure. (The state agricultural extension office can usually help with the determination of this date.)

Family Simuliidae

The Family *Simuliidae* consists of the **Buffalo Gnats** and the **Black Flies**. These small, hump-backed flies derive their name (Buffalo Fly) from this appearance rather than their choice of host. Often swarming in great numbers, especially during the spring, members of this family are recognized by their blackish color and clear wings. The eggs of this family are laid near or in swift running water. If laid in the water, the eggs are often attached to rocks. The aquatic larvae hatch and cling to the surface of either the rocks or other objects and feed on plant life. When grown, the larva spins a cocoon that is attached either to the substrate or another cocoon. After passing the pupal stage in this cocoon, the adult black fly likes heat and light, but like some of the other members of the Order *Diptera*, the Black Fly attacks either man or beast, causing painful, irritating swelling that lasts for days. Avoidance of successful attack by members of the Family *Simuliidae* on humans consists of the use of protective clothing and insect repellant as well as the location of camp sites on high ground away from concentrations of the pest. Smudges of bark, old wet wood, or grass in pastures and near stable doors can be effective for the control of the attack of these insects on beasts.

Family Tabanidae

The **Deer Flies** and **Horse Flies** of the Family *Tabanidae* prefer sunny, moist areas and associated are hot weather. This large family contains hundreds of species that are stout-bodied, one third to one inch long with large heads and a strong flying capability. While the males of the family feed on plants' sap and nectar, the females have piercing and sucking mouthparts to permit a diet of the blood of cattle, horses, deer, or other large mammals as well as a diet of plant juices. The eggs are often laid on the grass growing at the edge of a body of water. Living in the mud, under water, or in soft damp soil, the larvae develop into the adult fly. As a result of the voracious appetite of the female of the species for the blood of the host, attack by these pests can become very severe. Cows can experience curtailed milk production and weight loss if severely attacked, both from lost grazing time and the loss of blood.

Family Trypetidae

Although many small pests are commonly known as **Fruit Flies**, this name is generally reserved for members of the Family *Trypetidae*. Four particularly pesky members of this family deserve a few words of discussion.

The **Apple Maggot**, **Railroad Worm**, or **Blueberry Maggot**, *Rhagoletis pomenella* (Walsh), is a native pest whose activities are most serious in the northern sections of the United States. These flies have the interesting characteristic that some apples, such as thin-skinned and summer apples, are more vulnerable to this pest than other apples. The adults, slightly smaller than a house fly, have wings that are marked with heavy dark bands and a spread of approximately one half inch. Their black bodies are illustrated with light markings on the upper side of the abdomen. Feeding freely on liquids, this fly appears in the orchards of New England during early July, for example, to commence its major infestation of the year. The female fly thrusts eggs singly into holes punctured on the surface of the apple. This act of nature is accomplished with the aid of a sharp ovipositor contained on the tip of the abdomen and occurs some twenty days after the emergence of the mother. Within four to ten days, the eggs hatch to reveal one quarter inch long whitish maggots who begin their existence by tunneling through the fruit while feeding on the juices of its environment. As a result of this activity, the tunnel walls assume a brown color and become streaks through the fruit. After the larval stage, lasting from two to four weeks, this pest leaves the fruit, which has usually fallen by this time. The larvae enter the ground to pupate in a brownish cocoon that is buried one to six inches into the soil for the overwintering period. A few *Rhagoletis pomenella* will complete their life cycle sufficiently to produce a "crop" of late summer flies.

The **Mediterranean Fruit Fly**, *Ceratitis capitata* (Wiedeman), is particularly destructive to grapefruit, orange, peach, nectarine, Surinam cherry, and guava, found in tropical areas such as the Hawaiian Islands. Discovered in large areas of central Florida in 1929, this pest was effectively eradicated by 1930. This one-fifth inch long adult fly makes a hole in the skin of the fruit to enable the fly to insert eggs. The eggs hatch in three to twenty days to liberate larvae that feed on the pulp of the fruit, destroying the fruit in the process. Descending to the ground to pupate, in ten days the larvae has been superceded by the adult that commences to lay eggs in approximately a week. The production of several such generations are probable during each year.

The **Mexican Fruit Fly**, *Anastrepha ludens* (Loew), has a life history similar to the previously described Mediterranean Fruit Fly. A serious pest in Mexico and southern Texas, this insect attacks citrus and other fruits, especially the Mango.

Family Anthomyiidae

A native of Europe, the **Cabbage Maggot**, *Hylemya brassicae* (Bouche), is found throughout the northern United States and Canada. The adult emerges in the spring when the cabbage and other crucifers are planted. Eggs that hatch in three to seven days are laid on these plants near the surface of the soil. The emergent maggots attack the main root and stem of the plant below the surface of the soil. After gorging for approximately three weeks, the maggot enters the soil to pupate. The adult fly emerges in twelve to eighteen days. Two to four generations per year may be developed, with later generations subsisting on radishes or turnips. Overwintering is accomplished in the pupal stage in a brown, oval puparium.

Mining into the bulbs of onions, the **Onion Maggot**, *Hylemya antiqua* Meigen, is a very serious pest in a series of wet spring seasons and the decay of the bulbs is the result of its attack. The flies of this pest lay their eggs near the level of the soil on the spring emergent onions. Within a few days, one third inch long maggots emerge from the eggs to feed on the bulbs. The one quarter inch long fly begins laying the eggs for the next generation of pests about ten days after leaving the pupal stage. Two to three generations of Onion Maggots may be produced during the year, with the last generation overwintering either in onion refuse or in the ground.

Other members of the Family *Anthomyiidae* that are considered as pests include the **Raspberry Cane Maggot**, *Pegomya rubivora* (Coquillet) or *Hylemya rubivora*, and the **Seed Corn Maggot**, *Hylemya cilicrura* (Rondani) attacks the seeds of corn, beans, and peas, particularly during cold, wet seasons, causing a failure of the seeds to sprout.

Family Drosophilidae

The **Vinegar Fly**, *Drosophila melanogaster* (Meigen), is the only member of the **Pomance Flies** of this family that is of particular significance as a pest. This one-eighth-inch long, yellowish fly is commonly employed in genetic studies since its entire life cycle can be completed within two weeks. In such studies, the Vinegar Fly is often mislabeled and referred to as a "fruit fly". The tiny, white maggots feed either on the fermenting juices of decaying fruit or products that contain vinegar, such as pickles.

Family Muscidae

Certain of the members of the Family *Muscidae* are significant pests, both from the bothersome and economic points of view. These pests are commonly known as House Flies.

The **Common House Fly,** *Musca domestica* Linn., is commonly found about human dwellings and is believed to be a carrier of some human diseases, especially dysentery. The one-half-inch long adult fly has long, transparent wings. The blackish body is somewhat darker on the top and lighter on the sides. The head contains reddish brown eyes and mouth parts designed for lapping. Overwintering chiefly in the larval or pupal stage, the flies emerge with the onset of warm spring weather to lay hundreds of eggs. These eggs, usually laid in either manure or decaying matter, hatch from eight hours to three days after they are laid, depending upon the temperature of the environment. The white maggots feed on the substrate on which they were hatched for at least five days before pupating. After emerging from the pupal stage, the adult fly may begin egg laying within three to twenty-three days, producing many generations per year. As a result of the nature of the material on which the fly reproduces, this pest is considered a passive disease carrier. Any substrate on which the fly crawls may be the recipient of small remnants of other substrates previously trod upon, such as the aforementioned manure. In addition, the fly may disgorge some of its previous food and/or deficate on a substrate, such as human food. Given this total disregard for the rules of deportment, the fly is a bother as well as a sanitation hazard.

The **Stable Fly,** *Stomoxys calcitrans* (Linn.), resembles the Common House Fly except that the Stable Fly has piercing mouth parts. Commonly found in the vicinity of barns, stables, pastures, and marshes, this bloodsucking fly feeds upon livestock, causing pain to the host. During warm weather, this pest breeds in wet, fermenting vegetable matter, such as wet stacks of straw or seaweed piled along the sea shore.

The **Horn Fly,** *Siphona irritans* (Linne), is smaller and more slender than the Common House Fly, but has a similar appearance. Living for about seven weeks, the Horn Fly uses its piercing mouth parts once or twice per day to suck the blood from its cattle hosts. A cow that carries several thousand of these hitch hikers on its back may suffer a severe loss of blood. The Horn Fly generally leaves the back of its host only when disturbed or to lay its eggs in fresh cow manure. The larvae develop in this environment to overwinter as pupae or larvae in the dung or the ground beneath it.

Family Calliphoridae

The major pest-type members of the Family *Calliphoridae,* **Blow Flies**, is the **Screw Worm,** *Callitroga hominivorax* (Coquerel) or *Callitroga americana* (Cushing and Patton), that is a serious pest in the southern and southwestern states. The dark blue-green adult fly is much larger than the Common House Fly, has three black stripes on its back, and attacks warm-blooded animals. Laying its eggs on an open wound or a bloody spot on the body of the host, the larvae hatch in a few hours to feed on the flesh and, consequently, enlarge the injury. An untreated wound may result in an increase in infestation that eventually will cause the animal to die. The full grown maggot drops to the ground for the purpose of pupation, a process that is completed in several days. Several generations of Screw Worms are possible each year.

Family Tachinidae

The **Tachina Flies** of the Family *Tachinidae* comprise a most useful family of insects since the larvae are parasitic to many other species of insects. Although the adult Tachina Fly bears some resemblence to the Common House Fly, the abdomen of the former insect is frequently more stout and often is decorated with stiff bristles. Since the eggs of this family are laid either on caterpillars or on the plants that are used by the caterpillars as food, the hatched maggots bore into the caterpillars. Dining on the tissues of its host, the maggot generally causes the death of the caterpillar.

The **Red-Tailed Tachina Fly**, *Winthemia quadripustulata* (Fabricius), is a parasite of many caterpillars, including the Armyworm. A parasite of the Cane Borer Weevil, *Ceromasia sphenophori* (Villers), was introduced to the Hawaiian Islands in an attempt to control that pest.

Family Oestridae

The **Bot Flies** of the Family *Oestridae* are internal parasites of mammals during the early stages of their development. The large adult flies do not feed during this stage of their life cycle.

A native of Europe, the **Common Cattle Grub** or **Heel Fly**, *Hypoderma lineatum* (DeVilliers), was probably brought to the United States in early cattle shipments. In late spring and early summer, this fly fastens its eggs to the hairs of its host, especially in the area of the flank or the heel. The larvae, commonly called *Warbles*, bore through the skin and tunnel through the body of the host. These pests live around the internal organs and connective tissue of the body for approximately six months. During the winter and fall, the larvae migrate to the skin along the back, cutting an air hole through the skin of the animal upon arrival at this destination. The mature Warble grows rapidly, producing a swelling at its point of residence. After the larvae have attained maturity, they burrow through the air hole to the surface of the host and drop to the ground. Forming a puparium, the insect later emerges as an adult fly.

Other undersirable Bot Flies include the **Sheep Bot Fly**, *Oestrus ovis* Linne, whose larval stage infests the nasal cavaties and frontal sinuses of sheep, and the **Horse Bot Flies**, consisting of several members of the *Gasterophilus* species, whose larvae infest the stomachs of horses.

BIBLIOGRAPHY

1. Baerg, W. J., Introduction To Applied Entomology, Third Edition, Burgess Publishing Co., Minneapolis, Minnesota, 1951.
2. Borror, D. J. and Delong, D. M., An Introduction to The Study of Insects, Rinehart and Company, New York, 1954.
3. Comstock, J. H., An Introduction to Entomology, Ninth Edition, Comstock Publishing Company, Inc., Ithaca, New York, 1947.
4. English, L. L., Illinois Trees and Shrubs: Their Insect Enemies, Illinois Natural History Survey, Circular 47, Urbana, Ill.
5. Fernald, H. T. and Shepard, H. H., Applied Entomology, Fifth Edition, McGraw-Hill Book Co., Inc., New York, 1955.
6. Imms, A. D., Insect Natural History, The Blakiston Co., New York, 1951.
7. Isely, D., Methods of Insect Control, Part II, Burgess Publishing Co., Minneapolis, Minnesota, 1947.
8. Metcalf, C. L., Flint, W. P., and Metcalf, R. L., Destructive and Useful Insects, Third Edition, McGraw-Hill Book Co., Inc., New York, 1951.

VI

Insects and Diseases of Field Crops

The entry of the European Corn Borer led to the first common use of protective chemicals on corn. The small grains, wheat, oats, barley, and rye, are still subject to attacks of varying intensity from the **Hessian Fly**, **Chinch Bug**, **Green Bug**, and **Grasshopper**. However, the diseases of **Bunt**, and other **Smuts**, and of **Rust**, which at one time swept the country, are being held in check by the work of the plant breeders and by improvement in the chemicals that are used in the treatment of the seed. The rapid spread of the **Spotted Alfalfa Aphid** is causing serious concern, but progress in the use of chemicals and biological control techniques is promising. Cotton growers, despite the injury caused by the **Cotton Boll Weevil** and other insect pests, continue to produce surplus crops. Pest control on field crops continues to offer many serious problems for the grower and researcher alike, including coping with the effects of the continued use of powerful chemicals, but the future for safe, but effective, control is hopeful.

ALFALFA

Fusarium Wilt of alfalfa, caused by *Fusarium solani* is found in various scattered areas throughout North America. This pathogen attacks other plants in addition to its primary host, alfalfa. Infected alfalfa develops yellow or pale green leaves before the plant wilts. Roots of an infected plant develop a dark red or brown color in their internal tissues. Fusarium Wilt may also cause *damping off* of the seedlings if they are infected.

Bacterial Wilt, probably introduced from Turkestan, was identified in the United States in 1925. This disease, caused by *Corynebacterium insidiosum*, seems most prevalent in those years in which the plants suffer damage from the cold during the winter followed by a warm summer in which moisture in the soil is particularly abundant. Alfalfa infected by Bacterial Wilt suffers stunted growth and the plants are dwarfed as a consequence. The leaves of an infected plant curl and become yellow as the plant wilts, then dies. The root of the plant becomes dark reddish brown on the inside as the disease progresses.

Currently research workers are attempting to develop a strain of alfalfa that will be resistant to both of these diseases.

The **Spotted Alfalfa Aphid** is a recently developed, serious pest of certain of the clovers in addition to the alfalfa plants. This aphid is smaller than the Green Pea Aphid and is characterized by the four to six black spots on the back. To this time, the Spotted Alfalfa Aphid is found principally in the southwestern states, but this pest is spreading rapidly throughout the nation. The aphid is especially injurious to seedlings and the stand of the crop may be weakened or killed by an infestation of the pest. Mature plants are stunted and the foliage is covered with a honey dew that forms a substrate for a growth of a black fungus.

Among the control measures that are being advanced for use against the Spotted Alfalfa Aphid are the development of resistant strains of the plant and the importation of small parasitic wasps. These wasps are bred in great numbers for distribution to the infested areas. Chemical protection measures include the application of demeton (Systox), four ounces of the active ingredient, as an emulsion concentrate, per acre. This spray is applied to seedlings when the aphid population reaches one per plant, or to older plants when five aphids per leaf and honeydew are found. Demeton (Systox) acts both as a contact and a systemic insecticide, giving protection for several days. Systemic insecticides for seed treatment that have been developed are also promising in the control of this pest. Insecticides that have been substituted for demeton (Systox) include Malathion, Parathion, and Phosdrin (mevinphos).

The **Pea Aphid** attacks all peas as well as alfalfa and clover. Control of this pest is the same as for the Spotted Alfalfa Aphid except that for seed crops, one and one half pounds of active DDT has been applied before the bloom with success. Demeton (Systox) should be applied only once per cutting of the crop.

Grasshoppers can be controlled through the application of Dibrom (Xnaled) at a concentration of 0.75 pounds of active ingredient, as an emulsion concentrate, per acre. Aldrin, Chlordane, DDT, Dieldrin, or Toxaphene should not be used on plants or ensilage to be fed to poultry, dairy animals, or animals that are being finished for slaughter. Malathion may be substituted for the Dibrom (Xnaled) if applied at a rate of one pound of the active ingredient in the emulsion concentrate per acre. The minimum length of time from the last application of the control chemical until the day of harvest or feeding is four days for Dibrom (Xnaled) and seven days in the case of Malathion.

The **Alfalfa Weevil** is an immigrant from Europe that has been a destructive pest for many years in the western states and is now established in the east as well. Both the larvae and the adults feed on the plant and, where abundant, may weaken the stand of the crop.

Cultural control of the Alfalfa Weevil consists, in the west, of an early cutting of the first and second crops of alfalfa. The early cuttings are harvested when only a few blooms of the plant are showing but when a few shoots of the next crop have begun to appear. This clean cutting, and quick removal of the hay, leaves the weevil larvae, eggs, and pupae exposed to the sun and with little food for the larvae and the adults.

Chemical control of the western hay crop consists of the application of methoxychlor, one to one and one quarter pounds of active ingredient that is supplied as an emulsifiable concentrate OR the emulsifiable concentrate of parathion, at a rate of four ounces of active ingredients, per acre. The eastern hay crops respond to the application of either eight ounces of active ingredient Guthion in the emulsifiable concentrate OR one and one half pounds of active ingredient methoxychlor per acre of crop.

In addition to the cultural and chemical control techniques for the Alfalfa Weevil, a small parasitic wasp has been introduced into Ohio as well as the southern and northeastern states. Reportedly this black "weevil parasite", *Bathyplectes curculionis*, can kill eighty to ninety percent of the weevil larvae that occur early in the season. Parasitism is most prevalent during the first crop of alfalfa, and very little parasite activity is reported to occur during the second crop.

The **Alfalfa Caterpillar** can be controlled, chemically, by the application of carbaryl (Sevin), one pound of active ingredient, as a wettable powder; OR methoxychlor, ¾ to one pound of active ingredient, as an emulsion concentrate, per acre. The application of the spray usually is made when the alfalfa is one third grown. Commercial preparations of disease organisms are available for use against the Alfalfa Caterpillar.

Mites are often a serious pest of seed-alfalfa in the southwestern states. Their feeding causes a heavy shedding of bloom. Under California conditions applications of sulfur dusts at dosages ranging from fifteen to forty pounds per acre, depending on the severity of the attack, will control the Atlantic Mite and depress other species of this pest. Sulfur may cause petal burn to the flowers when it is applied during the summer when the temperatures are high (deOng, 1960) General control of Mites can be accomplished through the application of four ounces of active demeton (Systox) as found in the emulsifiable concentrate per acre of crop.

The group of sucking insects known as **Lygus Bugs** may be controlled by an application of toxaphene that is sold as an emulsifiable concentrate. The toxaphene is applied at a concentration of one and one half pounds of active ingredient per acre. Forage or ensilage made from plants that have been treated with toxaphene must not be fed to dairy animals, poultry, or animals being finished for slaughter.

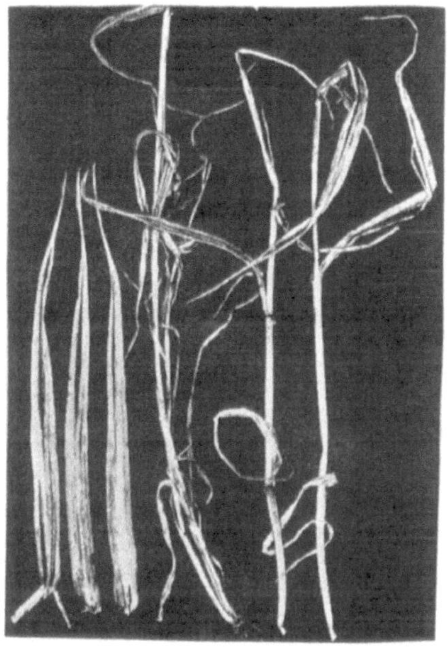

Figure VI-1. Barley stripe. (Illinois Natural History Survey.)

BARLEY

Stripe or **Blight** of Barley is a result of an infection of the plant by the fungus *Helminthosporium gramineum*. The entire plant is affected and the first symptoms are small yellow spots on the leaves of the seedlings. A few weeks before the barley heads, pale yellow streaks appear on the blade and sheath of the leaves. These stripes develop dark edges and the diseased tissue dies while splitting longitudinally. Plants that are infected by this disease are stunted, seldom head, and usually die before the time of the harvest. To control this disease, the barley seed should be treated with a fungicide that contains an organic mercury compound.

For the descriptions of the following diseases and the recommended means for their control, refer to the indicated sections of this chapter.

Ergot—see Rye, *Claviceps purpurea*

Anthracnose—see Wheat, *Colletotrichum graminicola*

Downy Mildew—see Wheat, *Sclerospora macrospora*

Powdery Mildew—see Wheat, *Erysiphe graminis hordei*

Stem Rust—see Wheat, *Puccinia graminis tritici* and *Puccinia graminis secalis*
Scab—see Wheat, *Gibberella zeae*
Septoria Leaf Spot—see Wheat, *Septoria passerini*
Scald—see Rye, *Rhynchosporium secalis*
Basal Glume Rot—see Wheat, *Pseudomonas atrofaciens*

Spot Blotch, *Helminthosporium sativum*, causes the formation of lesions on the seedlings in addition to **Crown Rot** and spots on older leaves and stems. The kernels that are infected with this disease have a brown discoloration. Treatment of the seed and the use of varieties of barley that are resistant to this disease are helpful in controlling outbreaks of the disease.

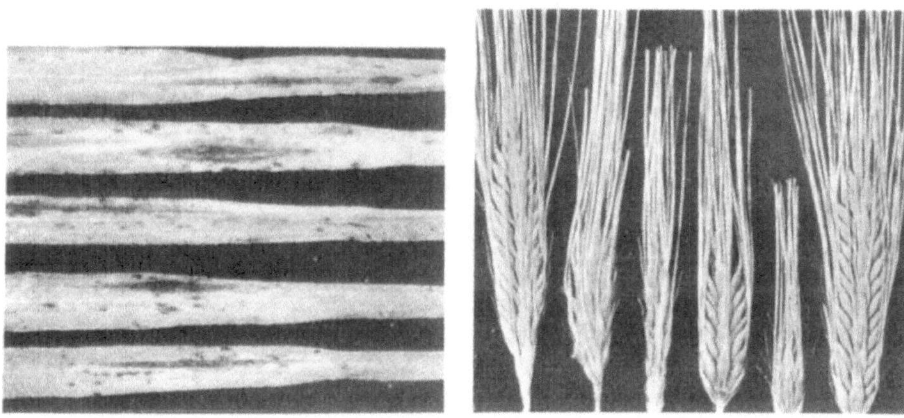

Figure VI-2a. Spot blotch on barley. **2b.** Scab on barley. Four center heads are infected with Scab, outer two are normal. (Illinois Natural History Survey.)

The **Bacterial Blight** of barley, *Xanthomonas translucens*, is similar to the Bacterial Blight of rye and Black Chaff of wheat. Small, water-soaked streaks and spots appear on the leaves and sheaths of an infected plant. These spots coalesce as the disease progresses, becoming translucent before turning yellowish brown. Severely infected leaves die back from the tips. The disease spots occasionally yield a milky exudate.

CLOVER

In the spring when the clover is two to six inches tall, the **Clover Leaf Weevil** may be controlled by the application of one pound of active ingredient per acre of either Malathion or the emulsifiable concentrate of methoxychlor.

The **Meadow Spittlebug** is a sucking insect that draws the sap from the stems of plants causing stunting and rotting of the crop. The common name of this pest is derived from its habit of covering itself with a froth. In the late spring when the first spittle case is seen, four ounces of active ingredient of methoxychlor should be applied to each acre for control of the pest. In the fall, control may be achieved by the application of one-and-one-half pounds of methoxychlor per acre in early September.

CORN

The **European Corn Borer** was first noted in the United States of America in 1917. By 1952 this pest could be found in thirty-seven states east of the Rocky Mountains, including the principal corn producing states. The losses of corn through the action of the European Corn Borer amounted to many million bushels of field corn annually during the period of maximum infestation. The degree to which the European Corn Borer is a pest is increased by the double as well as single brooding of the insect. The Borer can live on a great variety of wild and cultivated plants, but corn is the crop that is most widely attached. The European Corn Borer passes the winter as a full grown worm inside a tunnel in the stalk or ear of corn or other host. In May or early June, the insect pupates, emerging as an adult moth that flys from plant to plant, laying clusters of eggs. The eggs hatch within a few days and the Borers develop into mature larvae in August, whose feeding continues until checked by cold weather. In addition to the immediate damage caused by the European Corn Borer, the infection of the host by various fungi increases the extent of the injury (Chiang, et al., 1961).

Cultivation methods for the control of this pest include the cutting and shocking of the crop in the field, leaving very short stubs, and feeding the stalks to livestock. Shredding or cutting the plant in very short pieces may kill ninety-five percent of the larvae. If cleanly done, plowing the infested material under is helpful in the control of the pest, since the Borers crawl to the surface and are exposed to their natural enemies before other shelter can be found.

Chemical control of the insect consists of the application of three pounds of either fifty percent granules of toxaphene or fifty percent emulsion concentrate of DDT per acre. The first application should be made when three-quarters of the plants showing feeding in the whorls, repeated in seven days. Forage, including ensilage, that has been treated with either DDT or toxaphene must not be fed to either dairy animals or animals being finished for slaughter.

The European Corn Borer can be controlled by the use of chemicals even if the corn is to be used as ensilage by the use of either thirty pounds of five percent Carbaryl granules OR seven pounds of fourteen percent Diazinon granules per acre of crop. In addition, the following insecticides can be used for corn that is to be used for ensilage, green feed, or pasture: fifteen pounds of Sevin 10G per acre used until one day before harvest, OR seven pounds of Diazinon 14G used until ten days before harvest, OR two pounds of Sevin eighty percent wettable powder for ground sprays only. All of the preceding sprays are given in pounds of the "as supplied" granules or wettable powder per acre. If the corn is to be used only for grain, a single application of ten pounds of Endrin as two percent active granules per acre may be used at least forty-five days prior to the harvest.

The **Corn Earworm** is a widely distributed insect that has breeding habits similar to those of the European Corn Borer, with the exception that the Earworm produces three or more generations each year. Since Sweet Corn is more seriously damaged by this pest than is Field Corn, the control of this pest is described in the section dealing with Sweet Corn.

The **Chinch Bug** is a serious pest of both corn and sorghum plants throughout the corn-growing states. The adult insects leave the cultivated crop in the fall and fly to native grasses where they pass the winter. The eggs of this pest are laid on grass and other plants as well as in the grain fields where they continue until the crop ripens. The second generation of pests develops on the corn while they suck the juices of the crop. As the host plants are destroyed, the advancing army of bugs moves further into the field of corn.

Resistant strains of both corn and sorghum have been developed that provide a degree of protection against the inroads of this pest. Barriers of newly plowed earth that have the furrow coated with creosote of dinitro-*o*-cresol have been laid down in front of the hord of advancing bugs with some success. The Chinch Bug can also be controlled by the application of one half pound of fifty percent emulsion concentrate of dieldrin or four pounds of the fifty percent emulsion concentrate of toxaphene per acre. Either spray should be applied to the base of the plants where the bugs are concentrated.

The **Southern Corn Rootworm** is a bright green beetle that is six millimeters long and has twelve black spots on the covers of the wings. This eastern insect has extended its presence westward to Arizona in the United States of America. The eggs of the Rootworm are laid about the crown of the plant just beneath the surface of the soil. The young larvae migrate downward, feeding on the roots and bulbs of corn and other grains.

The Southern Corn Rootworm can be controlled by the application of either aldrin or heptachlor to the surface of the soil. If the chemical is applied to the rows or hills, or is applied at the time of planting, one pound of the active ingredient per acre is sufficient. If a broadcast application is used, or if the chemical is applied before planting and working into three to four inches of the soil, two pounds of the active ingredient should be used.

Figure VI-3. Left, roots of cornstalk damaged by corn rootworms, right, undamaged. (Shell Chemical Company.)

Two other varieties of **Corn Rootworm**, the **Northern** and the **Western** are also widely found and pose other problems in regard to their control by the use of insecticides. Specimens of the Northern Corn Rootworm with apparent resistance to chlorinated hydrocarbons were discovered near El Paso in central Illinois in 1962. In Nebraska, the Western Corn Rootworm has become very resistant to such chlorinated hydrocarbons as BHC, aldrin, and heptachlor. For these Rootworms, but not the Southern Corn Rootworm, crop rotation has been reported to be a helpful control measure. One of the following organic phosphate insecticides is recommended for the control of these resistant varieties of the Rootworm: ten pounds of ten percent Thimet granules, OR ten pounds of ten percent active ethyl parathion granules, OR seven pounds of fourteen percent active Diazinon granules per acre. A carbamate insecticide, BuxTen has been developed as a control agent for use against these chlorinated hydrocarbon-resistant Rootworms. The manufacturer recommends an application of seven-and-one-half to ten pounds of BuxTen per acre in a seven inch band in the top one half to one-and-one-half inches of soil, based on crop spacing of forty inch rows.

For the general control of the larvae of either of the three varieties of Corn Rootworm, one of the following insecticides can be applied in a seven inch band over the seed row during the time of planting, or on both sides of the row during the time of cultivation of the field: seven pounds of fourteen percent active Diazinon granules, OR ten pounds of ten percent active Thimet granules, OR ten pounds of stabilized, ten percent active ethyl parathion granules per acre of corn.

Cutworms, Wireworms, Corn Seed Beetles, Corn Seed Maggots, and **Ants** may deserve attention in any pest control program. When damage from the Cutworms is first noticed, an experienced operator should be used to apply either one pint of endrin (one sixth pound per gallon) or one half gallon of toxaphene (six pounds per gallon) to each acre of crop. The spray should be applied either over the row or it may be broadcast. For the control of the other pests, the seed should be treated according to the directions of the manufacturer with either dieldrin, aldrin, heptachlor, or lindane.

Many diseases of corn are troublesome, but the control techniques that are available are limited. Presently, about the only control method that exists consists of the selection of a specific disease-resistant hybrid. Unfortunately, no one specific hybrid is imbued with resistance to all diseases, although many breeders of corn hybrids are currently striving toward this end. The use of fungicides to control some diseases appears to be beneficial only in the case of seed corn in the control of seed rot. However, some of the more significant diseases are briefly described to indicate the extent of the problem of diseases in corn. Based upon a knowledge of the most serious disease, the local agent of the Agricultural Extension Service (in the United States of America) should be able to help in the selection of an appropriate disease-resistant hybrid.

Rots of Seed and Blights of Seedlings

Invasion of the newly planted seed by a disease organism can result in a variety of symptoms. Lesions may form on the sprout, small discolored spots may occur on the young root, or, most seriously, the embryo may be killed before sprouting can occur. Affected seedlings may have leaves that are yellowish and/or wilted.

Cold, wet soil is most conducive to seedling blights for two reasons. Seeds germinate more slowly under such conditions, allowing more time for infection. In addition, the pathogens that attack the seeds and seedlings flourish in the cold, wet environment. Seed that was infected with Ear Rot before it was harvested will not germinate properly when it is planted. Seed that had its shell cracked or damaged during harvest, storage, or in planting, is more susceptible to infection and disease. Because of these two considerations, seed must be carefully chosen and planted during a period of time when conditions favor germination and do not favor the development of the pathogens.

Several fungi are responsible for rots and blights of seeds and seedlings. The most common fungi are *Pythium graminicola* and *Pythium debaryanum*, while *Diplodia maydis, Gibberella zeae*, and *Penicillium oxalicum* are less frequently responsible for rots and blights of seeds and seedlings.

Rots of Roots and Stalks

Substantial losses of crop and potential profit for the grower of corn can be caused by rots of the stalks and roots of the crop. If the infection of the plant is extensive, the plant may be killed. However, often the plant is weakened and either falls over or is blown down but the loss may be significant in either case.

Diplodia Stalk Rot, caused by the fungus *Diplodia maydis* (D.Zeae), usually kills the stalk within ten days of the initial infection of the plant. The leaves of the infected plant suddenly turn greenish-gray and the plant reaches a condition in which it is easily crushed within a few days. The diseased stalk contains brown pith and has visible black dots, or fruiting bodies, immediately under the surface of the lower internodes. The fruiting bodies cannot easily be removed from the stalk.

Gibberella Stalk Rot, caused by *Gibberella zeae*, and *Fusarium Stalk Rot*, produced by *Fusarium moniliforme* or *Fusarium moniliforme* var. *subglutinans*, have similar symptoms. The leaves of an infected plant become grayish-green, and the lower stalks lose their firmness, becoming brown as the disease progresses. Inside the stalk, the pith is deteriorated and reddish at the height of the infection. *Pycnidia*, the fruiting bodies of the *Gibberella* pathogen, appear after the death of the plant. The Pycnidia are raised black dots on the plant that can be scraped off the stalk. Fusarium fungi produce powdery pink spores on the sheaves of the leaves at the nodes.

Charcoal Rot, a result of an infection by the fungus *Macrophomina phaseoli* (*Sclerotium bataticola*), is first characterized on the roots of the seedlings by brown, watersoaked lesions that become black as the disease progresses. The disease spreads to the lower parts of the plant, causing gray streaks on the internodes near the base of the plant. *Sclerotia*, minute black bodies, are present on the vascular strands within the shredded stalk.

Kernel and Ear Rots

Diplodia Ear Rot, caused by *Diplodia maydis* that is responsible for the Diplodia Stalk Rot, flourishes when the early summer is rather dry and rainfall is abundant from the time of silking until maturity of the crop. Young ears that are infected by *Diplodia maydis* become straw colored and early infection can result in an ear that is shrunken, lightweight, brownish, and rotted. If the corn is infected at a later point in its development, the damage may be confined within the husk, where a white mold grows between the kernels.

Gibberella Ear Rot, also known as **Red Ear Rot**, is the result of an infection by the same *Gibberella zeae* that causes the Gibberella Stalk Rot. This Ear Rot is characterized by a reddish or pinkish mold that begins at the tip of the ear and grows between the ear and the husk. An ear that is infected early in its development may not rot completely from the attack of this fungus. Hogs that eat corn that has been attacked by this disease frequently lose weight and vomit and some hogs die.

Fusarium Ear Rot is caused by the fungi *Fusarium moniliforme* and *Fusarium moniliforme* var. *subglutinans* that also cause the Fusarium Stalk Rot. The first sign of an infection by this disease is a pinkish discoloration of the caps of the kernels. A Fusarium infection often occurs when the kernel has been injured and is more prevalent during dry weather. Since no specific control has been devised for this, as well as the other diseases, the avoidance of susceptible varieties of the corn is the best control.

A less prevalent disease, the **Grey Ear Rot**, is caused by *Physalospora zeae* or *Macrophoma zeae*. Infected ears are lightweight and the kernels are held in tightly adhering husks that have a bleached color. A grayish-white mold develops on and between the kernels of the corn.

Cob Rot, or **Nigrospora Rot**, caused by *Nigrospora oryzae*, is a disease that enjoys fairly wide geographical distribution. Damaged or unhealthy corn is most susceptible to this disease. The lightweight, infected ears have loose kernels and cobs that are easily broken. Masses of black spores are frequently found in the pith of infected ears, and the kernels from infected cobs generally are lightly streaked.

Storage Rots and Molds

If the moisture content of stored corn, either shelled and stored in a bin, or left on the ear and stored in a crib, rises above twelve to thirteen percent, Storage Rots or Molds may develop. Molds of blue, green, white, reddish, tan, or black are only some of the hues that may be observed, especially at the top or center of the bin. The mold grows on and between the kernels, and cakes the kernels together. Different organisms are able to flourish in various concentrations of moisture.

When one of these organisms begins to grow, various species of *Penicillium* and *Aspergillus* are the most common. The growth process emits moisture and heat. Thus, if one fungus can begin to grow, it may be able to create an environment that is favorable to the growth of another variety.

Control measures for shelled corn begin with storage of the crop in a tight, clean bin after drying to twelve to thirteen percent moisture content. Several samples of the crop should be removed from the storage site periodically for determination of moisture content. In order to make a judgment concerning the "storeability" of the crop, the sample *highest* in moisture content should be used for the basis of the decision, rather than the average. If moldy corn is observed, it should be removed immediately to prevent further spread of the damage.

Other Diseases of Corn

Northern Corn Leaf Blight is most severe during seasons in which the weather is damp. This disease causes long, grayish-green lesions to form on the leaves, beginning with the leaves at the bottom of the plant. In damp weather, large numbers of spores are produced on the surface of the lesions and appear to be a black or gray fuzz on the surface of the infected area. Caused by the fungus *Helminthosporium turcicum* (*Trichometas-phaeria turcica*), the Leaf Blight also infects sorghum, Johnsongrass, Sudangrass, and several other grasses as well as corn.

Southern Corn Leaf Blight, which produces small lesions that are up to one-and-one-half inches long, is caused by the fungus *Helminthosporium maydis* (*Cochliobolus heterostrophus*). The lesions are tannish and occasionally possess a dark colored margin.

Common Smut is characterized by large silvery-white or greenish galls that rupture to disclose numerous sooty-black spores. Smut is extremely common, but not a great deal is known about the spread and other characteristics of this disease except that it is caused by the fungus *Ustilago maydis* (*U.zeae*).

Stewart's Disease or **Bacterial Wilt** is spread among corn plants by the Corn Flea Beetles who carry and deposit the disease as they feed. Sweet corn is particularly susceptible to this disease, which is also more prevalent after a mild winter that allows the survival of a large number of corn flea beetles. This wilt is caused by the bacteria *Xanthomonas stewartii* (*Bacterium stewartii*). Ears of corn that are infected with Bacterial Wilt have long, pale green or yellowish streaks that become brown and dry as the disease progresses. If the disease is severe, much of the leaf may become nonfunctional, causing a reduction in the yield of the crop.

Crazy Top, or **Downy Mildew** is not of great economic importance in most instances. This disease is usually prevalent when the soil was flooded between the time of germination and when the corn reaches ten inches in height. Instead of a normal tassel, a bushy mass of small leaves appears on the infected plant. These leaves are often strap-like and ears may be wanting on such diseased plants in addition to the stunted growth. Crazy Top is caused by the fungus *Sclerophthora macrospora* (*Sclerospora macrospora*).

Maize Dwarf Mosaic, caused by one or more strains of virus, is found in bottomlands. Young leaves that are infected by the Mosaic are marked by a mottled pattern of various shades of green or a pattern of flecks and stripes. The disease usually causes stunted or bushy growth of the plant, frequently the leaves become yellowish as the plant grows and, at the time of tasseling, the leaves often become red or purplish. In addition to infecting corn, The Maize Dwarf Mosaic viral disease is capable of attacking many grasses.

COTTON

Cotton Insects

The light yellow to gray **Boll Weevil** is approximately one fourth of an inch long. Depending on whether the Weevil dines "in the square" (flower bud) or in the boll (fruit), the size of this pest may vary. The Weevil damages the cotton plant by the shedding of attacked squares as well as direct damage to the fiber in the boll. The extent of damage caused by the Boll Weevil varies from year to year, with recent years of considerable damage being in the nineteen twenties and in nineteen fifty.

The cultural control of the Boll Weevil is based upon a number of factors, including: early seeding, seed treatment, resistant varieties, fertilization, and early destruction to the cotton stalks. Chemical defoliation of the cotton plant, if properly done, quickens the opening of the bolls and drives the Boll Weevil from the field (Gaines, 1952).

The Boll Weevil has succumbed to the chemicals shown in Table VI-1.

TABLE VI-1

*Amount of Active Ingredient**	*Chemical*	*Form*
0.25 - 0.50	aldren	dust or E.C.
0.30 - 0.50	benzene hexachloride	dust or E.C.
7 - 15	calcium arsenate	dust
1 - 2.5	carbaryl (Sevin)	dust or W.P.
0.25 - 0.50	Guthion	dust or E.C.
1 - 2	malathion	dust or E.C.
2 - 4	toxaphene	dust or E.C.

*pounds of active ingredient per acre
E.C. = Emulsifiable Concentrate
W. P. = Wettable Powder

The **Cotton Bollworm** is identical with the Corn Earworm, both technically referred to as *Heliothis obsoleta*. This pest attacks the squares and tender leaves of the cotton plant as well as boring into the bolls, cutting and destroying the fiber. Although both the egg and larval stage of the Cotton Bollworm have a number of natural enemies, most growers rely on chemical means to control this pest. Since control of the Cotton Bollworm is difficult after the insect has entered the boll, early control measures are necessary. In order to affect early control, the extent of the Bollworm population must be estimated while the larvae are quite small. Based upon the conditions in the environment and the estimate of the pest population, a decision between reliance upon natural control agents or the use of an insecticide must be made. If the insecticide is chosen, the first application of spray must be made before the larvae have attained a size that makes it difficult, if not impossible, to kill with an economical dosage of insecticide. Until the time when more information is available regarding the effect of various levels of Bollworm infestation on yield and quality of the cotton, the timing of the application of insecticide will remain a slowly progressing art. As an example of the type of investigation that will transform this art into a more-or-less reliable "science", Adkisson, et al. (1964) studied the affects of various levels of bollworm infestation on cotton grown under screen cages. In this study, more than two larvae per ten feet of row were required to cause significant losses of yield of the crop. In addition, the Bollworm infestations were not responsible for "spotted" grades, according to the evidence presented.

To chemically control the Cotton Bollworm, the following insecticides have been used: one to two-and-one-half pounds of carbaryl (Sevin) applied as either a dust or from a wettable powder, OR one to three pounds of DDT applied as either a dust OR from an emulsifiable concentrate spray, one pound of methyl parathion applied as a dust or spray from the emulsifiable concentrate, OR two to four pounds of Strobane as either a dust or spray from the emulsifiable concentrate. The dusts or sprays are applied to the foliage as needed and all concentrations are in pounds of active ingredients per acre of crop.

The **Cotton Aphid** may be controlled by the application of either four tenths to one half pound of benzene hexachloride, OR one half to one pound of carbophenothion, OR one half to one pound of malathion per acre. The insecticide concentration is based upon the amount of active ingredient and the chemical may be applied either as a dust or as a spray, in the latter case, the use of the emulsifiable concentrate form of the insecticide is in order.

The **Cotton Fleahopper** feeds by sucking the juices of the plant from the buds, squares, or tender leaves. This pest overwinters in the egg stage on weeds that serve as the host during the early spring before the Fleahopper flies to the cotton plants. Six to eight generations of this insect are produced each year that cause the leaves of the plant to become ragged and the small squares to turn brown and drop from the plant. The Fleahopper can be controlled by the application of 0.17 to one half pound of aldrin, OR one to two pounds of carbaryl (Sevin), OR one half to three pounds of DDT, OR one half to two and a half pounds of malathion, OR one pound of trichlorfon (Dipterex, Dylox) per acre. The above concentrations of active ingredients can be applied as either a dust or a spray from the wettable powder in the case of the trichlorfon or from the emulsifiable concentrate in the case of the other insecticides.

Figure VI-4. Aerial Application of Insecticide on Cotton Fields. (Shell Chemical Co.)

The **Cotton Leafworm** may be controlled by the application of seven to ten pounds of calcium arsenate, 0.25 to 0.375 pounds of Guthion, OR one to one and a quarter pounds of carbaryl (Sevin), OR one quarter pound of methyl parathion, two pounds of Strobane, OR two pounds of toxaphene. The concentration of the selected insecticide is based upon the active ingredient per acre which may be applied either as a dust or a spray which may be prepared form either the emulsifiable concentrate or the wettable powder.

The adult **Pink Bollworm** has a brown wing spread of approximately four-fifths of an inch. The eggs of this pest are deposited at the base of maturing bolls of cotton under the calyx; if the insect is particularly abundant, some eggs may be placed in the squares. The pinkish, mature larvae are about one half inch long. Seven or eight generations of this pest may be produced in the longer seasons of the more southern districts of the United States. Severely infested squares and small bolls are shed, reducing the yield of the crop. Larvae that feed in the boll cut through the lint to reach a seed, consume the contents of the seed, and bore through the lint to the next seed. Much of the lint may escape the damage that is caused by a light infestation of this pest, but a heavy infestation of the Pink Bollworm will render the boll useless (Curl and White, 1952).

A number of methods of treating the cotton seed are available for use to control infestations of the Pink Bollworm, including subjecting the seed to 150°F for over thirty minutes, fumigation at atmospheric pressure with methyl bromide and, fumigation with hydrocyanic acid gas under a vacuum. The lint may be passed between steel rollers, or exposed to vacuum fumigation together with compression pressure during the baling process (Curl and White, 1952). Alternatively, or additionally, one-and-one-half to two pounds of active ingredient carbaryl (Sevin) may be applied as either a dust or a spray that is prepared from the emulsifiable concentrate per acre of cotton.

Thrips may be controlled by the application of one of the following insecticides to the foliage of the cotton plant as necessary: 0.13 to 0.5 pounds of benzene hexachloride, OR 0.35 to one pound of carbaryl (Sevin), OR 0.25 to one pound of DDT. The given concentration of active ingredient may be applied either as a spray or dust to each acre of the field.

Spider Mites may be controlled through the use of either a dust or spray of one of the following chemicals (as usual concentrations are given in pounds of active ingredient to be applied per acre): 0.33 to one pound of aramite, 0.375 to one pound of carbophenothion, 0.125 to 0.38 pounds of demeton (Systox), 0.25 to 1.25 pounds of malathion, or twenty to forty pounds of sulfur (as a dust only).

Specific Safety instructions and restrictions for the use of insecticides on cotton

Benzene hexachloride *must not* be applied to cotton that is rotated with root crops.

Demeton (Systox) is highly toxic, even when compared to other insecticides, and should be applied only by trained and qualified personnel.

Livestock must not be grazed in fields of cotton that were treated with either aramite, benzene hexachloride, carbophenothion, or DDT.

Livestock must not be fed gin waste that is produced from cotton that was treated with aramite or demeton (Systox).

Cotton Diseases

Disease control in cotton is dependant upon good cultural practices. High quality seed that germinates well is the first step. The seed should be treated by one of the following methods to prevent contaminating the seedlings with seed-borne diseases. The seed should be treated with either Captan 75, Ceresan L, Ceresan M, Ortho LM, OR Panogen 15 at the rate of three ounces of chemical per hundredweight of seed if the seeds were machine delinted OR two ounces of agent per hundred pounds of seed if the seeds were acid delinted. The seeds that have been treated should be planted only if the soil temperature is either 70°F or greater or if the soil has been greater than 65°F at a depth of six inches for five consecutive days (the temperature should be taken at approximately 9:00 a.m.). In addition, the soil should have a pH of between 6.0 and 6.5, regulated with lime if necessary. A soil fungicide is required in addition to the treatment of the seed if seedling disease has been a problem. One of the following methods may be used for the application of the soil fungicide: 1) three pounds of ten percent Terraclor (PCNB) plus either Captan OR Terraclor Super X dust may be mixed with the seed per acre, if a planter box application is desired, OR 2) an in-furrow spray of one half gallon of Panterra emulsifiable concentrate, one gallon of Terraclor (PCNB) emulsifiable concentrate, OR one half gallon of Terraclor Super X emulsifiable concentrate may be applied, OR 3) in-furrow granules of ten pounds of Terraclor Super X granules (10% Terraclor - 2.5 Terrazole) OR fifteen pounds of the mixture of 6.5% Terraclor and 1.63% Terrazole may be used.

Specific Diseases of Cotton

One or more common soil-borne fungi cause a disease of seedling cotton that is commonly called **Soreshin**, or **post-emergence Damping Off**. Among the organisms that are responsible for the disease in cotton seedlings are various species of *Phythium, Rhizoctonia solani* Kuhn, and *Fusarium*. Often the seedling is infected before it emerges from the ground. The stems of diseased plants are dark reddish brown near the soil line. A seed that is infected when first planted may rot prior to the emergence of the seedling. Emerging seedlings that are infected will be stunted and often are unproductive. Infected seedlings will die if the infection is sufficiently severe.

Cool, wet growing seasons are conducive to Soreshin as is hard-packed soil. The control of this disease is aided by the adherence to good cultural practices and the preceding control techniques.

Root Rot, whose presence becomes evident at the fruiting and squaring time when the leaves of the cotton plant turn yellow, wilt, become brown and dry, is a fungus disease caused by *Phymatotrichum omnivorum* (Shear) Duggar. The roots of infected plants have reddish-brown lesions and support the growth of light tannish threads of fungus which bear brown sclerotia. The fungus is capable of penetrating the outer layers of the tissue of the root, after which it "rots" the roots. This soil borne disease attacks many species of plants, but spares corn, cereal grains, and grasses. Crop rotation, deep plowing, and generally good weed control programs are important in the control of Root Rot.

Verticillium Wilt, a soil borne fungus, is caused by *Verticillium albo-atrum*. The disease may attack the plant in any stage of its growth, but generally the symptoms of Verticillium Wilt become evident at the onset of fruiting. The leaves of diseased plants are stunted in their further growth. While the internal tissue of the stem develops brown stains, the leaves and young bolls of diseased plants drop to the ground.

Sandy soils are the favorite habitat of *Meloidogyne incognita*, the **Root Knot Nematode**. The above ground portion of diseased plants is stunted and may die, while the root system is smaller than that of a healthy plant, containing round, knot-like growths. Crop rotation is a beneficial control measure as is fumigation of the soil. Two to three gallons per acre of either ethylene dibromide, Fumazone, or Nemagon S1 is the recommended soil fumigation agent.

The fungus that is responsible for **Southwestern Cotton Rust**, *Puccinia stakmanii*, also attacks Grama grasses. After the beginning of summer showers, the disease becomes evident as reddish-orange lesions on the upper side of the cotton leaves, and black or rusty brown spots of the stems and leaves of grasses in the range lands next to the cotton fields. This rust overwinters on the infected grasses on adjacent rangelands, suggesting the obvious control measure of either heavy grazing of such range lands or burning of the grasses. The application of Zineb before the onslaught of an infection is a helpful preventative measure. Two pounds of Zineb plus a "sticker-spreader" agent as directed should be applied as a spray in forty gallons of water per acre at ten day intervals from late June until mid August.

A bacterial disease of cotton that results from infection by *Xanthomonas* is variously called **Leaf Spot**, **Angular Leaf Spot**, **Boll Blight**, and **Blackarm**. Leaf veins and areas of infected stems and branches at first appear to be water-soaked, but later become dark brown or black, retaining water-soaked edges. The Blight causes dark green, water-soaked spots that become black, to form on bolls that later often do not open. Among the control measures that are advanced to control this disease are crop rotation, the use of disease resistant seed, destruction of old, infected plants, and the use of fungicide-treated and acid delinted seed. In areas of New Mexico and district 6 of Texas that are known to be subject to infestations of this bacterial blight, the use of blight resistant seed Acala 1517BR-1 has been advocated while the use of Acala 1517C seed was recommended for use in soils that are known to be mildly infected with bacterial blight.

FORAGE GRASSES

Since **Cutworms** and **Armyworms** are reasonably similar pests from the standpoint of control considerations, the same insecticides may be used for both types of insects. The application of one-and-one-half pounds of carbaryl (Sevin), either as a wettable powder or as a suspension concentrate, OR one-and-one-quarter pounds of emulsifiable concentrate of malathion or one-and-one-half pounds of malathion dust may be used per acre for the control of either Cutworms or Armyworms.

Grasshoppers are best controlled while they are young. The best results are obtained when the borders of the pasture are treated with one of the following insecticides: one pound of carbaryl (Sevin) suspension concentrate, OR either one-and-one-quarter pounds of malathion emulsion concentrate OR one-and-one-half pounds of malathion dust, OR one pound of naled (Dibrom) emulsifiable concentrate per acre of pasture. If naled (Dibrom) is used for the control of grasshoppers, cattle must be removed from the treated pasture for four days after the application of the chemical.

Leafhoppers, Meadow Bugs, Spittlebugs, Plant Bugs, and other assorted insect pests can be controlled by the judicious application of either malathion or naled (Dibrom), generally at the same concentrations of insecticide that are used for the control of grasshoppers.

Most cereal grains and many forage grasses are attacked by **Anthracnose.** A plant that is severely attacked by this disease loses vigor, possesses shriveled seeds, and may die prematurely. *Colletotrichum graminicolum*, the fungus responsible for Anthracnose, is most active during periods of moist, warm (approximately 80°F) weather. Affected plants often possess bleached stems and rotted roots and crowns. Small light colored lesions with dark, reddish-brown borders develop on the leaves as the plant becomes diseased. The fruiting bodies of this fungus, or *acervuli*, are tiny raised black dots that may be found near the joints of the diseased plant or in the center of old, bleached lesions.

Anthracnose is a particular problem in crops grown on coarse, poor, sandy, dry, or infertile soil. Therefore, maintenance of the soil in a good state of fertility, especially having adequate concentrations of phosphorus and potassium in the soil, is important in the control of Anthracnose. The fields should be prepared by clearing plant debris and deep plowing. When forage grasses are planted, a mixture of grass varieties should be sown whenever feasible. The use of certified, disease-free seed of a resistant variety of grass is always a necessary portion of a disease control program. Mercuric based seed treatments have proven particularly effective in the control of Anthracnose. The control of weed grasses and a crop rotation plan including corn, soybeans, etc. are also helpful in the control of this disease.

Stripe Smut is a pest of many pasture and hay grasses as well as such favorite lawn grasses as Kentucky bluegrass. Plants that are infected with the Strip Smut fungus, *Ustilage striiformis*, commonly die during the hot, dry spells in the summer. Initially long, thin, gray stripes appear on plants infected with Stripe Smut. As the disease progresses, these lesions become black as the spores become visible. When the diseased plant tissue bursts, the dark, greasy spores are liberated from their host. The leaves of a diseased plant curl downward, wither, shred, and finally die.

A field that becomes severaly infested with Stripe Smut should be plowed deeply before planting a nonsusceptible crop for the next two years. Effective control of grass-type weeds is important in the control of this disease, so that a reservoir of the disease is not created in such weeds. Disease-free, certified seed of a well adapted variety of grass should be chosen for planting and the seed should be treated with either Captan or Thiram fungicide prior to planting to complete the control procedures.

PEANUTS

The grower of peanuts must contend with a great variety of insect and disease pests. Some of these pests will be discussed as well as possible control measures that can be used to minimize the damage that they would otherwise cause. As is always the case, chemicals are dangerous and must be used properly and with great care. Proper use of chemical control techniques always includes the use of no more than the recommended concentration and no less than the recommended interval between applications.

Seed and Seedling Diseases of the Peanut

A variety of organisms attack the seeds and seedlings of the peanut. Some of these organisms are soil-borne, whereas others are borne by the seed. A poor stand of seedlings may be the result of **pre-emergence Damping Off. Post-emergence Damping Off** is characterized by an area on the stem near the soil that is soft, yellowish, and has a water-soaked appearance. The stem becomes constricted, causing the plant to fall and die. The fungi alleged to be the cause of Damping Off in the seedlings include various species of *Botrytis, Fusarium, Phytophthora, Pythium*, and *Rhizoctonia*. The application of one of the following substances to each one hundred pounds of seed should help control seedling diseases: four ounces of Arasan 75, OR a four ounce mixture composed of two parts of Captan 75 and three parts of two percent Ceresan (by weight), OR three ounces of Semesan, OR four ounces of Yellow Cuprocide, or four ounces of a mixture of equal parts by weight of Captan 75 and two percent Ceresan. It is particularly important that the recommended dosage of two percent Ceresan is not exceeded.

The quality of the soil, local climate, types of soil disease that are prevalent in the immediate area, and other factors affect the amount and type of seed treatment that is most effective. The local extension agent is generally very helpful in the selection of the best treatment for any particular disease problem.

Root, Stem, Peg, and Pod Rot

Various soil-borne diseases cause root, stem, peg, or pod rot. The use of good cultural practices is the first step in control of any of these diseases. Deep plowing of infected weeds and litter from the crop, rotation of the crop with small grains, corn, or grasses, and good weed control are helpful measures in the control of these diseases. The choice between mechanical or chemical weed control is influenced by disease control considerations since the crown of the plant must not be covered during mechanical cultivations and injury to the plant incurred during mechanical cultivation provides an entrance for the organisms that cause these diseases. As a result of these considerations, the use of herbicides to control weeds generally causes less damage to the peanut plant and thus is generally preferred to mechanical cultivation.

Peanut Leafspot diseases are caused by several different organisms. Since the treatment of these diseases is not dependent upon the source of the infection, these diseases can be grouped together for the purpose of discussing recommended control procedures. *Leptosphaerulina arachidicola* causes a Leafspot that is manifested by symptoms that vary from the appearance of tiny spots, called "pepper spots", to leaves that appear scorched. The fungus *Cercospora arachidicola* produces small, pale spots on the underside of the leaflet. As the disease progresses, these spots become first yellow, then brown, often developing a yellow halo. Infection by *Cercospora arachidicola* produces the disease **Early Leafspot. Late Leafspot**, caused by the fungus *Cercospora personata*, is similar to **Early Leafspot**, except that the spots of the latter have no halos.

These Leafspot diseases can generally be controlled by the regular use of a crop rotation plan and deep plowing. When the spots first appear on the older leaves at the base of the plant, fifteen to twenty pounds of one of the following dusts can be applied per acre: sulfur, copper-sulfur mixture, or Maneb. If mites are a problem in the field, sulfur containing dusts are recommended. The dust should be re-applied every ten to fourteen days until two weeks prior to the harvest. Alternatively, one of the following sprays may be applied when the active ingredients are mixed with ten to fifteen gallons of water (per acre): one third to one half gallon Copoloid 6, OR one half gallon of CAL-COP 10, OR one pound of Cyprex wettable powder, OR one-and-one-half pounds of Maneb wettable powder, OR one to two pounds of Polyram wettable powder, OR one half to three quarters of a gallon of TC-90. As above, the sprays should be applied at ten to fourteen day intervals.

Insects of the Peanut

For the control of **Aphids**, either one half to three quarters of a pint of Parathion emulsifiable concentrate (four pounds of active ingredient per gallon of concentrate), OR one to one-and-one-half pints of malathion emulsifiable concentrate (five pounds per gallon of concentrate), OR fifteen to twenty pounds of five percent malathion dust. Malathion must not be applied within thirty days of the harvest of the peanuts. (All concentrations are per acre.)

Various varieties of **Caterpillars** may be controlled if the foliage of the peanut is treated with one of the following: fifteen to twenty pounds of ten percent Sevin dust, OR two to two-and-one-half pounds of eighty percent Sevin wettable powder, OR one half to one gallon of 4-2 toxaphene-DDT emulsifiable concentrate, OR ten to twenty pounds of 20-10 toxaphene-DDT D, per acre of peanuts.

At the first sign of damage to the peanut by either **Corn Earworms, Velvetbean Caterpillars**, or **Fall Army Worms**, one of the following sprays may be applied to the foliage (concentrations are stated per acre): fifteen to twenty pounds of ten percent DDT dust, OR one half to one gallon of DDT emulsifiable concentrate containing two pounds of DDT per gallon, OR fifteen to twenty pounds of ten percent active Sevin dust, OR two to two-and-one-half pounds of eighty percent active Sevin wettable powder, OR ten to twenty pounds of 20-10 toxaphene-DDT dust, OR one half to one gallon of 4-2 toxaphene-DDT emulsifiable concentrate. The use of Sevin or a toxaphene-DDT mixture may prove necessary to control the DDT-resistant strains of Corn Earworm that are evolving.

Although **Cutworms** are difficult to control, some success is achieved through the application of either one half to one gallon of 4-2 toxaphene-DDT emulsifiable concentrate, OR ten to twenty pounds of 20-10 toxaphene-DDT dust per acre to both the foliage of the plant and the soil directly under the plant.

To minimize the damage that is caused by **Leafhoppers** or **Thrips**, one of the following sprays may be applied to the foliage when required: ten pounds of ten percent DDT dust, OR one half gallon of two pounds per gallon DDT emulsifiable concentrate, OR twenty pounds of five percent Sevin dust, OR one-and-one-quarter pounds of eighty percent Sevin wettable powder, OR ten pounds of twenty percent active toxaphene dust, OR one third gallon of six pounds per gallon toxaphene emulsifiable concentrate per acre of crop.

The application of one of the following insecticides in a twelve inch band over the row immediately prior to "pegging time" will control infestations of the **Southern Corn Rootworm** (concentrations in pounds per acre): fourteen pounds of fourteen percent active Diazinon granules, OR twenty pounds of ten percent active parathion granules.

If **Thrips** are expected to be a problem, then application of one of the following at planting time proves to be an effective control procedure: three quarters to one-and-one-quarter pints of six pounds per gallon Di-Syston, OR six to ten pounds of ten percent active Di-Syston, OR three quarters to one-and-one-quarter pint of Thimet (six pounds per gallon), OR six to ten pounds of ten percent active Thimet granules. The smaller dosages should be used if the soil is either light or sandy. The granules of insecticide are applied in the row with the seed or the liquid is applied one to two inches to the side of the row of seed.

Twenty to thirty pounds of three percent Ethion dust OR one to two pints of four pound per gallon Ethion emulsifiable concentrate may be applied per acre to prevent damage by the **Spider Mite**.

The damage that is caused by the **Red-Necked Peanutworm** may be minimized by the application of a foliage spray that contains one of the following insecticides: ten pounds of twenty percent active toxaphene dust, OR one third gallon of toxaphene emulsifiable concentrate (six pounds of active ingredient per gallon of concentrate), OR ten pounds of ten percent active DDT dust, OR one half gallon of DDT emulsifiable concentrate (two pounds per gallon), per acre.

Two weeks prior to planting twenty pounds of ten percent active paration G OR one half gallon of parathion emulsifiable concentrate should be broadcasted on the surface of the soil and worked into the top three to four inches of the soil immediately after application to control either **Wireworms** or **White Grubs**.

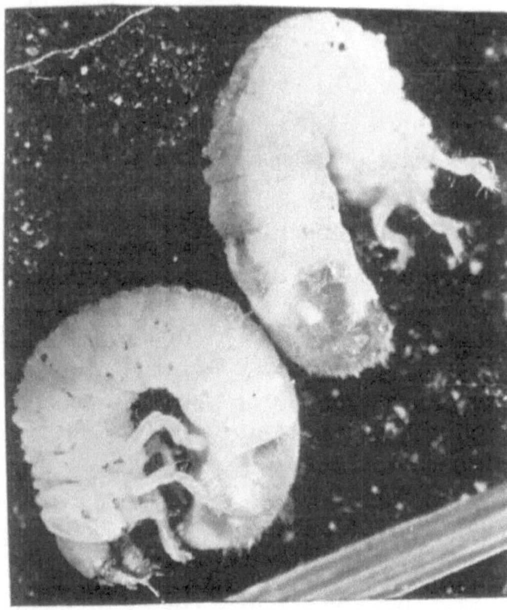

Figure VI-5. White Grubs. (Shell Chemical Company.)

General Considerations

Care must be taken to avoid the drift of the pesticide to adjoining fields, especially those that are used for either food crops or for grazing areas.

Plants that have been treated with either DDT, Cyprex, Ethion, maneb, Polyram, or toxaphene must not be fed to dairy animals, poultry, or animals that are to be slaughtered for their meat.

Seedling peanuts may be burned by Di-Syston, Sevin, or Thimet, but seem to recover in most cases.

Di-Syston, Ethion, parathion, and Thimet are particularly *highly toxic* to warm blooded animals and must be handled with extreme care and a great deal of caution must be exercised in their use.

OATS

The **Yellow Dwarf** disease of oats is caused by a virus and is also known as either **Red Leaf** or **Gray Spot**. The first symptoms of this disease are greenish-yellow blotches that appear on the tip of the leaf. These spots enlarge as the disease progresses, growing down the leaf and often turning bright or dark reddish color. The sections of the plant that are not either yellow or red have an unnaturally dark green color. Young plants that are infected by this disease may perish, but the infection of older plants may result in either reduced yields or the formation of no crop at all. Although no simple remedy for this disease has been identified, an effective insect control program is important since the disease is spread by several species of aphids.

Blue Dwarf is a virus disease vectored by the six-spotted leafhopper, *Macrosteles fascifrons*. Plants that are infected with this disease are stunted, uniformly dark blue-green, and maintain their green color longer than normal, healthy plants. This disease was first reported in the United States in 1951 and, while the disease is not yet widespread, no control measure has yet been found that controls **Blue Dwarf**.

Septoria Black Stem, a fungus disease that is caused by infection from *Leptosphaeria avenaria*, is widely spread throughout the oat raising regions of both the United States and Canada. Leaves that are infected with this disease, once known as **Septoria Leaf Blotch**, contain round or oval brown spots that contain small, black fruiting bodies. Infected stems develop black, necrotic spots, especially at the joints or internodes. Infected stems are often weak and break at the point of maximum infection. Even the hulls of the plant may become infected, exhibiting brown discolorations. This disease overwinters on infected oat refuse, suggesting the importance of both crop rotation and the avoidance of spreading manure that contains such refuse on fields that are to be sown with oats. Resistant strains of oats have been developed that should be used whenever practical or necessary for the control of Septoria Black Stem.

Scab, caused by the fungus *Gibberella zeae*, is called **Fusarium Blight** when it attacks the seedlings. Infected seedlings suffer stunted growth and may turn yellow before dying. Often the roots of such seedlings are reddish-brown and rotted. Later in the growing season, infected plants' roots and hulls may be partly covered with a pink mold. If the disease attacks a joint of the plant, the growth above the joint usually dies, becoming white. Control measures include an effective crop rotation plan, avoidance of top-dressings of manure that contains infected refuse, and the use of disease resistant varieties of oat seed whenever possible.

Helminthosporium Leaf Blotch is characterized by long reddish-brown spots that appear on the leaves soon after they emerge on the seedling. Diseased leaves serve as the source of infection for the healthy leaves. When the oat head becomes infected, the fungus enters the hulls and may even enter the kernel, the basal end of which turns brown. Diseased seeds are often not as heavy as fully developed, healthy seeds. As a result of their lighter weight, diseased seeds can be separated from healthy seeds by fanning or air separation processes. Treatment of oat seed with an appropriate fungicide is an effective measure in the control of *Helminthosporium avenae* infection in the emergent seedlings.

Anthracnose attacks most parts of the plant, including the leaves, joints, heads, roots and stems. This disease is often more severe on winter than spring varieties of oats. This disease, caused by the fungus *Colletotrichum graminicola*, is manifested on the oat by reddish, brown lesions that develop tiny black fruiting bodies after the leaf dies. When the crown or basal tissue is infected by this disease, it becomes bleached before turning brown and developing the characteristic fruiting bodies.

Figure VI-6a. Scab on oats. White spikelets are diseased; the darker ones are healthy. **6b.** Victoria blight of oats. (Illinois Natural History Survey.)

Basic general cultural practices constitute the first control measure for this disease. A proper crop rotation plan should be followed. Light grains should be fanned from the seed since the light seeds are most likely to come from an infected plant and be the source of ensuing disease. Whenever possible, the sowing of disease-free seed is also recommended.

Victoria Blight, at times a very destructive disease, is caused by the fungus *Helminthosporium victoriae*. Every stage of the development of the oat is susceptible to attack from this disease. Seedlings that are infected before they emerge from the soil are sickly, having narrow reddish-brown stripes that span the length of their grayish-blue leaves. Badly infected seedlings soon die. The presence of fruiting bodies causes the hulls of the seeds from infected plants to appear black from the black velvet-appearing bodies that also coat the sheaths of the leaves and the lower joints of the stem.

Oats that are planted in poorly drained soil, or during especially wet seasons, may be infected by the fungus disease known as **Downy Mildew**. The growth of diseased plants is stunted and the plants mature later than do healthy oats. The leaves of diseased plants possess a more upright stance and are thicker and shorter than those of a normal plant. Any heads that are produced by a plant so infected by *Sclerospora macrospora* are malformed, and viable seed is not produced by the diseased plant.

Good cultivation practices are important in the control of Downy Mildew. Disease-free seed that is planted in well drained areas helps prevent the disease. The elimination of weed grasses near grain fields removes a refuge for the disease. Manure that contains straw from diseased plants should not be spread on fields in which grain will be planted.

Stripe Blight, a bacterial disease that is usually of little consequence, is caused by *Pseudomonas striafaciens*. Although the leaf of the oat is the prime target of this disease, other parts of the plant may be attacked. When the leaves are initially attacked, small, sunken, water-soaked lesions appear on their surfaces. As the disease progresses, the lesions enlarge, become brownish, and yield a bacterial exudate in the form of droplets. General, good, clean, cultivation practices are helpful in the control of this disease.

Infestations of **Halo Blight** vary in severity from one year to the next, partly as a result of fluctuations in the weather. This bacterial disease, caused by infection with *Pseudomonas coronafaciens*, primarily affects the leaf portion of the plant. The initial manifestation of Halo Blight on the leaf is a small, light green, oval spot that is rather sunken. The center of the lesion turns brownish or gray with a yellow border as the disease destroys more tissue of the leaf. No bacterial exudate appears on these "haloed" spots. If infestations of this disease are severe, a resistant variety of oats should be selected if possible. In any case, good cultivation practices, including weed control, insect control, seed treatment, and crop rotation, are helpful in the control of Halo Blight as well as most other diseases.

RYE

Dark, grayish-blue, water-soaked spots that appear on the leaves of the plant are the first sign of an infection by *Rhynocosporium secalis* that causes the disease that is known as **Scald**. These first blotches grow pale and develop a yellowish border and greenish-yellow halo as the disease progresses. Cool weather is conducive to the growth of this disease that destroys the leaf of the plant. Crop rotation, clean cultural practices, and the use of disease-resistant varieties of Rye help to control Scald. The control of weeds is especially important in the control of Scald since some perennial grasses can act as reservoirs of this disease.

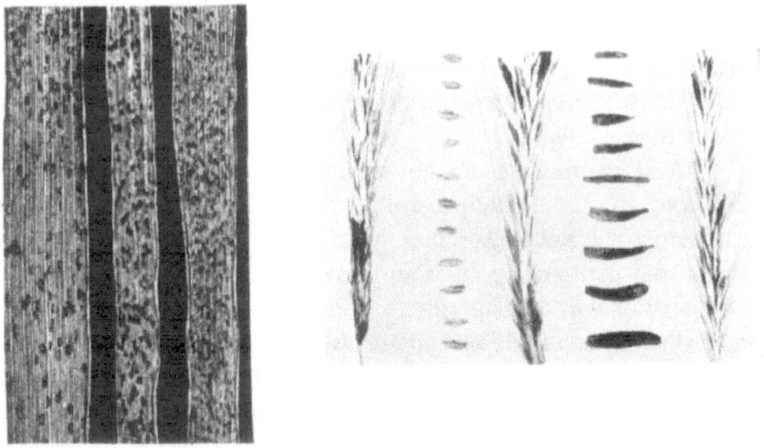

Figure VI-7a. Leaf rust of rye. **7b.** Ergot on rye. (Illinois Natural History Survey.)

The **Leaf Rust** of Rye, *Puccinia rubigovera secalis*, is similar in appearance to the leaf rusts that appear on other cereal grains. The small pustules on the leaves of the plant are oval and orange-brown. The life history of the disease is rather involved, and includes a stage in which dark-grayish pustules exist.

For the description of **Stem Rust** (*Puccinia graminis secalis*); **Spot Blotch** (*Helminthosporium sativum*); **Scab, Head Blight**, or **Fusarium Blight** (*Gibberella zeae*); **Anthracnose** (*Colletotrichum graminicola*) see the same diseases in the section "Wheat". The **Stem Smut,** *Urocystis occulta*, is somewhat similar in appearance to the Flag Smut of Wheat. For the control of any of the above diseases, the general observance, of clean cultivation practices, the use of a well-designed plan for the rotation of the crops, and the use of varieties of Rye that are resistant to the specific disease(s) that are of greatest concern are recommended.

Ergot, also known as **Spurred-Rye, Spur-Kernels**, or **Bright Kernels**, results from the infection of the plant by the fungus *Claviceps purpurea* (Fr.) Tul.. Flowers of the infected plant secrete a substance that is similar to "honeydew". Dark, purplish-black "sclerotium", or horn-like structures, develop in the place of the normal kernel. These "sclerotium" contain substances that are very toxic and are capable of causing serious symptoms in any man or beast that ingests the infected grain. Early symptoms include diarrhea and vomiting, rapid breathing, high pulse, and the abortion of pregnant animals. In the later stages of reaction to the ingestion of infected grain either spasms or gangrene occurs. Neither stages are pleasant. Careful attention is being given to the avoidance of this disease. Poisoning by Ergot has become less common now than was the case in former times; the practice of modern crop sanitation and milling methods have lessened the prevalence of this disease. Control of Ergot includes planting clean seed, cutting weed grasses before they "head out", and the use of crop rotation. Badly infested hay must not be fed to livestock, nor should livestock be allowed to graze either in infected pastures or roadsides.

SUGAR BEET

Insects of the Sugar Beet

For control of **Aphids** on seed crops only, one half pound of the gamma benzene hexachloride can be applied as either a dust or as an emulsion concentrate to each acre of the foliage of the sugar beet crop.

The control of **Armyworms** in the sugar beet is the same as the control of these pests for use on the small grains (see section on Small Grains).

The small (three to three-and-one half millimeters long) **Beet Leafhopper** carries the disease organisms that cause "curly top" of the beet and tomato. This insect is widely distributed in the western United States, breeding on both native plants and the Russian Thistle. One to four broods are produced each year depending on the location. In California the Beet Leafhopper overwinters as an adult on the Russian Thistle as well as on certain perennials. In the spring, the pest migrates from the region of the foothills to the area of the cultivated crops in the valley. Insects which have fed on a diseased plant transmit the disease to healthy plants as their feeding continues. Since the losses to the growers of sugarbeets from this disease and the insect that carries it have been severe, a concerted plan of control of this disease has been devleoped. The plan is based upon the location of the areas of heavy infestation in the fall of the year, followed by the application of insecticides from an aircraft. In California alone, the fall treatment may cover from 150,000 to 200,000 acres. The usual insecticide is either three quarters of a pound of malathion or one pound of DDT per gallon of diesel oil per acre of crop that is to be treated (Harper and Lockwood, 1961).

Diseases of the Sugar Beet

Curly Top is a virus disease that is most prevalent in the United States, west of the Rocky Mountains. This disease is characterized by the rolling up of the leaves of the beet. Infested beets exhibit an inordinately large amount of growth in the root section of the plant. This phase, of extraordinary growth of the root, is also known as **Wooly Root** or **Hairy Root**. As the disease advances, the leaves of the beet turn yellow and wilt. Finally, the infected plant dies.

The **Beet Leafhopper**, *Circulifer tenellus* (Baker), is the vector of the Sugar Beet **Curly Top**. Control of this insect through the use of chemicals as well as destruction of its alternate weed hosts is helpful in the elimination of the further spread of this disease. Natural predators have been imported from the Mediterranean area in the hope that they will aid in the control of the Beet Leafhopper. For effective control of this disease, virus resistant varieties of the sugar beet plant should be planted whenever possible.

Cercospora Leaf Spot, a fungus disease that results from the infection of the plant by *Cercospora beticola* Sacc., is destructive east of the Rocky Mountains in the United States. The first signs of infection by this disease are the emergence of dark bordered, brownish, small spots on the older leaves of the plant. These spots enlarge and turn gray as the disease advances. At last, the center of the spot falls from the plant, leaving a ragged leaf that eventually dies. The crown of the beet from an infected plant is rough. Cercospora Leaf Spot is similar to most fungal diseases in that it grows most readily during periods of warm, humid climate. Spraying the beets with a fungicide provides some protection from this disease in addition to the control that is achieved from the use of disease resistant varieties of the plant.

SUGARCANE

Insects of Sugarcane

The **Sugarcane Borer** is found in the gulf coast regions of Texas, Louisiana, and Mississippi as well as in the southern half of Florida. Sugarcane, sorghum, corn, rice, and a number of grasses are attacked by this pest. The adult moth is straw-colored and has a wingspread of approximately one inch. The female lays clusters of eggs that may contain over twenty-five eggs. The emergent larvae feed on the leaves of the host before boring into the stalks in which they burrow up and down. Four or five generations of Sugarcane Borers each year accomplish this patterned boring, ending with the last generation which overwinters in the trash and stubble of the sugarcane.

The damage that is caused by the Sugarcane Borer includes killing the eyes of the seed cane, the whorls of the young shoots, injury to the growing shoots, broken stalks, and loss in weight of sucrose that is held in the cane. Cultivation practices that may be used to control the Sugarcane Borer include the use of borer-free, resistant varieties of sugarcane, the early harvest of the cane, cutting it close to the ground, and cleaning all trash from the field before burning the refuse during the winter. The Sugarcane Borer can be controlled by the application of twelve pounds per acre of two percent endrin granules on a 30-60 mesh

RVM-AA attapulgite. Four applications of this spray should be made at fourteen day intervals. If the trimmings of the plant are to be fed to livestock, twelve to fifteen pounds of forty percent ryania dust should be applied to each acre of sugarcane (U.S.D.A. Leaflet 479, 1960).

TOBACCO

Control of Insects in the Plant Bed

NOTE: In this section the recommendations are for the application of the dosage to each one hundred square feet.

For the control of **Aphids**, parathion should be applied to the foliage of the plant. One of the following forms of parathion is satisfactory: one pound of a one percent dust, three to five gallons of a spray that is composed of either two-and-one-half fluid ounces of the forty-two percent active emulsifiable concentrate or one half pound of the fifteen percent wettable powder per twenty-five gallons of spray. The treatment of the plants should be conducted at ten day intervals, and five days before transplanting a mixture of parathion and TDE should be used. As always the high level of toxicity associated with the parathion indicates that a trained operator should be employed to apply the spray or dust.

Budworms, **Flea Beetles**, and **Hornworms** can be controlled by the application of TDE to the foliage of the plant as necessary. The TDE may be applied from a dust composed of one pound of the five percent active form of TDE, OR three to five gallons of a spray that is composed of either one half pint of the twenty-five percent active emulsifiable concentrate or one quarter pound of the fifty percent active wettable powder per twenty-five gallons of water.

The broadcast application of one half pound of a bait per one hundred square yards of plant bed or fifteen to twenty pounds per acre of field prior to transplanting can be used to control the **Cutworm**. Later applications of the bait can be made around the base of the plant. A recommended bait consists of two pounds of fifty percent active DDT, fifty pounds of coarse corn meal or wheat bran, and enough water to moisten the powder to the consistency of a paste.

White Grubs, such as the larvae of the Green June Beetle, may be controlled by the application of one half pound of fifty percent active wettable Diazinon powder per one hundred gallons of water. This dispersion may be applied at the rate of one gallon per square yard of plant bed from a sprinkler can or similarly suitable instrument.

To control **Mole Crickets**, broadcast the problem areas with one half pound per one hundred square yards or fifteen to twenty pounds per acre of a bait. This bait is composed of fifty pounds of cottonseed meal or corn meal, two pounds of forty or fifty percent active Chlordane wettable powder and enough water to moisten the powders into a paste.

Insects of the Tobacco

Aphids may be controlled by the use of a spray that is composed of either one to one-and-one-half pounds of active endosulfan (Thiodan) in the form of either a dust or a spray prepared from the emulsifiable concentrate, **OR** one to one-and-one-half pounds of malathion active ingredient, **OR** 0.15 to 0.35 pounds of active parathion dust or spray per acre of crop. The spray or dust is applied to the foliage of the plant as necessary for effective control of the Aphid.

The **Budworm** is a serious pest of the tobacco in the southern United States. The moths of this pest are greenish and have a wing span of one-and-one-half inches. The hind wings of the Budworm are silvery with a brown fringe. During the thirty days from the egg to adult stage, the larvae grow to a length of one-and-one-half inches at maturity while the color ranges from green to brown. The adult moth emerges from the pupa that is hidden in the soil and lays its eggs either in the seed head or on the under-side of the leaves of the plant. The larvae that hatch from these eggs bore into either the buds or the seed pods, destroying them in the process. The larva also cut small holes in the leaves of the plant, damaging the leaves for use as leaf tobacco (Allen and Lawson, 1962).

The Budworm may be controlled by the application of a commercial preparation of *Bacillus thuringiensis* as a bait consisting of one pound of the wettable powder per nineteen pounds of corn meal. This bait is applied to the buds of the plants when the eggs and small worms of the pest are evident. Alternatively, one of the following insecticides may be used for the control of this pest: fifteen to thirty pounds of five percent Sevin dust, **OR** one to two pounds of eighty percent Sevin sprayable powder, **OR** fifteen to thirty-five pounds of five percent TDE dust, **OR** two to four pints of twenty-five percent TDE emulsifiable concentrate, **OR** one to two pounds of fifty percent wettable TDE powder, **OR** fifteen to thirty-five pounds of four percent Thiodan dust, **OR** two to three pints of twenty-five percent Thiodan emulsifiable concentrate, **OR** ten pounds of a bait composed of two pounds of fifty percent DDT wettable powder per seventy-five pounds of corn meal. All of the above concentrations are per acre of crop to be treated, and the smaller dosages should be used when the tobacco plants are small, increasing the dosage to the recommended maximum for fully mature plants.

The adult **Hornworm** is a large grayish moth possessing a wing expanse of four to five inches. The mature larva of this insect are three to four inches long and green in color. The larva is armed with a long spine on the posterior end of the body. This pest is very destructive to the foliage of the tobacco plant.

A great deal of attention has been given to determine the value of natural control of the Hornworm by the use of the **Polistes Wasp**, a natural predator of this pest. However, damage to the crop has not been held to less than seventy-four percent if this technique of control is solely relied upon. Better control of the Hornworm was obtained by the combination of a light application of either endrin or TDE and the activities of the predator (Lawson, et al., 1961).

Alternatively, the application of a commercial preparation of *Bacillus thuringiensis* applied at a rate of two pounds of the wettable powder per acre has proven effective in the control of the Hornworm. This preparation may be applied either alone or as a bait to the tender young leaves. As a substitute for the above treatment, 0.5 to 0.75 pounds of active endosulfan (Thiodan) applied as either a dust or a spray based on the emulsifiable concentrate, OR one pound of active TDE may be applied as a sprayed emulsion, OR Sevin (see the section on the control of the Budworm), OR one-and-one-half to two-and-one-half pounds of twenty percent Guthion emulsifiable concentrate per acre of the crop.

The Thiodan and Guthion will control the Aphids in addition to the Hornworm or **Flea Beetle**, but the other materials must be used in combination with parathion for control of aphid pests. For control of the Hornworm with the maximum efficiency, the larger concentration of the Guthion should be employed.

One of the following should be broadcast on the surface of the soil and **immediately** disked into the upper six to nine inches of soil to discourage **White Grubs**: one gallon of twenty-five percent parathion emulsifiable concentrate, OR twenty pounds of ten percent parathion granules, OR forty pounds of five percent parathion granules, OR twelve to fifteen pounds of fifteen percent parathion wettable powder, OR one gallon of twenty-five percent Diazinon emulsifiable concentrate, OR forty pounds of the five percent Diazinon granules, OR fourteen pounds of the fourteen percent Diazinon granules per acre of crop to be treated.

For control of the **Mole Crickets**, see the section of the control of these pests in the Plant Bed.

If control of the **Wireworms** has been less effective than desired in the past, two ounces of the fifty percent Diazinon wettable powder per fifty gallons of transplant water can be used. Two-hundred-and-fifty to three hundred gallons of this mixture per acre is used to thoroughly saturate the soil around the plant. Fields that have histories of severe infestations merit a broadcast application of one of the following, per acre of crop land: Diazinon (one gallon of the twenty-five percent emulsifiable concentrate, fourteen pounds of the fourteen percent granules, OR forty pounds of the five percent granules), OR parathion (one gallon of the twenty-five percent emulsifiable concentrate, or twenty pounds of the ten percent granules, or forty pounds of the five percent granules, or twelve to fifteen pounds of the fifteen percent wettable powder). At least two weeks prior to planting, broadcast the selected insecticide on the soil and immediately disk it into the top six to nine inches of the soil. These insecticides should not be applied mixed with the fertilizer.

The grower has a choice of methods for the control of **Snails and Slugs**. Two pounds of one of the several commercial formulations that contain metaldehyde can be scattered on the ground around the plant. Alternatively, late in the afternoon, two to five pounds of parathion dust (one percent active) can be applied. The application of the parathion dust should be repeated in four to six days for maximum effectiveness.

For control of the **Vegetable Weevil**, two pounds of a five percent active TDE dust can be broadcasted.

Diseases of the Tobacco

The sudden wilting of the tobacco plant is one of the first symptoms of **Black Shank**. This wilting is accompanied by the discoloration of the roots. As the disease progresses, the roots become entirely discolored and the stem is blackened to a point above the level of the soil; then the plant dies. A cross section of the plant reveals blackened bark that encloses healthy-looking pith at the top of the plant, separated from the dead tissue by a watery, discolored zone. The soil-borne fungus that causes this disease, *Phytophthora parasitica*, often attacks plants in low, wet areas of the field. Black Shank is one of the most destructive diseases that attacks the tobacco crop. Crop rotation and the choice of disease resistant varieties of plants are important measures in the control of this disease. Disease-free seedlings must be planted to control the disease. Since irrigation or other waters may spread the disease, care in the selection and use of water is a final measure to be used in the control of Black Shank.

Damp weather enhances the growth of the bacterium *Pseudomonas angulata* that causes the **Angular Leafspot** of the Tobacco. Symptoms of this disease usually appear within four to five days after a rain. The bacteria enter the young leaves and produce irregularly shaped, water-soaked spots that turn black or dark brown as the disease progresses. These spots cause the leaf to appear as if it had been scorched. As the spots age, they may develop tan centers, while the borders remain dark. Severely infected leaves may be puckered or ragged.

This disease may be seed-borne and may overwinter on the roots of weeds as well as on the plant litter. Clean cultivation practices, including weed eradication, are effective in the control of Angular Leafspot. A farmer who uses his own seed should have the seed treated with silver nitrate to avoid the contamination of the resultant seedlings. To control Angular Leafspot in the plant bed, the bed, sideboards, and any contaminated equipment should be fumigated with methyl bromide. If Angular Leafspot occurs on plants in the bed, a spray containing streptomycin at the level of four hundred parts per million should be applied at the rate of five gallons per one hundred yards of bed. A field containing plants with wet leaves should not be worked to minimize the mechanical spread of the disease that is possible. If a field becomes badly infected, tobacco should not be grown for a period of at least two years.

Tobacco is just one of the crops that are attacked by the soil-inhabiting fungus *Rhizoctonia solani*. Beans, peanuts, cabbage, and cotton also are attacked by this fungus that causes the **Sore Shin** of tobacco. Tobacco is attacked both in the plant bed and in the field. The infected young plants possess a dark colored, decayed area near the soil line. This lesion girdles the stem and climbs up the plant until the plant drops. Few plants become infected in the field if uninfected seedlings are transplanted to the field. This situation is fortunate as no suitable measures exist to control Sore Shin except for the fumigation of the plant bed with methyl bromide.

Beginning at the base of the plant and progressing upward, the leaves of tobacco that are infected with **Southern Stem Rot** turn yellow, wilt, die, and turn brown. At the soil level, the stem is encircled by a brown, depressed, dry-rotted area. This depressed spot becomes covered with a white, cottony growth during periods of damp weather. Appearing on the spot later in the course of the disease are fruiting bodies of the fungus that are responsible for this disease, *Sclerotium rolfsii.* This soil-borne disease is fairly common in many tobacco-growing areas but seldom is responsible for extensive damage to the crops of tobacco. No chemical control for the Southern Stem Rot is currently available.

Granville Wilt, a bacterial disease, can be very damaging to tobacco crops. *Pseudomonas solanacearum*, the causative agent, also attacks many plants that are related to the tobacco, such as potato, tomato, pepper, eggplant, ground cherry, horsenettle, string bean and ragweed. One or two leaves of an infected plant wilt during the day, or the leaves may be only half wilted. Initially, the leaves may recover from the wilt during the night, but as the disease slowly progresses, the wilted leaves gradually lose their color, becoming yellow. In cases of severe infection, the entire plant wilts and soon dies. Infected plants develop a yellowish-brown discoloration within their vascular tissue. As the disease advances, these discolorations darken and may even become visible on the outer surface of the stem. An off-white material can be expressed from the root and stalk of a diseased plant.

Control methods begin with the utilization of clean cultivation practices, including the control of weeds and nematodes. Resistant varieties of tobacco should be planted whenever possible and only healthy stock should be transplanted from the seed beds. If Granville Wilt should develop in the field of tobacco, the area should be isolated as much as possible, for instance, soil from the infected area should not be borne into non-infected areas by cultivation practices. Finally, fields that are infected with the disease should not be planted with either tobacco or other crops that are susceptible to this infection for at least five years.

Tobacco Mosaic, a virus disease, is characterized by the presence of a light and dark green, mottled pattern on the infected leaves. Infected plants are often stunted. A healthy plant may become infected simply by being brought into casual contact with a diseased plant. Field workers can transmit the disease to healthy plants if they have first handled infected ones. Tobacco products, such as chewing tobacco, that are used by field workers may carry the disease to healthy plants if these products were manufactured from infected leaves. Control of the Tobacco Mosaic virus embraces the old adage that "an ounce of prevention is worth a pound of cure". The best control of this disease is its advoidance by the planting of disease-free plants in disease-free soil. Field workers should be prohibited from the use of tobacco products while working in or near the plant beds. Before workers enter the fields their hands should be decontaminated by dipping in milk. Susceptible crops and weeds, especially tomatoes and peppers, should not be grown in the plant beds. After the tobacco is harvested, the roots of the plants should be excavated and exposed to the top of the soil for a period of three weeks before they are deeply plowed under. Finally, a three year crop rotation plan should be followed, using crops that are not susceptible to the disease, such as grain, cotton, or peanuts.

WHEAT

Diseases of Wheat

Wheat, oats, and rye, as well as several other grasses, are attacked by *Xanthomonas translucens undulosum*, that bacterial agent that is responsible for the disease **Black Chaff**. Grain that is infected with this disease has sunken, dark-colored, long spots on its chaff. These lesions develop small yellow beads of ooze during damp weather. When the beads dry, they resemble yellow scales. Small water-soaked areas that become black appear when the leaves and stems of the plant are attacked by this bacterial agent. Crop rotation, general good cultivation practices, and the use of treated seed are recommended for the control of this disease. Organic mercury compounds are the most widely used seed treatment agents.

A discolored area at the base of the glume, the portion of the plant that contains the kernel, is the primary symptom of **Basal Glume Rot** that is caused by *Pseudomonas atrofaciens*. At the present time, no specific control measures are effective in the treatment or prevention of this bacterial disease. Adherence to the general cultural practices that are recommended for Black Chaff and other diseases will provide a small measure of control.

Yellow Mosaic, a virus disease, is noticeable in the spring. Large patches of wheat appear yellow or pale green. The leaves, sheaths, and chaff of infected wheat are mottled. The plants may be somewhat dwarfed as a result of an infection of Yellow Mosaic.

The infection of the plant with **Green Mosaic** virus causes the plant to assume a different appearance than is caused by the Yellow Mosaic virus. Excessive tillering which results in a bushy, dwarfed plant as well as an intense green color are characteristic of infections of **Green Mosaic**.

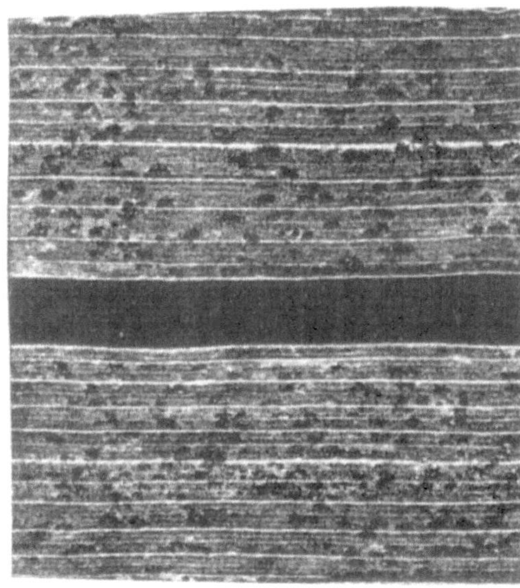

Figure VI-8a. Basal glume rot on wheat. Upper row of kernels are diseased. **8b.** Leaf rust on wheat. (Illinois Natural History Survey.)

Both Yellow and Green Mosaic diseases are able to remain in the soil for a period of time. Heavy soils hold the virus longer than do lighter soils. Wheat sown in the spring does not appear to be affected by this disease, but when control is necessary, the use of disease resistant varieties of seed is recommended whenever possible.

Hot, dry weather is the ideal climate for the development of **Streak Mosaic**. The area of Western Kansas is particularly troubled by infections of this disease. In the spring, the symptoms of this disease become evident, as the leaves become streaked and mottled. Later the leaves turn yellow and the growth of the plant is curtailed. At the normal time of maturity, the infected plant turns brown and dies.

Leaf Rust infections are noticeable from the small, raised, red-orange pustules that initially appear on the older leaves of the plant. Throughout the season, these lesions infect the new leaves. Near the time of harvest, a second type of spot, darker in color, appears on most parts of the plant. Both types of pustules remain until after the wheat is mature. The life history of Leaf Rust is far too complicated to be dealt with in the brief discussion offered here. The results of the infection and method for its control are very common for the control of many diseases; infection causes a reduction in the yield of the crop and infection can be controlled by the utilizaiton of disease resistant varieties of wheat.

Stem Rust is caused by the fungus *Puccinia graminis* and results in light-weight, poor quality grains being produced by the infected wheat. The eruptions of Stem Rust on the leaf of the plant are elongated and rather ragged. These brick red lesions initially appear on the wheat in late spring or early summer. At the time of harvest, black spores are produced from the lesions. The life history of this disease is very complex, involving the common barberry plant as an alternate host. Some varieties of wheat have been developed that possess a measure of resistance to Stem Rust. Early maturing varieties of wheat usually sustain less rust damage than later maturing varieties. Eradication of the common barberry (not to be confused with the Japanese Barberry) is very helpful in interrupting the life cycle of the rust, especially if the eradication measures are coordinated by the growers within a large, contiguous area.

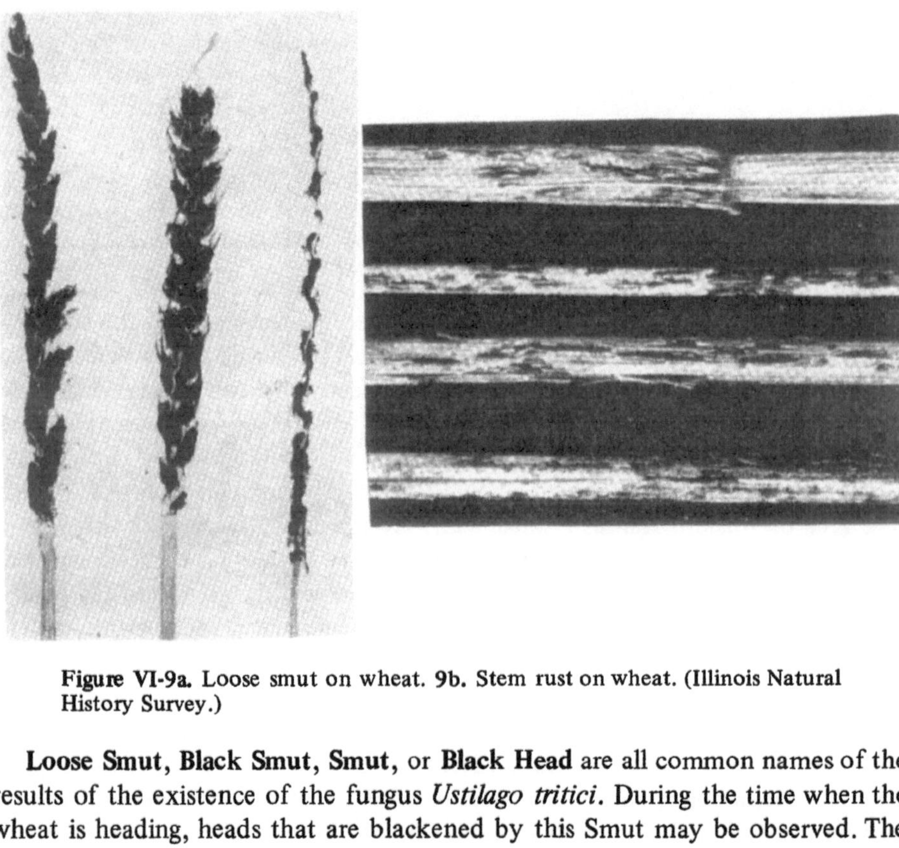

Figure VI-9a. Loose smut on wheat. **9b.** Stem rust on wheat. (Illinois Natural History Survey.)

Loose Smut, Black Smut, Smut, or **Black Head** are all common names of the results of the existence of the fungus *Ustilago tritici.* During the time when the wheat is heading, heads that are blackened by this Smut may be observed. The spores of this fungus are loosely attached to the plant and may be readily dislodged to leave only an inconspicuous, naked head. The fungus is carried within the diseased grains of wheat, rather than on the surface of the grain. For this reason, fungicides that are intended for contact treatment are ineffective in the control of this disease. The treatment of the wheat in a hot water bath is effective in killing the fungus, and not harming the wheat. This operation must be carried out carefully so as not to damage the ability of the grain to germinate.

Cool weather is the prime time for damage by the fungus *Septoria tritici*. This fungus causes a disease that is known as **Speckled Leaf Spot, Septoria Leaf Spot, Nebular Leaf Spot**, or **Leaf Spot.** Elongated, reddish-brown spots that have light colored centers as well as being dotted with many tiny black fruiting bodies, are characteristic of an infection of Speckled Leaf Spot. Seed should be sown early enough in the autumn so that the young plants have established strong growth before the weather becomes cool enough for the growth of the fungus. In any case, the use of good cultivation practices are an aid in the control of Speckled Leaf Spot.

Flag Smut, a potentially damaging fungal disease, results from infections by *Urocystis tritici*. The infected plants are often deformed and twisted and have long dark stripes on the leaves, leaf sheaths, and upper stem areas. The spores of the smut are contained within these long lesions. Damage that is caused by Flag Smut often is unnoticed since the diseased plants are frequently stunted and die prior to harvest. Disease resistant varieties of wheat should be sown in areas that are infected by this disease whenever possible. The treatment of the seed with a fungicide supplemented by adherence to good cultivation practices is helpful in the control of this disease.

Stinking Smut, also known as **Bunt**, is caused by the fungus *Tilletia foetida* and attacks rye as well as wheat. The head of an infected plant is more slender at the time of blossoming than are the heads of normal, healthy plants. However, by the time of harvest, the diseased heads appear to be plumper than the normal heads. The ball of smut develops in the place of the kernel of wheat and consists of an oily, dark brown, foul-smelling powder. The foul, fishy odor is capable of permeating healthy wheat, rendering the healthy grain unfit for milling.

The fungus that causes **Scab**, *Gibberella zeae*, attacks wheat and most other grains, and many grasses. The severity of this disease varies from year to year depending upon the weather since the disease flourishes during warm, moist seasons. This fungus is capable of causing Seedling Blight, Stem Blight, Crown Rot, Root Rot, and Head Blight, or Scab. Grain that is from plants that have been infected with Scab is chaffy and light, containing substances that are toxic to man and to animals which have similar digestive systems (especially hogs). Diseased spikelets are pale yellow instead of green at the dough stage. During damp weather a pinkish color appears at the base of the spikelet or along the edges of the chaff. Severely infected kernels are shrunken as well as discolored and have a seed coat that is rough, flaky, and sometimes covered by small, dark, fruiting bodies. To control infestations of Scab, clean cultivation practices should be employed, including the avoidance of top dressings that contain diseased straw and the use of clean seed for planting when the temperature of the soil has descended to 60°F or below. Of course, the seed should be of a disease resistant variety of wheat whenever possible.

Figure VI-10. Scab on wheat: The head and the nine kernels on the far left are uninfected, the other kernels and heads show varying degrees of infection. (Illinois Natural History Survey.)

Low, poorly drained areas are ideal sites for the development of the fungus that causes **Downy Mildew,** *Sclerospora macrospora*, the causative agent of this disease, makes some plants to tiller excessively, dwarfed plants, and the early death of other plants. Thickened twisted leaves and abnormal growth of the plant are often seen when this disease is prevalent. If the dieease becomes a significant problem, susceptible crops should not be planted for a period of several years.

The fungus that causes **Powdery Mildew,** *Erysiphe graminis tritici*, also attacks other grains and grasses besides wheat. Infection usually occurs when the air is moist. The growth of the plants is dense, and the plants do not suffer from a deficiency of nitrogen. The leaves of the plant are the primary site of attack for this disease and the first symptoms are the appearance of small, pale gray spots on the upper surface of the leaves. Later, these spots appear as if flour had been sprinkled on the leaves. When the wheat ripens, these spots develop dark fruiting bodies. Some varieties of wheat that are resistant to Powdery Mildew are available for use when necessary, but crop rotation is a simple and effective measure in the control of this disease.

Figure VI-11. Downy mildew on wheat. (Illinois Natural History Survey.)

The fungus, *Helminthosporium sativum*, attacks many grasses and some grains causing a disease that is called **Spot Blotch**. All stages of the growth of the plant are susceptible to **Spot Blotch**, including the seedlings, roots, crowns, leaves, and stems. Affected parts of the plant become rotted and the diseased plant is dwarfed and rather reddish-brown. Diseased plants occur in patches in the field of healthy wheat. Cankers at the nodes of the plant are dark and often velvety. Infected leaves develop dark spots and blotches. Good cultivation practices, crop rotation, and the treatment of the seed with mercuric fungicides are helpful in the control of Spot Blotch. Wild grasses should also be controlled since they can serve as a reservoir of the disease.

The blades and sheaths of the leaves of wheat are the sites of infection of *Helminthosporium tritici vulgaris*, the fungus responsible for **Yellow Leaf Spot**. This disease, known to occur only on wheat, is characterized by elongated, brownish-yellow spots. These spots fade a bit as the disease progresses and develop a yellow border. In the presence of a severe infection, nearly all the leaf will become yellow. For the control of Yellow Leaf Spot, the recommendations that are advanced to control Spot Blotch should be followed.

The fungus *Colletotrichum graminicola* that causes **Anthracnose** attacks most parts of the wheat plant. The diseased plant tissue of the crown and at the base of the stems becomes bleached, followed by browning as the infection becomes more severe. Diseased joints are the site of brownish or purplish water-soaked spots. When the wheat is nearly mature, the fruiting bodies of the fungus appear on the stems, leaf sheaths, and, occasionally, other parts of the plant. These fruiting bodies are elevated, extremely small, somewhat elongated, and black. Severely infected plants are not vigorous and produce shriveled grain. Control methods similar to those described for Spot Blotch are generally effective in the control of Anthracnose. In addition, the proper balance of potassium and phosphorus in the soil is desirable for the control of this disease.

SORGHUM

Diseases of Sorghum

Many diseases of sorghum can be controlled through the treatment of the seed before it is planted. Treatment of the seed with one of the following should afford protection against seed rots, seedling blights, and kernel smuts: Arasan, Captan, Phygon, Spergon, or Thiram. The use of one of these agents in the treatment of the seed will increase the yields of sorghum for both forage and grain.

Charcoal Rot, resulting from infection by *Macrophomina phaseoli*, is troublesome on light, sandy soils when conditions of drought persists in the late summer. In 1953, the varieties of Redbine 60 and Weskan sorghum were crossed to produce, by 1959, the disease resistant sorghum New Mexico 31. Since this strain possess many desirable characteristics in addition to Charcoal Rot resistance, it has been suggested as a base for future hybrids.

In most areas of the United States in which untreated seed is planted, **Covered Kernel Smut** is a common disease. Infected plants possess a normal appearance with the exception of the smutted heads of the crop. These heads are infected with *Sphacelotheca sorghi*, the fungus that is responsible for Covered Kernel Smut. Infected heads suffer the destruction of their kernels and the replacement of the kernels by masses of spores of the Smut. These spores are dark brown, powdery, and covered by a membrane. At the time of harvest, the spores are released, clinging to healthy seed. When these seeds are planted in the spring, the Smut spores germinate and infest the seedlings of the sorghum. Covered Kernel Smut can be controlled by treatment of the sorghum with a fungicide that contains Captan, Chloranil, Dichlone, or Thiram.

The heads of sorghum that are infected with **Head Smut** are replaced by a large gall of smut before the heads have even been able to emerge from the boot. Inside these galls are dark colored, powdery smut spores as well as long, thin, broom-like structures. Various elements of the weather are instrumental in scattering the spores to the soil and plant litter. In the spring, these residual spores of *Sphacelotheca reiliana* infect the sorghum seedlings, but the infection is not evident until the sorghum heads later in the year. Control of Head Smut is achieved by treatment of the sorghum seed in the same manner as is recommended for the control of the Covered Kernel Smut.

Sphacelotheca cruenta, the culprit that is responsible for **Loose Kernel Smut**, attacks all types of sorghum. Long, pointed smut galls replace the kernels in heads that are infected by this disease. When the galls mature, their membrane covering ruptures and the dark colored, powdery spores are released. Remaining on the plant is the *columella*, a long, curving, pointed appendage. Infection that results in the Loose Kernel Smut is usually unobservable until the plant heads. Infected plants head earlier than do normal plants, are stunted, and often produce an abnormal number of side branches. Measures to control the Loose Kernel Smut are similar to those that are recommended for the control of Covered Kernel Smut.

SOYBEAN

Soybeans arrived in the United States of America from their native home in the orient before 1804. During the 1920's, several people attempted to promote the soybean from a long-standing laboratory curiosity to a useful crop. In 1936 a USDA Soybean Laboratory was established to coordinate research and development activities connected with the soybean throughout the United States and Canada. These early research units helped develop improved varieties of the soybean, investigated disease control, and determined potential uses for the crop in the creation of both edible and industrial products. From this research and development, the soybean emerged as a valuable addition to the crop rotation sequence and a plant that yields useful oil and meal after processing. The soybean is currently the second most important cash crop for the farmer in the United States, with a value of over $2.5 billion per year.

Insect Control for the Soybean

The control of the three most serious insect pests of the soybean, the **Bean Leaf Beetle**, the **Green Cloverworm**, and the **Mexican Bean Beetle**, are briefly and succinctly summarized in table VI-1.

TABLE VI-1

INSECT PEST	*INSECTICIDE DOSAGE+*	*AGENT*	*RESTRICTIONS**
Bean Leaf Beetle	1.5 lb.	80% Sevin WP	none
Green Cloverworm	1.5 lb.	80% Sevin WP	none
Mexican Bean Beetle	1.5 pt.	57% malathion EC	one day
	OR 3.0 lb.	50% methoxychlor WP	seven days
	OR 1.5 lb.	80% Sevin WP	none

+Quantity per acre
*Days that must elapse between the last application of insecticide and the commencement of the harvest
WP = Wettable Powder
EC = Emulsifiable Concentrate

Disease Control for the Soybean

The presence of seedling blights or seed rots in the crop of soybeans indicates that the seed that will be planted next season should be treated with either Arasan, Captan, or Thiram to control these diseases.

Bacterial Blight, *Pseudomonas glycinea* Coerper, is first evident as tiny, brownish or yellow, water-soaked spots. After these spots turn brown and become numerous along the vein of the leaf, the leaves assume a ragged appearance. Wet, cool weather is favorable for the growth of Bacterial Blight. Good crop rotation combined with deep fall plowing that buries infected debris is helpful in the control of this disease. High quality seed is a second control measure that should not be overlooked.

Bacterial Pustule, *Xanthomonas phaseoli* var. *sojensis* (Hedges) Starr and Burkholder, initially appears as small, yellow spots with brownish-red, wrinkled centers on the leaves of the plant. This disease, which is most rampant during warm, damp weather, usually initially attacks the bottom leaves of the plant. Damaged leaves may become ragged and drop from the plant. If the infection is severe, brown spots may even develop on the pods of soybeans. To control this disease, the following techniques are recommended: provide for at least a one year crop rotation schedule with cotton, grain or grasses and use a good quality of seed (the recommended varieties, except Jackson, are resistant to Bacterial Pustule).

Downy Mildew is characterized by yellow-green spots that grow dark brown and irregularly shaped on the upper surface of the leaf as the disease progresses. The underside of the infected leaf sports a grayish mildew. This fungus disease is caused by *Peronospora manshurica* (Naoum.) Syd.

Pod and **Stem Blight** usually attacks the stems and pods of the almost mature plants. The affected parts of the plant exhibit black, small, raised fruiting bodies. The standard practices of deep plowing, use of high quality seed, and conventional crop rotation are effective in the control of Pod and Stem Blights.

Purple Stain is a seed- and soil-borne disease that is caused by the fungus *Cercospora Kikuchii*. This disease causes pinkish-purplish discolorations on the soybean seeds, often cracking them as well. Hampton 266, Hardee and Hill are varieties of soybean that are particularly resistant to Purple Stain, but most recommended varieties of soybean are somewhat resistant to this disease.

Root Knot is caused by the gall-producing nematode *Meloidogyne spp*. Plants that are severely infested with galls on their roots are yellow, stunted, and may be wilted. Crop rotation helps control this nematode, especially when a disease resistant seed, such as Bragg, Hardee, or Jackson, is also used.

Southern Blight often is prevalent when soybeans are planted following either peanuts or other soybeans. This soil- and debris-borne fungus, *Sclerotium rolfsii* causes the plants to die suddenly. Near the soil line, the stems become rotted and are covered with mats of thick white fungus growth if the plants are infected by Southern Blight. The fungus growth contains very small brown bodies.

First noticed in central Illinois in 1944, **Brown Stem Rot** has become a serious disease. Cool weather in late July and early August enhances the growth of *Cephalosporium gregatum* Allington and Chamberlain that is responsible for Brown Stem Rot. This soil-borne disease enters the plant through the lower portion and the roots and then moves slowly up the plant. The external symptoms of an infection by this disease appear on the leaves in late August or early September. The stem of the infected plant contains a brown stain. The tissue of the leaf becomes brown between the veins; as the disease progresses, the leaf withers. Since a satisfactorily resistant variety of seed has not been developed, crop rotation is the only established measure for the control of Brown Stem.

Phytophthora Rot and **Stem Rot** are fungal diseases that are caused by *Phytophthora sojae* Kaufmann and Gerdemann. This disease was first reported in central and northern Illinois in 1955, but appears to be a problem only during wet seasons at the present time. Seedlings that are infected by this disease may shrivel and die, whereas older, infected plants develop brown areas on their stems, wilt, and die. This fungus also rots the roots and turns them dark brown. Resistant varieties of seed include Blackhawk, Clark 63, Lindarin 63, Harosoy 63, and Hawkeye 63.

Sclerotinia Stem Rot is characterized by large, black bodies, or *sclerotia*, that develop with a white mass of fungal growth on the lower part of the stem of the plant. No control technique is available that is effective against the cause of this disease, the fungus *Sclerotinia sclerotiorum* (Lib.)de Bary.

Bud Blight, caused by the *Tobacco Ring Spot Virus*, is able to attack plants throughout their growing season. When the young plants are infected, their terminal bud curls and becomes dry. Brown areas may be noted on other parts of the plant growth and the plant is unable to produce seed. Later infections may result in stunted pods that fall prematurely, or purple spotted pods that don't drop from the plant. No measures are available for the control of this seed-borne virus.

Brown Spot, caused by the fungus *Septoria glycines* Hemmi, is manifested as brownish-red spots on the first two leaves that are produced by the plant in the spring. The disease progresses up the plant, infecting the new leaves as they open and leaving the older leaves to turn yellow and fall from the plant. The lower half of an infected plant may be defoliated as a result of a severe case of Brown Spot. This disease grows most rapidly during warm, humid weather. The fungus overwinters on the debris on infected plants and crop rotation is, at present, the only effective measure for the control of Brown Spot.

Wildfire, a result of an infection by the seed-borne bacterium *Pseudomonas tabaci* Wolf and Foster, is a concern in the southern United States. This disease is characterized by brown areas of dead tissue of the leaf that are bordered with yellow haloes. Wildfire is often found on beans that are already infected with Bacterial Pustule. To control outbreaks of Wildfire, seed that was raised in an uninfected field should be used.

VII

Insects and Diseases of Orchards and Vineyards

At the beginning of the twentieth century, orchardists were faced with a series of severe problems in their efforts to produce marketable fruit. A constant, and frequently losing, battle was waged against insect pests and fungal diseases.

Wormy apples were the rule as a result of the Codling Moth. Apple Scab and Mildew were a constant threat. Defoliating Caterpillars, Canker Worms, Leaf Rollers, and Tent Caterpillars weakened trees and stunted their growth. Plant mites defoliated the trees prematurely. Pear blight, a bacterial disease of both the apple and the pear, caused heavy losses in their trees and killed hundreds of acres of trees despite severe pruning of the infected limbs. Fruit Flies attacked the apple and the cherry. Aphids abounded, curling the leaves and twisting the fruit. Peaches, plums and apples were injured by the curculio beetle. Scale insects infected the fruit and weakened the tree. Brown rot destroyed many peaches and plums, both on the tree and after the harvest. Citrus groves were invaded by Mited, Mealy-bugs, and Scale insects that threatened their very existence.

Since fruit standards were not high, generally low quality fruit was marketed. The consumer bore the loss of trimming out the wormy and scabby portions. The loss to the consumer might have easily amounted to thirty percent of the total weight of the fruit, materially increasing the cost of the fruit. In an effort to produce acceptable fruit in commercial quantities, the grower turned to the use of chemicals.

About 1890, orchardists began trying *Paris Green* (a copper-arsenic insecticidal compound) to control the Codling Moth. Paris Green previously had proven effective in checking the invasion of the Colorado Potato Beetle that was sweeping the country. The available supply of Paris Green was impure, frequently containing sufficient free arsenic to injure the tree. Later legal standards of purity for the insecticide were established that did much to improve the quality of the product. With the development of lead and calcium arsenates, better control of the pests in orchards was achieved. Even with these improvements, some districts required such heavy, repeated applications that the trees were injured and the soil became so saturated with arsenic that great blocks of orchards had to be pulled and the land planted to arsenic resistant crops, such as rye and barley.

Figure VII-1. Scab shown on leaves of ornamental crap apple. (Illinois Natural History Survey.)

For scale insects and the winter eggs of Aphids and Mites, a lime-sulfur solution and crude fractions of petroleum oil were commonly used at the turn of the century. Citrus growers found that an emulsion of kerosene was ineffective protection against scale insects and, in desperation, turned to fumigation of their trees under cloth tents with hydrocyanic acid. This custom, although occasionally causing injury, was followed for several years until the development of refined petroleum oil fractions that could be used as foliage sprays. This practice, coupled with the introduction of a numbers of parasites and predators, have held the insects and mites under control for many years.

Concern over the safer use of insecticides was shown in a study begun in 1915 by the United States Department of Agriculture (USDA) on the amount of arsenic on sprayed fruit and vegetables when harvested. Up to this time, the subject of public health has apparently received little attention. Added interest in the subject was provided by reports from England of illnesses caused by eating apples from the United States and Canada. These apples, were alleged to have a larger amount of arsenic that the British legal tolerance of 0.01 grains of arsenic trioxide per pound of food. In 1927, the Federal Food and Drug Administration established the legal tolerance of 0.025 grains. This limit is now expressed as 7 ppm (parts per million) of combined lead per pound of fruit. Any excess of lead arsenate may be removed by washing the fruit with 1.5 percent solution of hydrochloric acid followed by rinsing with clean water. This practice has become so common that seizures of fruit for exceeding the legal tolerance of arsenic are now rare.

From 1940 until the present, the development of insecticides such as DDT, Guthion, and Sevin have further improved the position of the orchardist in his battle against pests. Codling Moth losses of from ten to twenty-five percent of the crop have been cut to the range of one to five percent. Plant mites have been checked by the addition of an acaricide to the sprays used on the Codling Moth. A new impetus has been given by these developments to the production of high quality commercial fruit. New and safer insecticides are being developed to safeguard both the public health and the health of beneficial insects.

The fact that insecticides are chemicals designed to kill should never be forgotten by the user. All insecticides, especially the more broadly toxic ones, should be used with careful observance of the directions, at minimum dosages and with as few repeated applications as is consistent with good practice for control of the pest. Observance of these suggestions tends to hold the residue of the pesticide on the crop within the legal tolerances, reduces the danger to predators of the pest, and lowers the probability of building a dangerous residue of the pesticide in the soil.

THE APPLE

Insect Pests of the Apple and their Control

The Rosy or Purple Aphid, *Anuraphis roseus* Baker, is one of the more serious pests of apples. These aphids feed on the young leaves of the tree (apple, pear, quince, or hawthorn), the growing tips of the tree, and the young fruit. The aphid can be controlled by the use of dormant sprays or foliage sprays. The dormant sprays generally consist of either an emulsion of a petroleum fraction or emulsifiable oil alone, or in combination with a dinitro compound. Such a dormant spray may contain two-and-one-half gallons of emulsified oil, or two gallons of emulsifiable oil plus one half pound of dinitrocresol per one hundred gallons of water. If the dormant spray has not held the aphids in check, a foliage spray may be needed. Typical foliage sprays are comprised of the following amounts of "active" ingredients per one hundred gallons of water:

1 pound 25% wettable	*Diazinion* powder
2 pounds 25% wettable	*Malathion* powder
¼ pound	Benezene Hexachloride
¼ pound	*Demeton*
¼ pound	*Diazinon*
½ pound	*Endosulfan*
¼ pound	*Lindane*
¼ pound	*Parathion*
0.12 pounds	*TEPP.*

The Wooly Aphid, *Eriosoma lanigerum* Hausman, is occasionally a serious pest of apple trees, especially with nursery stock. Control of the Wooly Aphid may be accomplished by the use of the foliage sprays recommended for the other aphids above.

The Apple Maggot *Rhagaletis pomonella* Walsh is a fruit fly native to the eastern and central parts of the United States and Canada. Three to five applications of one of the following sprays, spaced ten days apart, are generally applied after the flies begin to appear:

0.37 pounds of Azinphosmethyl per 100 gallons of water

OR

0.25 pounds of Azinphosmethyl plus
0.25 pounds of Carbaryl per 100 gallons of water

OR

1.00 pound of Methoxychlor per 100 gallons of water.

Later in the growing season, the application of many pesticides must be curtailed to prevent the presence of a residue from the spray as a contaminating agent on the fruit. The following sprays may be applied later in the season than the previous pesticides:

0.50 pounds of Diazinon per 100 gallons of water

OR

3.00 pounds of Lead Arsenate per 100 gallons of water.

Glass, et al. (1961) have shown that the half grown fly larvae are killed by storage of thirty-three or more days at 31 - 33°F, the standard temperature for apple storage. The gas concentration is similar to normal air, while the relative humidity ranges from ninety to ninety-five percent throughout the storage period. In a controlled atmosphere at 38°F, all fly larvae were killed in less than sixty days. In the controlled atmosphere storage of the McIntosh apple, the carbon dioxide content was raised from two to eight percent while the oxygen content was reduced from twenty-one to eight percent.

The Apple Maggot is said to have increased in numbers in the New Brunswick district of Canada since the introduction of the mild organic fungicides. In earlier years combinations of calcium arsenate, and either Bordeaux or lead arsenate alone, or in combination with lime-sulfur solutions had given satisfactory control of the Apple Maggot. Maxwell (1961) who studied both new and old control formulae in both field and laboratory reported the following results: Diazinon and Guthion were effective in laboratory and field tests with the exception of one field report concerning the failure of Diazinon. Present recommendations suggest the combination of lead arsenate and DDT for late varieties of apples, with the use of DDT dependent upon recent regulations concerning (in some areas abolishing) the use of DDT.

Fumigation experiments were conducted by Sanford (1962a, 1962b) concerning the control of the apple maggot in several varieties of apples. In the first series of tests, bulk quantities of apples were fumigated under a plastic tarpaulin with ethylene dibromide (six ounces per 1,000 cubic feet) for two hours at a temperature of 55°F. The Washington Red Gravenstein and McIntosh Red apples were used in the tests and all larvae of the apple maggot were killed in four of the five tests. Fumigation at temperatures above 60°F is suggested as a substitute for the six week, low temperature treatment now required to insure apples free of live maggots to meet the regulations of importing countries that have a zero tolerance of the Apple Maggot.

Further fumigation experiments were conducted by Sanford (1962b) with methyl bromide, ethylene dibromide, and ethylene chlorobromide. Crimson Gravenstein, McIntosh, Red Golden Russet, and Red Delicious apples were treated at temperatures near 60°F. The experiments were designed to determine the dosages necessary to kill the Apple Maggot, and the injury to the fruit that might occur from such treatments. Methyl bromide applied at dosages that were lethal to the apple maggot caused injury to Gravenstein and McIntosh apples. The application of ethylene dibromide or ethylene chlorobromide at a concentration of four ounces per one thousand cubic feet resulted in high mortality of the Apple Maggot without injury to the fruit.

The Codling Moth, *Carpocapsa pomonella* Linné, is a particular pest to the apple and pear, but there are increasing reports of the infestation of the English (persian) Walnut and frequent injury to apricots grown under California conditions. The proper timing of the second to the fourth Codling Moth spray is dependent upon the discovery of moths in bait traps. As is generally the case, the choice of an insecticide is based on a combination of effectiveness, residue tolerance, and the effect of the spray on bees and other beneficial insects. Ryania, a plant derivative, is favored in some districts, based on its greater safety to parasites (deOng, 1960), but this agent has yielded unsatisfactory control when used in the Pacific Northwestern United States. A commonly used insecticide formulation for preparation of the four Codling Moth sprays each year includes two pounds of fifty percent wettable DDT powder plus two pounds of twenty-five percent wettable powder per one hundred gallons of water (lighter dosages should be used outside of the Pacific Northwest). Parathion is not to be used on McIntosh apples and related varieties. If this spray does not provide acceptable results, the use of either Sevin at one pound of fifty percent wettable powder, or Guthion at two pounds of twenty-five percent wettable powder per one hundred gallons of water will avoid the buildup of residue in early sprays. Diazinon, at a concentration of two pounds of twenty-five percent wettable powder, is used in the control of both Aphids and the Codling Moth in midsummer. Hoyt (1961) found that frequent applications of Sevin may tend to build up large populations of mites, and should not be used until four weeks after petal fall. Kelthane, applied at a rate of six ounces of active material (either as an emulsifiable compound or a wettable powder) per one hundred gallons of water will check such outbreaks of mites. Parathion and Guthion should not be used in small home plantings since they should be applied only by trained operators.

Falcon (1967) and co-workers have reported a 98% reduction in pupation of the codling moth when trees were sprayed with a viral treatment. The virus they used appeared to be selective in its action to the larval stage of the codling moth, not harming any other insects.

The defoliating Red-Banded Leaf Roller has been controlled by the application of either two pounds of fifty percent wettable TDE (DDD) powder OR a combination of three pounds each of lead arsenate and hydrated lime per one hundred gallons of water. Both of these sprays are applied either at petal fall OR at the first codling moth spray. A second application may be applied at the fifth or sixth cover spray if necessary. The following sprays can also be used to control the leaf roller: one pound of carbaryl applied during the fifth and sixth cover sprays, OR 0.37 pounds of Azinphosmethyl applied at petal fall and the first cover spray, OR either 0.75 pounds of Malathion OR 0.3 pounds of Parathion applied at petal fall, the fifth, or the sixth cover spray (all concentrations given in pounds of active ingredients per one hundred gallons of water).

The Leaf Hoppers (Family *Cicadellidae*, Order *Homoptera*) are small sucking insects that normally feed and breed on grass. In some districts, these insects have become serious pests of the apple, russeting the leaves when the insect infestation becomes sufficiently severe. The application of either one pound of fifty percent wettable DDT powder, OR one pound of fifty percent wettable Malathion powder, OR 0.25 pounds of active Azinphosmethyl per one hundred gallons of water has proven effective in controlling such outbreaks of Leaf Hoppers.

The Japanese Beetle, *Popillia japonica* Newman, is another defoliating pest that can be controlled by any of the following sprays, applied in pounds of active ingredient per one hundred gallons of water: a) one pound of DDT, OR one pound of Carbaryl, OR 1½ pounds of Methoxychlor, b) 0.3 pounds of Parathion, OR 0.5 pounds of Malathion, OR 0.1 pounds of Rotenone. Whereas all sprays should be applied when the Beetles appear, the three "a" sprays should be repeated in ten to fourteen days (if necessary) while the "b" sprays can be repeated in seven to ten days. Of the few parasites that have been introduced from Asia to aid in controlling the Japanese Beetle, the most promising appears to be the Spring Tiphia, *Tiphia vernalis* Roh.. Polivka (1960) found that the beetle could be controlled for four years or more with very light dosages of Chlordane (one pound of active material per acre) OR Aldrin (three pounds of active material per acre) applied as a soil treatment.

The Clover or Brown Mite, *Bryobia praetiosa* Koch, the European Red Mite, *Metatetranychus ulmi* Koch, and the Two-Spotted Mite, or Two-Spotted Spider Mite, *Tetranychus bimaculatus* Harvey, can be controlled by the oil based dormant sprays described under the Aphids, providing that the dinitro compounds are not used. If these dormant sprays have not held the mites in check, then early summer applications of the following sprays can be used:

Kelthane (six oz of active material per one hundred gallons of water)
Sulphenone (one pound of active material)
Tetradifon (0.12 - 0.25 pounds of active material)
Binapacryl (0.19 pounds of active material)
Demeton (0.25 pounds of active material).

Whereas the Demeton may be applied at pink cover or petal fall, the Tetradifon or Binapacryl may be applied at petal fall or later covers if necessary.

The Plum Curculio Weevil *Conotrachelus nenuphar* Herbst, occurs in the eastern United States and adjoining provinces of Canada. The following sprays may be applied at petal fall and once or twice afterwards at seven to ten day intervals: one pound of twenty-five percent wettable Guthion powder, three pounds each of lead arsenate and hydrated lime, three pounds of fifty percent wettable Methoxychlor powder or 0.37 pounds of active Azinphosmethyl, all per one hundred gallons of water.

A number of closely related species of Scale Insects attack the apple, the most important of which is the San Jose Scale (*Aspidiotus pernicioesus* Comstock). Introduced accidentally into California (c. 1873) from northeastern China, its natural enemies, which held the scale in check in China, were not present in its new environment, allowing the San Jose Scale to spread rapidly across the United States. Native parasites soon adapted themselves to a diet of San Jose Scale, and these, together with introduced parasites, have reduced the scale to minor importance in many districts (Flanders, 1960). The scale remains important as a result of the variety of host plants to which it avails itself. Although found primarily in apple and peach orchards, it is also found on a long list of stone fruits, walnut, and many other cultivated trees and shrubs. The San Jose scale attacks the trunk, limbs, and fruit of bearing trees, causing severe injury to the tree if left unchecked. Anthon (1960) described the difficulties of controlling the San Jose Scale on stone fruits in Washington. Dormant sprays consisting of emulsions of naphthenic petroleum (three gallons per one hundred gallons of water) caused injury to stone fruits while Parathion applied in the delayed dormant stage is not as effective as it once was. Combinations of one gallon of paraffinic (Supreme) petroleum with the following two compounds have yielded good control of the San Jose Scale: one pound of twenty-five percent emulsion Parathion OR two pounds of wettable Malathion. In cases of severe infestation, the amount of Supreme oil was increased to two gallons per one hundred gallons of water. Two basic types of petroleum oil are used on plants as insecticides: the naphthenic oil of the Pacific coast and the paraffinic oil of the Eastern States. The latter, because of its somewhat greater efficacy as an insecticide, is known as "Superior" oil. Chapman, et al. (1962) presented the following list of specifications for a petroleum fraction including the following: SUS viscosity at 100°F of sixty seconds, and API gravity, unsulfonated residue, pour point and distillation range. The European Red Mite found on the apple trees of the Red Delicious variety was used as the test insect for this work.

Diseases of the Apple and their Control

Fire Blight, a disease of various fruit trees and ornamentals, is caused by the bacterium *Erwinia amylovora*. Beginning with the tips, the leaves on infected branches become brown. A heavily infected tree appears to have been burned. The blooms of a tree infected with Blossom Blight become water-soaked, shrivel, and turn dark. **Leaf Blight** occurs when the bacteria spread from the blossoms to the leaves, which become dark and appear scorched, but remain on the tree. **Twig Blight** usually is the result of the exudate of the bacteria being splashed by the rain. The less common Fruit Blight causes the young fruit to become water-soaked and brown, later shrivelling and turning black. Cankers develop on the blighted areas of a tree attacked by blight and the bark attains the appearance of being water-soaked while darkening, becoming sunken and dry. A canker that girdles a portion of the tree causes the death of the girdled portion.

The bacteria that are responsible for these infections are inactive during cold weather and overwinter in the cankers. In the spring, the bacteria become active and the cankers ooze. Rain and insects spread this exudate and transmit the infection to new areas. During warm, humid periods, the bacteria multiply causing the greatest problem.

Control Considerations:

1. Strains of the Red Delicious Apple possess considerable resistance to the attack of these bacteria, while Jonathon and Yellow Transparent trees are particularly susceptible.

2. Prune and burn the blighted twigs during the winter or early spring. This pruning should be conducted one half foot below the extent of the infection, if the limbs are removed in the summer, a one foot margin should be allowed to insure that all of the infected area is removed.

3. Spray approximately three times during the blossom period with an antibiotic substance, such as streptomycin (Trade Names include: Agri-strep

Type D, Agri-mycin 17, and Ortho Streptomycin), or a one to three parts per hundred Bordeaux mixture. The latter spray may be used in the household orchard and is sprayed during the period of full bloom.

4. The control of local insects is a secondary means of control of the blight since the insects spread the disease.

Figure VII-2a. Fire Blight infection on a crap-apple tree. **2b.** Close-up of a branch conker on a crab apple tree caused by Fire Blight. (Illinois Natural History Survey.)

The fungus *Botryosphaeria ribis* Gros. & Dug. causes small, brown, circular lesions, known as **Botryosphaeria Rot**, that are enclosed in a circular red area. The infected flesh of the fruit is soft and cupped, but no fungal fruiting bodies are visible. Fruit that appears healthy at the time of harvest may rot in storage as a result of this disease. This disease is often called "White Rot" when it infects red-skinned fruit. The disease can infect twigs or branches of the tree, causing cankers in the process. Most prevalent in the central and southern United States, this disease primarily attacks the Golden Delicious, Duchess, Transparent, Rome, Gallia Beauty, and Northwestern Greening varieties of the apple. Botryosphaeria Rot will also infect avocado and rose and may cause Cane Blight in the gooseberry and current.

Control Consideration:

Cankered twigs, limbs, or canes must be pruned and destroyed. Fungicides should be sprayed throughout the summer, especially late in the growing season and during warm, moist periods throughout the season.

Powdery Mildew, caused by the fungus *Podosphaera leucotricha* (E. & E.) Salm., covers twigs, leaves, and blossoms with a pearly-gray or white velvety mold. The terminal bud of the twig that is infected is usually killed, the infected twig dwarfed, and the infected leaves narrowed and rolled inward. In the summertime, older infections grow tannish-gray and sometimes become imbedded with tiny, black fruiting bodies of the fungus. Infected fruit becomes either etched or russeted.

Control Considerations:

Prune the infected branches. A fungicide should be applied from early spring until the growth ceases.

Apple Bitter Rot is a fungal disease caused by *Glomerella cingulata* (Stone) Spauld.. Jonathan, Golden Delicious, Northwestern Greening, Yellow Newton, and Grinus Golden trees are most susceptible to attack by this fungus. The fungus penetrates the skin of the apple and causes circular, light brown spots to form underneath the skin. When these spots grow to approximately one half inch in diameter, concentric rings of sticky beads of pink spores that later become dark colored appear. In early stages of a Bitter Rot infection, the flesh of the apple is watery and the decayed areas have the shape of a cone whose tip is toward the center of the apple. The most favorable climactic conditions for the growth of the fungus are rain and seventy to eighty degree (F) temperatures. Most infected apples drop from the tree, but a few remain, to shrivel into hard nobs called "mummies". These mummies spread the infection from one season to another.

Control Considerations:

Remove and burn all infected branches and cankered areas of the tree. An old tree that appears to be a reservoir for the disease may have to be removed completely.

Spray with a fungicide at the following times: June 10-15, July 1, July 15-20, and the first week in August. A spray that is composed of either Captan or Ferbam (two lbs per one hundred gallons of water) OR one containing Dichlone (three-quarters to one pound per one hundred gallons of water) has proven effective in controlling Apple Bitter Rot.

NOTE: Since such sprays leave a residue, the directions of the manufacturer should be carefully followed.

Black Rot, or **Frogeye Leafspot**, is caused by *Sphaeropsis malorum* Pk. in the "imperfect stage" and *Physalospora obtusa* (Schw.) Cooke in the "perfect stage". In the foliage phase, Frogeye Leafspot manifests itself as small purple spots on the leaves soon after the unfolding of the leaves. These tiny spots grow and become dark, sharply outlined, round spots with purplish edges. The fungus spreads throughout the growing season and can infect twigs, bark, and fruit. When attacking the fruit of the tree, this disease is known as Black Rot. The infection appears as the fruit nears maturity or while it is in storage. Lesions appear as small brown spots that expand, darken, and turn black. The infected area, that frequently appears at the calyx end of the fruit, does not become sunken, but appears as concentric circles of black and brown on the surface of the skin. Later black eruptions occur on other parts of the skin and the apple becomes mummified.

Control Considerations:

Prune and destroy all infected branches and mummified fruit. Much infection occurs early in the growing season from the pre-pink through the petal fall stage of seasonal growth.

Spray with Phaltan or Captan prior to the petal fall to control a Black Rot infection.

Cedar Apple Rust, the fungus *Gymosporangium juniperi-Virginianae* Schw. overwinters on the red cedar and other species of juniper, giving rise to the name of the disease that is commonly known. Spores from the galls formed by this fungus infect the fruit and leaves of apple trees. The upper surface of the leaves of an infected branch develop tiny pale yellow spots. As these spots grow, they become orange, red bordered, and centered with small black pustules. Under these spots of infection, blister-like growths develop in the tissue of the leaf. At a later stage of the infection, cup-shaped structures appear in these blisters. The fruit sustains similar lesions as those described that infest the leaves.

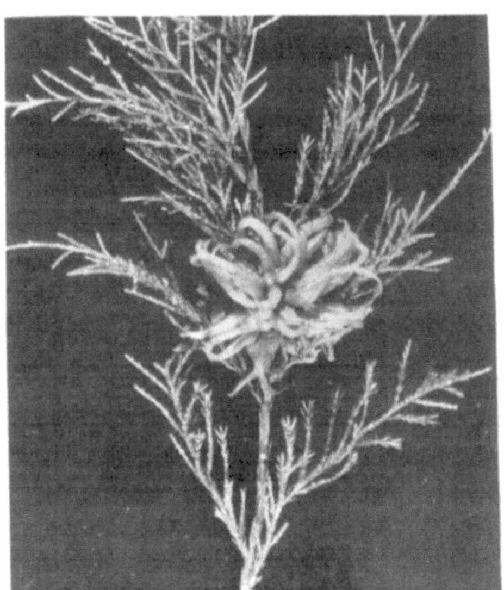

Figure VII-3. Red Cedar infected with Cedar-Apple Rust. (Illinois Natural History Survey.)

Control Considerations:

This disease is little problem if cedar trees are absent from a one half mile radius around the orchard, but destruction of such trees within the critical radius is often impractical.

A cover spray that contains ¾ pound of Ferbam (76 percent) added to the conventional Captan sprays that are applied from prepink to the first cover spray has been recommended for the control of the Cedar Apple Rust.

THE APRICOT—*See Peach*

THE AVOCADO

Insect Control Spray Program

The following spray program has been recommended for insect control of the Avocado.

December, January, February

During these months a miticide should be applied for the control of the Red Mite. If a Mite population count indicates at least six mites per leaf, one of the following control measures is in order:

1) A sulfur dust is applied at the rate of four to eight ounces of sulfur per tree or forty pounds of sulfur per acre,
2) A spray containing an emulsion of three quarts of ninety to ninety-two percent oil per one hundred gallons of water,
3) A spray containing ten pounds of a wettable sulfur powder per one hundred gallons of water should be applied in such a way as to thoroughly cover the leaves.

January, February, March

Insects that infest the blossoms of the avocado should be controlled during this period.

The Avocado Tree Girdler, *Heilipus apiatus* Oliv., eats the buds of the avocado and occasionally feeds upon the blossoms and twigs of the tree. If these beetles become populus enough to warrant the use of control measures, the following may be used in one hundred gallons of water to form the spray:

1) two pounds benzene hexachloride (10% gamma isomer),

OR

2) one pound Lindane (25% gamma isomer wettable powder),

OR

3) one pound Parathion (15% wettable powder).

Webbing Worms, *Argyrotaenia amatana* (Dyar) may be controlled by a spray containing three pounds of lead arsenate per one hundred gallons of water.

Either one of the following two sprays have been used for the control of Blossom Anomala, *Anomala undulata* Mels.,:

1) two pounds* benzene hexachloride (10% gamma isomer wettable powder),

OR

2) one pound* Parathion wettable powder.

May, June, July

During this period of the year, measures to control Scales and Mealybugs, in addition to the larvae of the Avocado Tree Girdler, are applied.

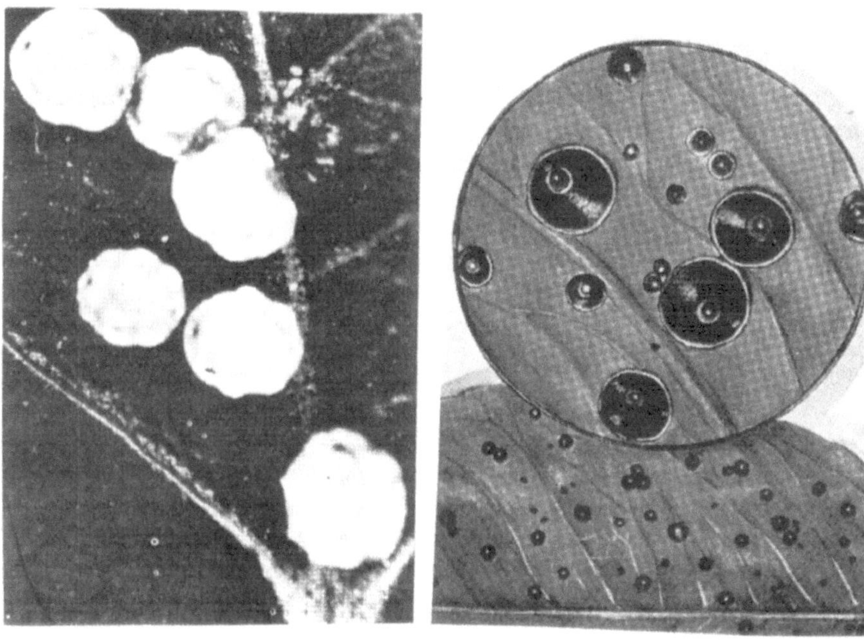

Figure VII-4a. Florida Wax Scale. **4b.** Florida Red Scale. (James E. Brogdon, Florida Cooperative Extension Service.)

The Florida Wax Scale, *Ceroplastes floridensis* Comst., may be controlled by the application of a spray that contains an emulsion of one-and-one-half pounds of fifteen percent wettable Parathion powder PLUS one--and-one-third gallons of emulsive oil per one hundred gallons of water.

The Florida Red Scale, *Chrysomphalus aonidium* (Linn.), seems easily controlled by a spray that contains an emulsion of one pound of fifteen percent active wettable Parathion powder and one gallon of emulsive oil per one hundred gallons of water.

"Spot" spraying of either four pounds of twenty-five percent active wettable Malathion powder or one-and-one-half to two pounds of fifteen

*per one hundred gallons of water

percent wettable Parathion powder per one hundred gallons of water may be used to control the Citrus Mealybug, *Pseudococus citri* (Risso), as well as other mealybugs.

If the larvae of the Avocado Tree Girdler, *Heilipus apiatus* Oliv., are feeding on the inner bark of the trees, a reddish excrement may be observed near the burrow hole. These larvae must be individually removed from the tree before the application of a tree wound paint to the site of the burrow.

August, September, October

During this period of the growing season, the control of Thrips and Fruit Scarring Worms is called for.

Greenhouse Thrips, *Heliothrips haemorrhoidalis* (Bouche), often feed upon the large fruits and may cause discoloration, russetting, and cracks on the fruit. The control of these pests can be achieved by the use of a spray that consists of one of the following active ingredients dispersed in one hundred gallons of water:

1. 1 pound Parathion (15% wettable powder),

OR

2. 3 pounds Malathion (25% wettable powder),

OR

3. 1 pound Lindane (25% wettable powder),

OR

4. 1 pint nicotine sulfate (40% liquid concentrate).

Any of several fruit scarring worms may become a problem during this portion of the season, such as the larvae of the two Tortricid Moths, *Argyrotaenia amatana* (Dyar) and *Platynota stultana* Wlsn.. If the presence of these pests begins to damage the crop, a spray composed of either three pounds of lead arsenate or one pound of fifteen percent active wettable Parathion powder dispersed in one hundred gallons of water should be considered for application.

Caution:

The use of any of the above sprays should be carefully controlled. Since Parathion is extremely dangerous in the hands of an inexperienced applicator, it should not be used on home plantings, and should only be applied by a thoroughly trained operator.

Sulfur should not be applied within four weeks of the application of one of the oil sprays.

The use of oil sprays should be discontinued when the weather is hot or the soil is dry.

Disease of the Avocado and their Control

The small feeder roots of the tree are attacked by **Phytophthora Root Rot**. Also known as the "Cinnamon Fungus", this disease is caused by *Phytophthora* *cinnamomi* Rand. Paler and smaller than usual, leaves from trees that are infected by this disease frequently wilt and then drop from the branch. The growth of an infected tree, needless to say, is arrested and often the branches "die back". This fungus attacks many other plants besides the avocado and is particularly active during periods when the temperature lies between 50 and 90°F. Control measures for use against this disease include planting trees in either uninfected soil or soil that is fumigated. Either uninfected seed or seed that has been subjected to treatment with hot water (120° to 130°F for at least one half hour) should be used. Resistant varieties of avocados are available that lessen the chance of infection. Lastly, the avoidance of overwatering during irrigation can curtail the development of the disease as well as its virulency.

BRAMBLES AND BERRIES

Spray Schedule for the Control of Insects and Diseases of the Blueberry

The following sprays are recommended for the control of the specified pests of the Blueberry. The dosage of the active ingredient listed below should be dispersed in one hundred gallons of water before application as a spray.

First Cover Spray—applied at the late bloom stage of the blueberry. This spray consists of two pounds of twenty-five percent wettable Malathion powder PLUS two pounds of seventy-six percent wettable Ferbam powder. While the Ferbam is used to control Botryosphaeria, the Malathion is present for the control of Curculio, Maggot, and Fruit Worms.

Second Cover Spray—applied seven days after the application of the First Cover Spray. The Second Cover Spray is composed of two pounds of twenty-five percent wettable Malathion powder PLUS two pounds of fifty percent wettable Sevin powder.

Additional Cover Sprays—applied at ten day intervals after the Second Cover Spray. Generally, two applications of the Second Cover Spray are necessary for effective pest control. Additional spray applications may be required for the control of maggots. Although Malathion should not be applied within twenty-four hours of the harvest, no such restrictions are noted for Sevin at the present time. As usual the manufacturers' directions and the results of recent research should guide the determination of specific control techniques.

Pests and Diseases of Brambles

General Considerations:

Many diseases of the Brambles, like those of other plants, cannot be "cured" by the simple application of some spray that contains chemicals. Diseases, like pests, are best controlled by the avoidance of those practices that encourage the development of the disease. The first step in the avoidance of disease consists of the purchase of disease resistant varieties, whenever possible, from reputable suppliers. Planting any of the brambles in a site that was previously planted with potatoes, tomatoes, eggplant, or related plants is undesirable since Verticilium Wilt may be present in the soil. When nursery stock is planted, any cane stubs or "handles" should be removed since they are often infected with Anthracnose. Patches of wild or neglected brambles should be destroyed if they are near domesticated plantings to reduce the transference of diseases to the new plants. Since red raspberries may harbor the virus disease, Mosaic, without exhibiting any visible symptoms, they should not be planted or maintained within five hundred feet of black raspberries, which are not resistant to Mosiac. After the plants are established, good cultural practices are important for the maintenance of healthy, productive plants. Any canes that don't leaf-out in a normal manner in the spring should be cut and burned. Old canes should be removed and destroyed after each harvest, in addition to the destruction of any canes that appear to be diseased. In the following paragraphs, some of the more important pests of the brambles will be discussed and the proper control of both the insects and diseases is the last, important aspect of the proper care of these plants.

Diseases of the Brambles

Anthracnose, resulting from an infection that is caused by the fungus *Elsinoe veneta*, appears on the new growth of the canes as a raised, purplish spot. As the growth of the cane continues, the infected spots become enlarged and develop a sunken, grayish center and a raised purple edge. Late in the summer, a white or grayish growth that is speckled with dark fruiting bodies develops on the growth that was formed during the current season. Anthracnose may also infect the leaves of the plant, giving them purple edged spots. Even the fruit of the plant may be infected by this disease. Since the disease overwinters in the old canes, these should be burned regularly. The adherence to a reasonable and sound schedule for spraying the plants is helpful in preventing the disease from infecting new cane growth.

Crown Gall, characterized by lumps or irregular growths on the roots and the lower parts of the plant, is the result of an infection by the bacterium *Erwinia tumefaciens* (Sm. & Towns.) Dows. This disease is able to survive in the soil and often enters a plant through a lesion that was made by an insect. Disease-free stock should be planted and maintained by an effective program for the application of insecticides to control this disease.

Orange Rust spreads through the whole plant once entrance to the plant is gained. This disease manifests itself in weak, spineless growth; yellowish, small leaves; and, on the underside of the leaves, swollen spots that contain red-orange spores. The canes will no longer produce blossoms the year following an infection by this fungus. Orange Rust does not attack red raspberries, only the purple and black members of this family. Neither pruning nor fungicides are effective measures for the control of this disease. The removal and incineration of the infected plants and their replacement with disease-free stock are the major solutions for cases of Orange Rust. The destruction of weeds and wild berries in the vicinity of the domesticated plants may be helpful in the control of this disease.

Black raspberries are especially susceptible to the disease **Verticillum Wilt.** Beginning at the base of the plant, leaves on infected canes become yellow, shrivel, and drop, while the canes become blue and die. The fungus responsible for Verticillum Wilt is borne in the soil, therefore it is important to chose disease-free soil in which to make the plantings. The avoidance of sites that have been used for tomatoes, potatoes, eggplant or peppers within the previous three years is an essential measure for the control of this disease.

Spur Blight is found particularly on red raspberries. This disease, also caused by a fungus, is first represented by purple or brown areas on or near the buds of the plant. These areas darken and the buds shrivel and fail to produce fruit. Sprays that contain fungicides are helpful in any program to control this disease.

Black and purple raspberries are more susceptible to the virus that causes **Raspberry Mosaic** than are red raspberries. The growth of the canes of infected plants is stunted while the leaves are abnormally small and have large green blisters that are edged with yellow. The tips of badly infected canes, especially on the black or purple varieties of the raspberry, may darken and die. The fruit that is produced by an infected plant is greatly inferior in its quality. The control measures for this disease are presented in the General Considerations section of the presentation.

Leaf Curl, another virus disease of the raspberries, is manifested by crinkled and curled leaves. On diseased plants, the canes become progressively shorter each year while becoming brittle at the same time. The combination of good cultivation practices and especially good insect control are essential in efforts to control Leaf Curl.

Cane Blight, apparent by the dark brown cankers on the plant, is caused by a fungus that frequently enters the raspberry cane through wounds in the canes. Canes that are infected with this disease turn gray in the summer and the lateral shoots of infected canes often die. The avoidance of pruning at least three days prior to rain and the incineration of any infected canes appear to be effective measures that can be used in the control of Cane Blight.

Spray Program for the Control of Insects and Diseases of the Brambles

The following active ingredients have been recommended for the control of insects and diseases that attack members of the Brambles. All active ingredients listed are applied as sprays after being dispersed in one hundred gallons of water.

Delayed Dormant Spray—applied in the early spring when the leaf buds have one half inch of green showing at their tips. This spray consists of ten gallons of liquid limesulfur.

New Cane Spray—applied when the new canes have grown eight to twelve inches and the first blossom buds are apparent. This spray may contain either two pounds of seventy-six percent wettable Ferbam powder* OR two pounds of fifty percent wettable Captan powder.

Pre-bloom Spray—applied after the New Cane Spray and just before the blossoms open in the spring. This spray is a repeat application of the New Cane Spray.

Pre-harvest Spray—applied ten days prior to harvest. This spray consists of two pounds of fifty percent wettable Captan powder.

Post-harvest Spray—applied after the old canes have been removed, this spray is a repeat application of the New Cane Spray.

Special Spray—applied for the control of mites, especially during June and July in areas such as Illinois. The spray consists of one-and-one-half pounds of 18.5% wettable Kelthane AP powder and is applied as needed, subject to the directions of the manufacturer.

Insects and Diseases of Berries and their Control

Current and Gooseberry

The Current Borer can be controlled through the application of two pounds of fifteen percent wettable Parathion powder per acre. This powder is dispersed in a convenient amount of water before application as a spray.

The application of one percent Rotenone dust is recommended for the control of the Gooseberry Fruitworm (Koehler, et al., 1962).

Diseases of the Strawberry

Botrytis cinerea Fr. causes the disease known as **Botrylis Gray Mold** that infects strawberries. A blossom blight, evidenced by wilted or discolored petals, is often the first symptom of this disease. In later stages of the infection, the berries become infected with the gray, fuzzy mold. Spores of this mold may be carried by the wind to infect new patches of berries. The loss of berries as a result of this disease is greatest when damp weather prevails during the blossom and fruiting periods.

The fungus that is responsible for the disease **Strawberry Leaf Spot** is named *Mycosphaerella fragariae* (Schw.) Lind in the perfect stage and *Ramularia tulasnei* Sacc. in the imperfect stage. This disease is most often manifested on the leaves by deep purple lesions that develop light colored centers. Strawberry Leaf Spot spreads to new growth most readily during damp weather. If the infection becomes severe, the strawberry plants may suffer defoliation.

Strawberry Leaf Scorch is another fungus disease that is similar to the Strawberry Leaf Spot, except that the former is caused by *Diplocarpon earliana*.

Red Stele is caused by the soil-inhabiting fungus *Phytophthora frageriae*

*NOTE: Ferbam must not be applied within forty days of the harvest.

Hickman. An infected plant develops a red stain in the center of its root. As the infection progresses, the root becomes rotted. The presence of this disease may be first noticed by the stunted growth of the plant and its general lack of an appearance of health.

Insects of the Strawberry and their Control

The Strawberry Crown Borer can be best controlled through the use of mechanical means. Since the adult beetle cannot fly and doesn't commence layings eggs until March, infected plants should be dug in February. The adults should be physically removed from the plants and crushed. New plants should be set at least three hundred yards from the old bed to forestall the continuance of the infestation.

The Cyclamen Mite may be controlled by the application of a spray that consists of four pounds of endosulfan (Thiodan) active ingredient (in either wettable powder or emulsifiable concentrate) per one hundred gallons of water. This spray should be applied at high pressure to the foliage at a rate of eight hundred gallons of spray per acre. The application of the spray should be made either at pre-blossom, post-harvest periods or during the year in which new plants are set out.

The Two-Spotted Mite can be controlled by the application of a spray containing 0.8 pounds of active Kelthane, either in the wettable powder, emulsifiable concentrate, or dispersed form per acre to the foliage of the plants.

Spray Program for the Control of Insects and Diseases of the Strawberry

In addition to two general sprays, any one of several sprays may be needed to control specific pests, such as the two insect sprays listed in the preceding section.

Pre-planting treatment of the soil—applied when plants are set out for control of white grubs. The soil should be treated with two-and-one-half pounds of ten percent Chlordane in either a granular or dust form per one thousand square feet of soil OR four ounces of a liquid emulsion that contains eight pounds of Chlordane per gallon.

Blossom bud spray—applied when the blossom buds have first appeared to control leaf diseases, fruit rot, weevils, sawflies, plant bugs, and spider mites. The spray contains two tablespoons of fifty percent active wettable Methoxychlor powder PLUS two tablespoons of twenty-five percent active wettable Malathion powder PLUS two tablespoons of fifty percent active wettable Captan powder per gallon of water.

Miscellaneous sprays—applied if needed from the time of bloom until within three days of the harvest.

Malathion—according to the manufacturer's directions, either the emulsifiable concentrate or the wettable powder may be dispersed in water and applied to control Leafhoppers, Aphids, and Strawberry Leaf Rollers. This spray may be used as needed, with the normal amount of discretion, but not within three days of the harvest.

Dicofol (Kelthane)—may be used according to the directions of the manufacturer to control Cyclamen Mites and Spider Mites. This material may be used as required, but not within two days of the harvest.

Dinocap (Karathane)—may be used as needed, but not within twenty-one days of the harvest to control Powdery Mildew.

Phenyl Mercury Acetate—applied at a rate of one half teaspoon of the ten percent solution per gallon of water. In the late fall when the plants are dormant, this spray is applied two weeks before the application of the winter mulch to control leaf diseases as well as Fruit Rot.

THE CHERRY

Insects of the Cherry and their Control

The **Black cherry aphid,** *Myzus cerasi*, overwinters in the egg stage on the cherry, attacking the leaves in the early spring forcing them to curl. In most situations, the sweet cherry is the preferred host of the aphid which feeds on the cherry leaves until midsummer when it migrates to other hosts. At the pre-bloom or petal fall stages, the tree may be sprayed with ¼ pound of Demeton in one hundred gallons of water. When the aphids appear, any of the following sprays may prove effective (all per 100 gallons of water):

Diazinon	0.5 pounds
Endosulfan	0.3 pounds
Malathion*	0.5 pounds
Parathion	0.25 pounds
TEPP	0.12 pounds

These sprays may be reapplied in ten to fourteen days, or later in the summer, if further control is necessary.

The **Lesser Peach Tree Borer,** *Synanthedon pictipes* Grote and Robinson, is of increasing importance on the Atlantic Coast to growers of peaches and cherries. Bobb (1961) reported similar results in the control of the Lesser Peach Tree Borer and the Peach Tree Borer, *Sanninoidea exitiosa*, Say, with the use of Parathion (three pounds of fifteen percent wettable powder per one hundred gallons of water). The first application was made in early May and a second application one month later. A compressed air sprayer was used to apply the spray to the rough bark of the trunk and main branches of the tree to avoid the contamination of the fruit. Other methods of controlling the Lesser Peach Tree

*Home orchards only.

Borer include the use of one of the following chemicals dispersed in one hundred gallons of water: two pounds of Malathion (fifty percent wettable powder) applied four times to the trunk and large limbs at three week intervals, 0.75 pounds of Endosulfan, or 0.5 pounds of Azinphosmethyl. In Georgia, for example, the trees are generally sprayed twice, in April-May to control the first brood, and twice in August-September to take care of the second brood.

The Cherry Fruit Fly, *Rhagoletis cingulata* Loew, is common in the Eastern United States and has become established on the Pacific Coast. Beginning when the first adult flies are trapped, one half pound of Diazinon per one hundred gallons of water may be applied every seven days for two to four applications. Alternatively, any one of the following may be dispersed in one hundred gallons of water and sprayed two to three times at ten day intervals:

Methoxychlor	1.00 pounds
Perthane	1.00 pounds
Parathion	0.25 pounds
Rotenone*	0.12 pounds

Mites should be controlled by the application of either a dormant oil composed of three gallons of emulsifiable oil per one hundred gallons of water or a foliage, postharvest, spray consisting of either 1½ pounds of 18½ percent wettable Kelthane powder or one pound of twenty-five percent wettable Ovex powder per one hundred gallons of water.

Scale Insects that are not controlled by the dormant spray used for control of mites should be sprayed three weeks before, and immediately after, the harvest with one pound of twenty-five percent wettable Parathion powder per one hundred gallons of water.

Diseases of the Cherry and their Control

Brown Rot, caused by the fungus *Monilina fruticola* (Winter) Honey, is a very destructive disease that attacks sweet cherries as well as other fruits. This fungus blights the blossoms, reducing the amount of fruit that is set, kills the twigs, occasionally causes limb cankers, and may rot the fruit both on the tree and in storage. An infected bloom becomes tan or gray with the flowers becoming covered with gray spores within a few days. The infected flowers are a potential source of infection for developing fruits and young twigs. Growing along a twig, the fungus may form a canker on a larger branch. Such cankers are often sunken with gray spore tufts on their surface. The leaves of a blighted twig become bleached, later turning brown as the twig dies. The presence of brown rot on a fruit is recognized initially as a circular, small brown spot on the nearly mature fruit. In warm, humid weather, the rot progresses quickly through the fruit, with the infected area becoming smooth, circular, light brown, soft, and unsunken.

*Rotenone may be applied at seven to ten day intervals.

Infected fruit may either drop to the ground or remain on the tree to become mummified. A mummy may drop to the ground and become partly covered to develop fruiting bodies that mature about the time that the trees blossom in the spring. These fruiting bodies may then release spores that infect the blossoms for a renewal of the disease cycle. Although the fungus does directly attack ripe peaches, the infection usually enters the fruit through a wound or lesion. The removal of the mummified fruit and cultivation of the surrounding area immediately prior to the blooming of the tree are effective control measures. Tree pruning conducted such that the air may circulate freely through the branches to dry the fruit during the rainy seasons is also recommended practice. A spray comprised of two pounds of fifty percent Captan to one hundred gallons of water may prove effective in cases where preventative measures were ineffective. The control of the insects whose injury to the fruit may provide an opening for the entrance of the fungal infection is, of course, recommended.

Cherry Leaf Spot, the most important fungal disease of cherries, is caused by *Coccomyces hiemalis* Higgens. In the spring, the fungus attacks the young leaves, producing small dark bluish spots that later turn brown or reddish-brown. The centers of such spots may fall from the leaves. During damp weather, whitish or pinkish growth appears on the underside of the leaf in the middle of the spots. Frequently infected leaves will turn yellow and drop prematurely. Such premature defoliation should not be confused with that caused by the virus yellows disease. The leaf-spot fungus overwinters on the fallen leaves and is available as a source of infection for the following year. An effective mechanical method of control consists of plowing the infected leaves. A ground spray of one half pound of seventy-five percent Dodine plus one pound of fifty percent Captan per one hundred gallons of water repeated two or more times at ten-day intervals has proven to be an effective control technique. After the harvest a spray comprised of either one-and-one-half to two pounds of Ferbam or a Bordeaux mixture of four pounds of copper sulfate and eight pounds of hydrated lime per one hundred gallons of water may be used.

The virus disease **Necrotic Ring Spot** can reduce cherry crops by as much as forty percent. This disease manifests itself on young leaves by small, light green spots bordered with dark rings. When the leaves mature, the spots fall out, giving the leaf a shredded appearance.

The viral disease **Cherry Green Ring Mottle** causes a chlorotic constriction along the veins of the leaf, yellowing the leaf and marking it with dark green blotches. The surface of the infected fruit is marked with necrotic pitting and possesses a bitter taste. Infected trees must be destroyed and replaced by virus-free trees, as no other control technique is presently available.

Cherry Yellows is the most serious viral disease that attacks the cherry tree. Most prominent in cool weather just after petal fall, Cherry Yellows produces a yellow and green mottling of the leaves of the tree. The primary loss from an infection of Cherry Yellows is through reduced yields of the fruit. As is the case of the Cherry Green Ring Mottle, the only known control is the removal of infected trees followed by replacement with virus-free stock.

CITRUS TREES

Citrus Insects and their Control

The control of pests in groves of citrus trees varies depending upon the district and therefore primary consideration should always be given to the locally accepted practices.

Aphids may be controlled by either a nicotine spray or dust, provided the spray or dust is not applied within three days of the harvest. A nicotine dust consists of four percent nicotine as nicotine sulfate and is applied at a rate of fifty to sixty pounds per acre. The nicotine sulfate spray is comprised of ¾ pint of forty percent nicotine plus 0.4 pounds of a casein spreader per hundred gallons of water. A Rotenone spray has also been used with good effects that is composed of one pint of rotenone extract containing 2.5 percent Rotenone dispersed in one hundred gallons of water. In January through April one pint of meta-Systox-R may be applied to nonbearing trees after it is dispersed in one hundred gallons of water. A Malathion spray, applied as directed on the label of the container, may also be used for the control of aphids.

The Citrus Red Mite, or Purple Mite (*Panonychus citri*) can be controlled when infesting Grapefruit or Oranges (only in emergency cases in the case of Navel Oranges) by the application of a petroleum oil emulsion. Such an emulsion consists of 1.6 to 1.8 gallons of emulsifiable refined petroleum oil, or two gallons of such a petroleum oil emulsion, per one hundred gallons of water. In August and September a medium viscosity grade of petroleum oil should be used, while a lower viscosity oil should be applied in later September and October. A similar emulsion, using the medium viscosity oil, may be used to control the infestation of Citrus Red Mites on lemons in April and May or September through December. Such a spray will also control the six-spotted mite and many of the scale insects that may infest the lemon.

Various other species of Mites may be controlled by either one pint of wettable Kelthane powder or two pints of emulsifiable Kelthane concentrate per hundred gallons of water. This spray should be applied at least seven days before the crop is picked. A spray based on two pints of twenty-five percent emulsifiable Ethion concentrate per hundred gallons of water has also been recommended.

The California Red Scale (*Aonidiella aurantii*), Black Scale (*Saisetia oleae*), and other miscellaneous Scale Insects may be controlled by some form of a petroleum oil emulsion. Grapefruit and oranges (except Navel unless an emergency situation exists) should be treated with light-medium grade refined petroleum oil in water at the rate of 1.6 to 1.8 gallons of oil per hundred gallons of water. This spray should be applied from August through October. A similar spray, using medium grade oil, can be used on lemons in April and May and/or September through October. A combination of petroleum oil and Parathion can be applied to Valencia oranges (not to Navel oranges) or grapefruit as an emergency treatment only. Such an emergency treatment consists of 1.6 to 1.8 gallons of emulsifiable petroleum oil, light or medium grade, plus 0.4 pounds of twenty-five percent wettable Parathion per hundred gallons of water. This spray should be applied from August through October. Lemons may also be treated with this petroleum oil–Parathion spray in April or May, or from September through December.

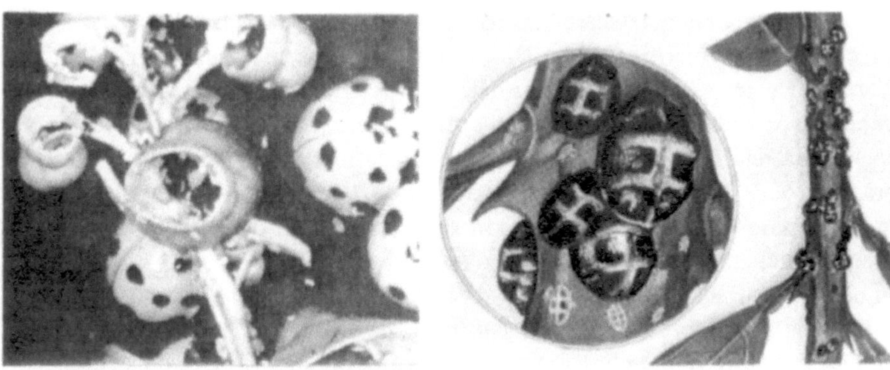

Figure VII-5a. Young citrus leaves severely curled by aphids. Note aphids and ladybeetles which feed on aphids. **5b.** Black Scale. (James E. Brogdon, Florida Extension Service.)

Citrus Thrips, *Scirtothrips citri*, may be controlled by the use of one hundred pounds of dusting-grade Sulfur per acre. Such a dust is usually applied three times: in mid-March, at petal-fall, and two to four weeks after petal fall. An alternative dust consists of a dusting-grade of Sulfur, at least eighty-five percent pure, that is, applied with two percent DDT at a rate of one hundred pounds per acre. This alternative dust should be applied at petal-fall and, if required, two to four weeks later. This latter dust must not be used during hot weather or near the time of picking.

Citrus Diseases

The fungal disease **Scab**, caused by *Elsinoe fawcetti* Bitancourt & Jenkins, is particularly troublesome in coastal areas. Attacking young trees, Scab forms raised, translucent, light brown spots on both fruit and leaves. The fungus is very active during humid, spring weather suggesting the importance of the application of a fungicide containing spray at the time of petal fall.

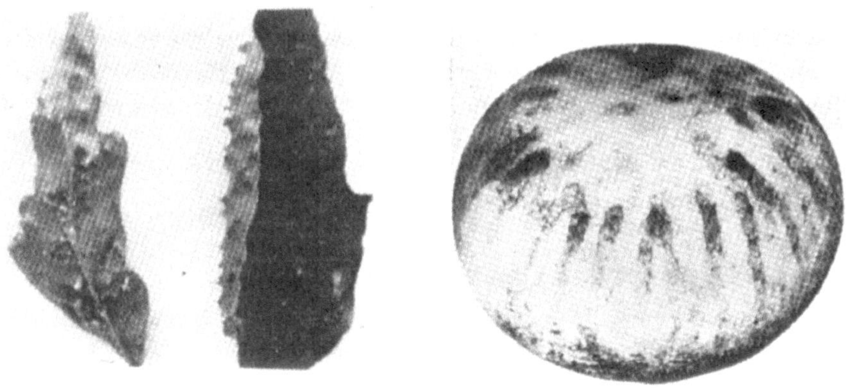

Figure VII-6a. Scab on young citrus leaves. **6b.** Sooty mold on grapefruit. (James E. Brogdon, Florida Cooperative Extension Service.)

Green Mold or **Blue Mold** often develop on water-soaked areas of the rind of citrus fruits. The areas infected by Green Mold become enlarged and covered with green spores surrounded by a white margin. As the infection progresses, the fruit becomes soft, shrivelled, and covered with green spores. The causitive agent of Green Mold *Penicillium digitatum* Sacc. enters the fruit through a wound in the skin. Blue mold is similar in its affect to green mold, except that the former is caused by *Penicillium italicum* Wehmer which is capable of penetrating the rind on undamaged fruit. Similar methods of control are effective in the reduction of losses due to either Green or Blue mold. Since fruit must be handled carefully to minimize abrasions, the fruit is often wrapped to prevent infection. While protective fungicidal dips are helpful, sanitary storage conditions are imperative for the control of these diseases.

Melanose, caused by the fungus *Diaporthe citri* Fawc. Wolf, attacks immature foliage as well as the fruit and twigs of older trees. Infection may be recognized by the raised, brown spots that appear on infected leaves, imparting a texture similar to that of sandpaper to the infected areas. Since Melanose grows in dead wood, removal of such wood from the trees provides a measure of control. The spores are most abundant in the early spring, especially during damp weather. The application of a copper spray to new growth soon after the shedding of the flowers is quite effective.

Sooty Mold is a fungus that may grow abundantly enough to blacken an entire tree. This fungus may be controlled through the control of insects such as aphids, immature whiteflies, and some soft scales. Such insects excrete a syrup-like, sweet substance known as "honeydew". Since the Sooty Mold uses the Honeydew as a growth medium, control of the honeydew producing insects results in the control of the Sooty Mold.

Spreading Decline is a condition caused by the invasion of the tiny feeder roots of the citrus tree by the burrowing nematode *Radopholus similes* (Cobb) Thorne. The damaged roots deteriorate and result in stunted trees whose productivity declines. Eradication of the disease is accomplished by treatment of the area with a heavy application of a nematocide after the infected trees and surrounding healthy trees have been pulled. The area should be planted with non-susceptible crops for two or more years to complete the treatment.

GRAPES

Diseases of Grapes and their Control

Old diseased canes, leaves, and dried berries serve as overwintering sites for diseases. Severe pruning of the vines during the winter to leave only the trunk and the healthy, strong canes from the growth of the previous year are an indicated control technique. All the material removed from the vines must be burned to destroy any disease organisms that may be harbored in the pruned vines. Before the new growth begins in the spring, the ground around the base of the vines should be cultivated. Since dried leaves and grapes can act as hosts for disease organisms, it is important that these be buried and thus rendered ineffectual by such cultivation practices. A spray that consists of either DN 289 or Elgetol, applied at a rate of two quarts per one hundred gallons of water, may be applied in the spring before the buds on the vines have opened. Such a spray is recommended for the control of scale insects as well as the diseases **Dead Arm** and **Anthracnose**. The first cover spray should be applied when the new growth is approximately two to four inches long. This cover spray may consist of either one pound of DDT and one-and-one-half pounds of twenty-five percent wettable Malathion or one-and-one-quarter pounds of Guthion (25% wettable) plus two pounds of Ferbam (76% wettable) per hundred gallons of water. A second cover spray, consisting of the same formulation as the first cover spray, should be applied immediately prior to the blooming of the grapes. After the bloom and when the berries have set, a third cover spray should be applied. The third cover spray is again identical to the first two sprays with the exception that two pounds of fifty percent wettable Captan should be substituted for the Ferbam. Ten days after the third cover spray, a fourth cover spray, identical in composition to the third, should be applied. Two pounds of fifty percent wettable Captan per hundred gallons of water should be applied as a fifth cover spray after the passing of another ten days. The use of Malathion or Guthion in addition to the DDT are recommended since the Grape Mealy Bug will not be controlled by the use of DDT alone. An insoluble copper spray, applied according to the directions of the manufacturer, may be used to control powdery mildew, but should be only used when necessary since a reduction in the yield of the grapes is frequently encountered. Bordeaux mixture has been very effective against diseases of grapes, but may reduce the yield of the vineyard. Since this mixture is incompatible with many insecticides, it must be used without other agents.

The use of fungicides is recommended to control fungal attack through prevention of infections rather than reliance on the fungicide to eradicate extensive infections. Since wind and rain are effective agents for the dissemination of fungi, a fungicide applied a day or two prior to a rain can result in added control.

The disease **Black Rot** is caused by the fungus *Guignardia bidwellii* (Ellis) Viala and Ravaz. Overwintering on infected canes, berries, and leaves, this fungus attacks the new growth of the plant on the leaves in the spring. The initial circular, reddish lesions on the leaves increase, developing dark margins while the centers turn to reddish-brown to tannish-gray. Lesions on the stems or veins of the leaves are generally dark purple or black. When the berries become half grown, the fungus may become noticeable as a whitish spot soon surrounded by an expanding band of brown. Such a spot may enlarge until the entire berry becomes rotten, at which point the berry dries, shrivels, and blackens.

Downy Mildew, characterized by pale yellow translucent spots on the upper surface of the leaf, is a result of the attack of the fungus *Plasmopara viticola* (B. & C.) Berl. and DeT.. The initially pale yellow spots become brown as the infection progresses. The underside of an infected leaf exhibits the white, velvety fungal growth for which the disease is named. The fungus overwinters as spores on the dead leaves. The spring rains are instrumental in spreading these spores to new areas of growth on the vine. The stunted, infected shoots are usually covered with the characteristically white fungal growth while infected berries become brown, wrinkled, and frequently drop from the vine.

Only the leaves of the grape are attacked by the fungus *Pseudopeziza tracheiphila* MT. to cause the disease **Rotbrenner**. The pale lesions on the leaf become first yellow, then brown and necrotic as the disease progresses. Severely infected leaves dry and fall from the vine. This fungus overwinters in the ground on infected leaves to discharge spores during the rainy spring period. Rotbrenner is thought to be a native of Europe and resembles sunscald in many instances.

The fungal disease **Powdery Mildew** is first noticeable as a powdery, grayish-white growth on the newest growing tips of the plant. Severely infected leaves often roll upward during warm, dry weather, while infected berries are spotted and misshapen. Powdery Mildew is caused by the fungus *Uncinula necator* (Schw.) Burr. which develops most rapidly during warm, humid weather.

Insects of the Grape and their Control

The Aphid may be controlled through the application of one pint of forty percent nicotine plus one quarter pound of casein spreader OR one pound of fifteen percent wettable Parathion powder per one hundred gallons of water.

The Grape Berry Moth occurs in many of the eastern states, but is especially troublesome in the northeastern states. The greenish larvae reach three-eighths of an inch in length at their maturity. The larvae of the first brood attack the bloom and the young clusters of berries. Mature larvae pupate within a fold of a leaf during June and July. The second brood usually drops to the ground when mature to pupate in the litter on the soil.

Control Considerations:

To avoid heavy DDT residues, both Taschenberg, et al. (1960) and Cox (1961) reported favorable results from the substitution of Sevin (carbaryl) at a concentration of two pounds of fifty percent wettable powder per one hundred gallons of water in comparison to common DDT control programs. The reduction of the concentration of the Sevin from two pounds to one pound resulted in slightly inferior control. Copper sprays that contain lime should not be used in conjunction with such a Sevin (carbaryl) spray. Since the use of DDT is discouraged or prohibited in many areas, substitutes such as this should be thoroughly investigated.

Either Methoxychlor (two pounds of fifty percent wettable powder) or DDT (one-and-one-half pounds of fifty percent wettable powder) per one hundred gallons of water are reported to be effective in the control of the Grape Berry Moth. Where the moth is abundant, the application of such sprays should be as follows: 1. just before or after the grape bloom, 2. ten to fifteen days after the bloom, 3. thirty-five to forty-five days after bloom, 4. ten to fifteen days after the spray #3. To avoid the DDT residue on the fruit, the last spray should be at least forty days prior to the harvest. For this reason, Methoxychlor (at the concentration recommended above) should be substituted for the DDT in the third and fourth sprays.

One pound of fifteen percent wettable Parathion powder per one hundred gallons of water may also be used instead of the DDT sprays in sprays three and four listed above. "Stickers" may be used in the earliest sprays as recommended by the manufacturer, but such applications are not recommended for the later sprays.

The control of the Leaf Folder should be exercised in the manner recommended for the Grape Berry Moth (above).

An application of one-and-one-half pounds of fifty percent wettable DDT powder diluted in one hundred gallons of water has been successfully used to control infestations of the Grape Leafhopper. Such a spray should be applied at petal-fall and seven to fourteen days after the first application.

The Grape Mealybug can be controlled through the use of a Parathion spray (one pound of fifteen percent wettable powder in one hundred gallons of water) that is applied either just before or just after the bloom and repeated in ten days if necessary.

When four or more leaves per vine show the spotting that is indicative of an infestation of Mites, an application of either one-and-one-half pounds of Kelthane (18½ percent wettable powder) or one pound of Parathion (twenty-five percent wettable powder) per one hundred gallons of water should control the infestation.

General Cultivation and Spray Program for the Control of Insects and Diseases that Infest the Grape

First Cover Spray

When the new growth is approximately two to four inches long, a spray composed of one pound of DDT (fifty percent wettable powder) plus one-and-one-half pounds of Malthion (twenty-five pereent wettable powder) per one hundred gallons of water should be applied. Alternatively, the first cover spray may be composed of a mixture of one-and-one-quarter pounds of Guthion (twenty-five percent wettable powder) and two pounds of Ferbam (seventy-six percent wettable powder) per one hundred gallons of water.

Second Cover Spray

Just before the bloom of the grape occurs, a second spray should be applied. The second cover spray is formulated similarly to the first cover spray (see above).

Third Cover Spray

After the bloom, when the berries have set, the formulation used for the first cover spray should be applied to the grapes once more, except for the substitution of two pounds of Captan (fifty percent wettable powder) for the Ferbam.

Fourth Cover Spray

Ten days after the third cover spray, the formulation of that spray should be applied to the vines.

Fifth Cover Spray

Two pounds of Captan (fifty percent wettable powder) per one hundred gallons of water should be applied ten days after the application of the fourth cover spray.

Notes:

If DDT were the only insecticide used in the cover spray, the Grape Mealy Bug would be able to develop. For this reason, malathion or guthion are added to the spray. An insoluble copper spray, used according to the directions supplied by the manufacturer may be used to control Powdery Mildew, but should not be used unless necessary since such sprays tend to reduce the yield of the grape. Bordeaux mixture is very effective against diseases of the grape, but may reduce yields and is incompatible with most insecticides that are used on the grapes. For this reason, Bordeaux mixture must be applied by itself.

Fungicides are considered more of an "ounce of prevention" than a "pound of cure". The grower is advised to prevent infections or to control them before they are extensive. Wind and rain are notorious agents that spread fungal infections, therefore the application of a fungicide a day or two in advance of rain, will do much good.

THE MANGO

The Ambrosia beetle, family *Scolytidae*, and the Lesser Snow Scale, *Pinnaspis strachani* (Cooley), are common pests of the branches and trunk of the Mango. The Ambrosia Beetle may be controlled by the application of one pound of Lindane (gamma-isomer concentrate wettable powder) per one hundred gallons of water. The Lesser Snow Scale responds to the application of either one gallon of emulsive oil or one-and-one-half pounds of Parathion (fifteen percent wettable powder) in one hundred gallons of water.

The leaves of the Mango are attacked by a number of insects. Infestations of Red-Banded Thrips occasionally become sufficiently severe to warrant treatment. An application of Dieldrin, either one pound of the twenty-five percent, one-half pound of the fifty percent wettable powder, or one pint of the emulsion concentrate (prepared at one-and-one-half pounds per gallon) per one hundred gallons of water is recommended to control these Thrips. This spray is quite versatile since the Dieldrin is compatible with sulfur, Parathion, copper, or carbamate fungicides.

The leaves are often attacked by various scale insects. Among the soft scales that attack the leaves are the following: Acuminate *Coccus acuminatus* (Sign.), Florida Wax *Ceroplastes floridensis* Comst., Mango Shield *Coccus mangiferae* (Green), and Pyriform *Protopulvinaria pyriformis* CK11. The armored scales sometimes attack the twigs as well as the leaves, and include in their number the following representatives: Dictyospermum *Chrysomphalus dictyospermi* (Morg.) and the Florida Red *Chrysomphalus aonidium* (L.). The Mango Shield, Pyriform, and Acuminate Scales may be controlled by spraying with one of the following suspended in one hundred gallons of water:

one pound emulsive oil
one pound Parathion (15% wettable powder)
3-4 pounds malathion (25% wettable powder)

The Florida Wax Scale should be controlled by a spray containing one-and-one-half pounds of Parathion (15% wettable powder) and one-and-one-third gallons of emulsive oil per one hundred gallons of water that is applied during December. The recommended procedure for the control of Armored Scales includes the application of either one-and-one-half to two pounds of Parathion (15% wettable powder) or one to one-and-one-half gallons of emulsifiable oil per one hundred gallons of water.

Mites often infest the young fruit, blossoms, and leaves of the Mango. Such an infestation can cause the leaves and young fruit to fall from the tree. Sulfur, either dusted on the trees at the rate of one half to eight ounces per tree, or sprayed at the rate of eight pounds of wettable sulfur per one hundred gallons of water, is effective in the control of some mites. Other types of mites may need a spray of three to four quarts of emulsifiable oil per one hundred gallons of water to effect their control. If the sprays that contain oil are applied to the trees during dry, hot (over 90°F), or unusually cool weather and the condition persists after the application of the spray, the trees may become "shocked" and their leaves may fall.

Occasionally the Banded Cucumber Beetle, *Diabrotica balteata* Lec., damages the young leaves at the site of their new growth. An application of a spray that contains Lindane (one pound of twenty-five percent wettable powder per one hundred gallons of water) is recommended for their control.

The following insects have been seen, from time to time, infesting various parts and stages of the blossom of the mango tree. None of these insects seems to have become of sufficient economic importance to warrant specific control measures: the Blossom Anomala *Anomala undulata* Mels., *Argyrotaenia amatana* (Dyar), Florida Flower Thrips *Frankliniella cephalica* (D. L. Crawford), and the Cotton Square Borer *Strymon melinus* (Hbn.).

The Citrus Mealybug, *Pseudococcus citri* (Risso), can infect the fruits of the mango. This pest can be removed from the surface of the fruit by a highly pressurized stream of water or can be controlled by an application of a spray of Parathion (one pound of 15% wettable powder) or Malathion (four pounds of 25% wettable powder) dispersed in one hundred gallons of water. Occasionally the larvae of the Papaya Fruit Fly infest the fruit of the mango, for their control see "Papaya".

THE PAPAYA

The Papaya Whitefly is an unwelcome visitor in the Papaya Orchard. The "crawler" stage of this pest sucks juices from the plant and, adding insult to injury, the insect excretes a honey-dew-like substance that becomes a substrate for the Sooty Mold. These Whiteflies may be controlled with sulfur dusts or a spray that is composed of eight pounds of wettable sulfur per one hundred gallons of water (or four tablespoons per gallon). Such a spray or dust should be applied when the adult fly population becomes large.

The Papaya Webworm, also known as the Fruit Cluster Worm, lives in a web that it spins along the stem of the plant and among the fruits. Not only does the Webworm injure the fruit and the plant, but these injuries allow the entrance of the fungal disease Anthracnose. A spray of Chlordane or DDT at the rate of two pounds of fifty percent wettable powder per one hundred gallons of water (or two tablespoons per gallon) has been found effective in the control of this pest.

The spray program may be begun at the first incidence of the Webworm or at the time of fruit setting and concluded thirty days prior to harvest.

The Papaya Fruit Fly possesses a very long ovipositor with which it thrusts its eggs into the seed cavity of the fruit of the Papaya. The larvae of this pest feed in the fruit and then burrow out of the fruit to drop to the ground for the two to four week pupation period. The emerging adults seek their mates and additional fruit on which to propagate. Control of this insect must be enacted before the eggs are laid to be successful. If only a few trees are involved in the infestation, the fruits of those trees may be individually wrapped in paper. If this method is impractical, the application of a spray that contains four pounds of fifty percent wettable DDT powder plus eight pounds of wettable sulfur dispersed in one hundred gallons of water at two to three week intervals until thirty days prior to the harvest has proven effective.

THE PEACH

Disease Control for the Peach

Peaches are subject to destructive attack by a number of diseases. The following important diseases are described prior to the presentation of a complete spray program.

Several fungae are capable of causing the disease known as **Brown Rot**. East of the Rocky Mountains the fungus *Monilinia fructicola* (Wint.) Honey causes this infection, while west of the mountains *Monilinia laxa* (Aderh. and Ruhl.) Honey is the culprit, especially in the Pacific Coastal area and the North Central states. Brown Rot is probably the most serious disease of peaches in the United States of America. The disease overwinters in mummified fruit to be carried, as spores, to new growth in the spring. The existence of humid weather during blossom time will cause the blooms to be damaged and the amount of fruit that is set to be reduced if the disease is present. This fungus also may cause cankers when it infects the twigs and branches of the tree. Pruning of infected parts of the tree and fungicide-containing sprays are effective means of controlling this disease.

Bacterial Spot, *Xanthomonas pruni* (E. F. Smith) Dows., appears at any time during the growing season. This disease attacks the underside of the leaves of the tree and produces small, pale, roundish areas. These areas grow larger and become deep purple or blackish as the disease progresses. In advanced cases of infection, the leaves may fade to a yellow-green color. If the fruit is infected, small brown spots are the first signs of such infection. As the fruit grows, the skin cracks and pits around these spots. A pale green circle often envelops the brown spot. Twigs infected with this disease are marked by dark purple, watery cankers on their surfaces. The twig cankers become sunken in time and are the overwintering sites of the Bacterial Spot. Warm, rainy weather accompanied by high winds is the ideal situation for the growth and spread of this disease.

Peach Cankers is a term applied to diseases that are caused by three different fungi: *Valsa cincta* Fr., *Valsa leucostoma* (Pers.) Fr., and *Fusicoccum amygdali* Delacr.. The small oval cankers that are characteristic of this disease may be observed on young twigs or branches during either the spring or the fall. Leaf scars and winter injury are the sites that are used for the entry of *Fusicoccum amygdali*. The exercise of care so as to prevent injury to the trees during the winter is one good measure for the control of this fungus. **Perennial Canker** is caused by *Valsa cincta* and *Valsa Leucostoma*. These fungi seem to invade dead or injured parts of the tree. A pale amber gum that exudes at the point of infection is a symptom of the presence of these unwelcome visitors. Gradually this gum becomes dark brown as the necrosis that lies under the gum grows larger and the tissue of the inner bark collapses. The bark around old cankers becomes shaggy and covered with a black fungus, while the cankers enlarge each year. These fungi also may cause a "die-back" of young twigs and the result of the attack is a gum-encrusted twig canker. These perennial canker organisms exhibit their greatest growth when the temperature is between 65° and 70°F, causing spring and fall to be the optimum periods for infection. Although the Perennial Canker is difficult to control, good cultivation practices are beneficial. Disease-free trees should not be placed in close proximity to older, infected trees to avoid further spread of the disease. The use of acceptable insecticidal programs will reduce insect-caused injury to the trees and block one means of entry that the fungus has. Spring pruning should be delayed until the new growth begins so that the injury caused by the pruning will heal quickly, discouraging entrance of the fungi. The elimination of fertilization past late in June will encourage the tree to enter dormancy earlier in the fall and reduce the chances of infection during the fall period of fungal activity. Lastly, the infected areas on trees can be scraped clean and a disinfectant can be applied to the injured area.

Peach Leaf Curl, resulting from infection by the fungus *Taphrina deformans* (Berkeley) Tul., is most severe during years in which the weather is wet. Infected new leaves become red, misshappen, thickened, silvery, and curled, finally dropping from the tree in June. If the twigs of a tree become infected, they develop greenish-yellow swollen areas. Infected fruits or flowers commonly drop from the tree. The application of an appropriate fungicide before the buds of the tree have swollen in the spring is generally effective in the control of the Peach Leaf Curl.

Peach Scab, manifested by small greenish-black spots on the fruit, is also known as either Black Spot or Freckles. A severe infection by the fungus that causes this disease, *Cladosporium carpophilum* Thumen, may cause the fruit to crack open. Round, tannish spots that have dark bluish-gray borders appear on twigs that have been infected. Infected leaves are marred by spots that ultimately drop from the leaf, leaving holes in the host. Peach Scab is generally easily controlled by adherence to a well prescribed program for the application of fungicides.

Virus Diseases

Peach Yellows causes the infected fruit to ripen much sooner than normal, destroying the good peach flavor and streaking the fruit with exaggerated coloration. Thin shoots crop up numerously along the large branches of trees that are infected by this disease. Leaves on the spindly branches of infected trees are narrow, misshapen, and have red spots on their surface.

X-Disease first manifests itself with purplish-red, water soaked areas on the leaves. The infected leaves roll upward, become frayed, and drop from the tree. Infected trees often lose all leaves with the exception of those at the tips of the branches.

Little Peach Infection often begins in one specific area of the tree. These infected areas have leathery leaves that are overcrowded, droopy, and have curled edges. Frequently such leaves grow pale as the season progresses. The size of the fruit on an infected tree is severely stunted and the flavor of such fruit is destroyed.

Phony Disease has occurred throughout much of the peach growing areas of the southern United States. This disease does not kill the trees, but renders their productive capacity ineffectual. Infected trees exhibit dwarfing of the terminal growth of the twigs. Diseased trees leaf out earlier in the spring and become dormant later in the autumn than do normal, healthy trees. The leaves are a darker green and the growth is thicker on infected trees than on healthy trees. Fruits from diseased trees are smaller, fewer, and more deeply colored than the fruits of normal trees. This disease is common among wild plum trees, so these trees should be destroyed in the vicinity of the peach, plum, almond, and apricot orchards, since all of these trees are susceptible to the Phony Disease. Vectors of this disease are Leafhoppers, especially *Oncometopia orbona* and *Homalodisca coagulata*. These insects, often called Sharpshooters, breed upon such plants as cotton, corn, sunflowers, and okra, as well as weeds such as ragweed, coffeeweed, and lambsquarters.

Peach viral diseases, such as the Phony Disease, are "treated" by the destruction of the infected trees and adherence to a rigorous program for the application of insecticides to control the vectors of the disease. Since wild plum and chokecherry trees, as well as cultivated plum trees, can be a reservoir of infection, these trees should not be maintained in the vicinity of the peach orchards.

Spray Program for the Control of Insects and Diseases of the Peach

(Note: all quantities per one hundred gallons of water)

Dormant Spray—applied in the late fall or early spring before the buds have swelled.

1. Bordeaux Mixture:
   8 pounds fresh spray lime
   8 pounds copper sulfate
   used especially for control of Peach Leaf Curl
   OR
2. Bordeaux Mixture:
   3 pounds fresh spray lime
   2 pounds copper sulfate
   used especially for the control of Peach Leaf Curl, Fusiococcum, Bacterial Spot, Mites, Scale Insects
   OR
3. Ferbam
   2 pounds of 76% wettable powder
   used especially for the control of Leaf Curl

Pre-Bloom Spray—applied when the blossom buds are full pink and before the blossoms have opened. Either of the two following sprays are used to control Brown Rot and Plant Bugs.

1. Captan or Phygon plus
   DDT or Dieldrin
   OR
2. Sulfur or Phygon plus
   DDT or Dieldrin

Bloom Spray—applied when the blooms begin to open. One of the following fungicides are used to control Blossom Blight. Note: Do NOT use insecticides during the blossom sprays to avoid killing the bees.

1. Captan
   OR
2. Phygon
   OR
3. Sulfur

Note: Captan and Phygon are more effective against Blossom Blight than is sulfur. Repeat the Blossom Spray every three to four days if the Brown Rot is particularly troublesome.

Petal Fall Spray—applied when approximately two thirds of the petals have fallen. This spray is used for the control of the following: Powdery Mildew, Brown Rot, Bacterial Spot, Peach Scab, Plant Bugs, Stink Bugs, Adult Oriental Fruit Moths, Curculio, Mites, and Scales.

1. Captan plus Sulfur plus Parathion
   OR
2. Parathion plus Sulfur
   OR

3. Captan plus Guthion

Sprays #1 and #2 are especially useful for the control of the Powdery Mildew. If Curculio and other insects are a particular problem, Dieldrin may be added to any of the above mixtures, provided that bees are not busy in the orchard. Tedion or Ovex may be added to the sprays for the control of mites if such specific control appears to be warranted.

Shuck-Split Spray—a repeat application of the Petal Fall Spray may be recommended for control of the same pests.

Shuck-Off Spray—a further repeat application of the Petal Fall Spray for cases in which the infestation warrants a third spray.

Cover Sprays—the First Cover Spray is applied approximately seven days after the application of the Shuck-Off Spray and repeat applications of the Cover Spray are made at seven to fourteen day intervals until the start of the Pre-Harvest Sprays. The Cover Sprays are intended to control Peach Scab, Powdery Mildew, Brown Rot, Sooty Mold, Bacterial Spot, Mites, Scales, Stink Bugs, Oriental Moth, Plant Bugs, Curculio, Japanese Beetle, Leaf Roller, and the Lesser Peach Tree Borer.

1. Captan plus Sulfur plus Parathion

OR

2. Captan plus Guthion

OR

3. Captan plus Parathion plus Trithion

If the Powdery Mildew is a problem, the use of Mixture #1 is especially good for a mixed apple-peach orchard. If Mites are a significant pest, the use of either Mixture #3 or the addition of Ovex or Tedion to Mixture #1 or #2 is recommended.

Pre-Harvest Sprays—these sprays are applied beginning about four weeks before the harvest and repeated at seven day intervals until one day before the harvest if the weather conditions are favorable for the spread of Brown Rot. The Pre-Harvest Sprays are intended to control Brown Rot, Peach Scab, Japanese Beetle, Oriental Fruit Moth, Curculio, Grasshoppers, and Rose Chafer. Note: Parathion containing sprays should NOT be applied later than fourteen days prior to the harvest.

1. Parathion plus Captan

OR

2. Parathion plus Wettable Sulfur

In the fourteen days prior to the harvest, Sevin may be substituted for the Parathion since Sevin may be applied up to one day prior to the harvest. Malathion may be used as a substitute for the Parathion in the above sprays, but, like Parathion, Malathion should not be applied within fourteen days of the harvest. Captan is preferable to Sulfur since the Captan imparts a better color and keeping quality to the harvested product.

Harvest Spray—applied immediately before or during the harvest to help control fruit rots.

1. Captan

   OR

2. Sulfur

The Captan is preferred to the use of Sulfur since the former imparts a better color and keeping quality to the peach.

Post Harvest Sprays

Several Post Harvest Sprays may be applied for the completion of an insect and disease control program. Generally the Post Harvest Sprays are tailored to the control of a particular pest or group of pests, and a few of the more common of these specific-purpose sprays will be presented.

Peach Canker—application of a phenyl mercury fungicide according to the directions of the manufacturer for the control of Apple Scab. When used on young trees, the concentration of the spray should be cut in half.

Peach Leaf Curl—application of a Bordeaux Mixture is generally preferred, but the use of Ferbam, Phygon, DN, or a liquid lime sulfur have been used with success.

Peach Tree Borers—application of a spray containing either Thiodan or DDT has been used.

Lesser Peach Tree Borer—application of a spray containing either Thiodan, Parathion, or Guthion has been recommended.

Insect Pests of the Peach and their Control

Various species of Aphids may be controlled by the application of a spray that contains either nicotine sulfate (¾ pint of a forty percent solution) plus soap or hydrated lime, OR Malathion (one pound of a fifty percent wettable powder) in one hundred gallons of water.

For the control of the Lesser Peach Borer, see the section on Cherries.

The Peach Twig Borer was introduced from Europe and has spread throughout the United States. This insect is a particular pest in Californian conditions where three annual broods of the insect are common. The Peach Twig Borer overwinters as the larval form in tiny cells beneath the bark and especially in the crotches of small twigs. As the buds open in the spring, they are attacked by the young larvae who also mine the young twigs. These actions result in the destruction of many of the buds. As the larvae mature, they become dark pink or brown in color. The second brood of moths deposit their eggs on the twigs and fruit, feeding mostly in the developing fruit. An unchecked third brood is capable of causing very serious injury to the crop. The Peach Twig Borer can be controlled by the application of one of the following sprays:

1. **A Dormant Spray** containing
   a. 2 pounds Parathion (25% wettable powder)

      PLUS

3 gallons oil emulsion

OR

2.5 gallons emulsive oil

OR

b. 2 pounds Diazinon (25% wettable powder)

per one hundred gallons of water

OR

2. **A Petal Fall Spray** containing

3-4 pounds basic lead arsenate

OR

2 pounds endosulfan (Thiodan) (25% wettable powder)

OR

2 pounds DDT (50% wettable powder)

per one hundred gallons of water

This spray may be repeated in ten days if necessary. Do not make more than two applications of the endosulfan (Thiodan) while the fruit remains on the tree. Do not allow animals to graze in treated orchards. Do not make more than one application of the DDT spray within six weeks of the harvest.

Under conditions that are prevalent in California, two applications of insecticides are necessary for the control of the Peach Tree Borer; the first in mid-May and the second in mid-July. One of the following sprays should be applied with a hand-gun to the trunk of the tree from the juncture of the main scaffold limbs to the soil line. A coarse spray under low pressure without a drifting mist should be used since the fruit must not be contaminated by the spray.

6 pounds Dieldrin (50% wettable powder)

OR

2 pounds Parathion (15% wettable powder)

PLUS

2 pounds DDT (50% wettable powder)

OR

2 pounds endosulfan (Thiodan) (25% wettable powder)

per one hundred gallons of water

Under New York conditions (Smith, 1961), the applications should be made about July 12 and during the first part of August. This investigator reported the Thiodan (endosulfan) spray to be the most effective spray that was evaluated. The same precautionary warning not to contaminate the fruit was stressed for all of these materials. A pre-planting treatment consisting of immersion of the roots and crown of the trees in a suspension of endosulfan (Thiodan) at ten pounds of fifty percent wettable powder per one hundred gallons of water provided a high degree of control of the Peach Tree Borer during the second growing season. Dieldrin and endosulfan were also effective at the five pound concentration, while DDT was found to be effective at higher concentrations. None of these treatments proved phytotoxic, but the emulsifiable concentration was superior to the wettable powders (Smith, 1963). Paradichlorobenzene may be used as a soil treatment for the control of the Peach Tree Borer. The recommended dosages range from 0.25 to 1.5 ounces depending upon the age of the tree that is to be treated. The crystals of this chemical are placed in a shallow trench around the crown of the tree, without touching the bark, and covered with fine soil for a few weeks. The treatment should be made in late August or early September in the northern United States and up to late October in the southern areas. The mounds that are formed by the covering of fine soil should be removed from the young trees after three or four weeks, while they may remain longer in the case of more mature trees.

For the control of the Japanese Beetle, see the section on Apples.

The Oriental Fruit Moth can be controlled by the application of two pounds of fifty percent wettable DDT powder dispersed in one hundred gallons of water. For other control techniques, see the section on the control of the Peach Twig Borer.

Dormant Mites may be controlled by the application of either three gallons of an emulsion of oil or two gallons of Superior Oil per one hundred gallons of water. This spray should not be applied until the rains have moistened the soil. Summer Mites have been controlled by a spray that contains either two pounds of 18.5 percent wettable Kelthane powder or one pound of twenty-five percent wettable tetradifon powder per one hundred gallons of water.

The application of a spray that contains either of the following active ingredients per one hundred gallons of water can be used in the control of the Plum Curculio:

1 pound Guthion (25% wettable powder),
1 pound Malathion (50% wettable powder),
3 pounds methoxychlor (50% wettable powder).

This spray should be applied at petal fall, with a second application ten days after the first spraying, with succeeding sprays at similar intervals as needed. For further control measures for use on the Plum Curculio, see Apple.

The application of one pound of fifteen percent wettable Parathion power in one hundred gallons of water can be used for the control of Scale Insects. For further control techniques, see the discussion for the Peach Twig Borer.

The group of insects known as Stink Bugs breed on grass and other plants, but migrate to the peach as the grasses dry. The Stink Bugs feed on the young fruit of the peach and cause the peach to twist out of shape. To control this pest, apply either one pound of fifty percent wettable Guthion powder or two pounds of fifty percent wettable carbaryl (Sevin) per one hundred gallons of water.

THE PEAR

Control of Insects of the Pear

The application of one pound of fifty percent wettable di-nitrocresol powder per one hundred gallons of water as a dormant spray is used for the control of dormant Aphids.

At petal fall, the application of one of the following active ingredients per one hundred gallons of water is effective in a program for the control of the Codling Moth:

3 pounds lead arsenate
PLUS
3 pounds hydrated lime,
OR
1 pound Guthion (25% wettable powder),
OR
1 pound carbaryl (Sevin) (50% wettable powder).

Mealybugs cause smutting of the fruit and can be controlled when the Pear tree is at the cluster bud stage by the application of two pounds of twenty-five percent wettable either Parathion or Diazinon powder per one hundred gallons of water.

Dormant Scale Insects can be controlled by the application of either four gallons of petroleum emulsive oil or five gallons of an oil emulsion per one hundred gallons of water.

Clover and European Red Mites can be controlled in either the dormant or delayed dormant stage by the application of either three gallons of a petroleum oil emulsion or two gallons of an emulsion of Superior Oil per one hundred gallons of water. The summer mites should be controlled by the application of eight ounces of fifty percent wettable ovex (Ovotran) powder or 0.4 pounds of chlorbenside wettable powder per one hundred gallons of water.

The Pear Psylla is a very important insect in the northeastern states and has been recently found on the Pacific Coast. If uncontrolled, this insect may cause defoliation of the tree and consequent loss of the crop. The adults overwinter under the scales of the bark on the tree as well as in the litter on the ground. Emerging early in the spring, the adults deposit their eggs in the bud cluster. The young nymphs feed on the peticle of the leaf and the stem of the fruit. Since the Pear Psylla reaches maturity in about a month, three to four generations of the pest may be produced each season. Attacked fruit is both stained and smutted. (Mills, et al., 1954) Application of either three gallons of petroleum oil or two gallons of Superior Oil per one hundred gallons of water is a control measure for the Pear Psylla. At petal fall and during the summer, apply one-quarter to three-eighths pound of wettable Guthion powder active ingredient or 0.1 pounds of active Rotenone plus two quarts of foligae oil per one hundred gallons of water.

The Pear Slug can be controlled by either a lead arsenate spray or a DDT spray as directed for the control of the Codling Moth.

As buds open in the green stage, the application of two pounds of DDT (50% wettable powder) per one hundred gallons of water may be used for the control of Pear Thrips.

Diseases of the Pear and their Control

Fire Blight a bacterial disease caused by *Erwinia amylovora*, inflicts a blight on leaves, blossoms, and twigs. The Fire Blight girdles branches and is capable even of killing the tree. The diseased parts of the tree turn dark and appear to have been scorched by fire. Infected branches have a water-soaked appearance initially that later becomes dry, sunken, and dark. This disease overwinters in cankers on the infected part of the tree. In the spring, the bacteria are spread by rain and insects to infect the blossoms and twigs. The bacteria rapidly multiply in humid, warm weather, increasing the extent of the infection. The control of the Fire Blight may take several forms. The obvious suggestion to plant varieties of the pear that are resistant to the disease is helpful only as a preventative measure. The application of nitrate fertilizers in the spring is not generally recommended since the resultant heavy growth of the tree is susceptible to further infection. Infected tissue should be pruned from the tree and removed and wounds should be disinfected with a one to one thousand solution of bi-chloride of mercury. Streptomycin is the most effective material to use in a spray for the control of Fire Blight, however it is generally not used since it is both expensive and difficult to apply. Zineb or Bordeaux Mixture give some measure of protection against this disease if they are sprayed during the blossom time of the pear. Since the disease is spread by both rain and insects, and little can be done about the former, control of the latter is an aid in the control of Fire Blight.

Stoney Pit, a disease that causes pears to be deformed by gnarled and hard spots in the flesh, is caused by a virus. The virus that causes Stoney Pit is transmitted by grafting diseased stock onto healthy trees. Since no chemical remedy for the disease is presently available, the use of disease-free grafting stock cannot be emphasized too heavily.

A Spray Program for Pears and other Home Grown Fruits

Early Spray—applied one to two times per week from the green tip through the bloom stage.
2 pounds Captan (50% wettable powder) per one hundred gallons of water.

Spring Spray—applied once per week from petal fall until approximately June 5 (in Michigan).

2 pounds Captan (50% wettable powder)
PLUS
3 pounds Methoxychlor (50% wettable powder)
PLUS
2 pounds Malathion (25% wettable powder) per one hundred gallons of water.

Summer Spray—applied every two weeks from approximately June 10 (one week after the last Spring Spray) until one week prior to harvest.

2 pounds Captan (50% wettable powder)
PLUS
2 pounds Sevin (50% wettable powder)
PLUS
2 pounds Kelthane* (18.5% wettable powder) per hundred gallons of water.

*Note: the Kelthane should only be used three times per season, or once a month.

THE PECAN

Pecan Scab, caused by the fungus *Fusicladium effusum* Wint., is one of the most serious diseases of pecans. This fungus attacks the tissues of the leaves, nuts, and twigs when they are in the stage of rapid growth. The Pecan Scab is manifested as dark spots on the leaves and nuts. On the nuts, the spots of infection become sunken. This disease is usually most severe in cool, wet seasons. Some varieties of the pecan tree are more susceptible to damage from the Pecan Scab than are other varieties. The Stuart variety appears to possess some resistance to the disease. Du-Ter and Cyprex fungicides are generally recommended for use in controlling Pecan Scab.

Powdery Mildew, a disease caused by the fungus *Microsphaera alni* Wint., sometimes becomes a serious pest in periods of either heavy rainfall or high humidity. A powdery white covering on the nut and leaves is an indication of the presence of this disease.

Crown Gall is manifested by galls, or lumpish growths, at the bud union on the trunk near the soil line or on wounds on the lower limbs or on the roots. Infested trees exhibit reduced vigor and may be blown over by a strong wind if the disease has progressed to a serious state in the tree. The bacteria, *Agrobacterium tumefaciens* (E.F.Sm. & Town), enters the tree through wounds in the bark. Disease-free trees must be used as new plantings in orchards as a control measure against this disease. The galls from infected trees should be removed and burned. The wounds should be painted with a compound that is composed of one part of creosote to three parts of coal tar. Since the disease can be carried to healthy trees via cultivation machinery, the area around infected trees should not be deeply cultivated.

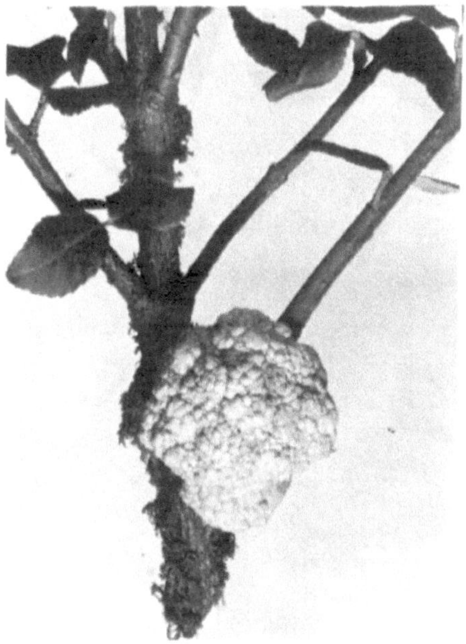

Figure VII-7. Crown Gall (Illinois Natural History Survey.)

Rosette, a nutritional disease, results from a zinc deficiency in the tree. The first symptoms of this disease are the yellowing and crinkling of the leaves at the top of the tree. As the disease progresses, the terminal growth dies and the tree becomes less productive. Rosette is easily remedied by the application of zinc sulfate, preferably in February, to the soil at the base of the tree. Approximately ten pounds of zinc sulfate should be used for each tree, depending upon the size of the tree. Alternatively, the zinc sulfate may be added to a Bordeaux Mixture at the rate of four pounds of zinc sulfate per one hundred gallons of the spray that contains the Bordeaux Mixture. This spray can be used during the regular spraying program to correct the zinc deficiency of the trees.

Mouse Ear, or **Little Leaf**, is a disease that is caused by a deficiency of manganese in the tree. The deficiency of this element causes the leaves of the tree to be stunted in size and blunted in shape. The condition of diseased trees can be improved by either a spray that contains one to two percent manganese sulfate or the addition of two to four pounds of manganese sulfate to the soil around a mature tree (correspondingly less of the chemical should be used for less mature trees).

Brown Leaf Spot, *Cercospora fusca* Rand, is a rather prevalent disease, but is rarely serious in an orchard of otherwise healthy, vigorous trees. Reddish-brown splotches on leaflets are an indication of the presence of the disease. A regular program of spraying should control this disease with little problem.

Figure VII-8. A Branch of a Pecan Tree Exhibiting Resette–a result of Zinc Deficiency. (Illinois Natural History Survey.)

Pecan Leaf Blotch is caused by *Mycosphaerella* dendroides (Cke.). The infection begins when small olive green, velvety spots first occur on the underside of the leaves. Later yellow spots develop on the upper surface of the leaves. Fruiting bodies appear in mid-summer and they join to turn the leaf black and shiny with a blotched appearance as the season progresses. This disease is more often a problem among nursery stock or unkempt trees than among healthy, mature trees.

Pink Mold, the fungus *Cephalothecium roseum* Corda, appears in late summer and early fall as a pink growth on the shucks of the nuts, especially on those that were injured by either mechanical means or Scab. The fungus is able to penetrate the shell of thin-shelled varieties of the pecan. If entrance through the shell is achieved, the fungus enters the kernel, decaying it. A rancid odor often is emitted from such affected nuts. Pink Mold is usually controlled by the spray program that is used to control Scab.

Dieback, a fungus disease that is caused by *Botryosphaeria berengeriana* DeNot, affects both twigs and limbs, causing terminal growth to abruptly wilt and die. Long ruptures in the diseased bark are imbedded with black crusts of the fungus. A diseased twig has a water-soaked appearance at its base, while infected portions of the branches become dry and sunken as the disease progresses. The most effective control method is to prune and burn the infected wood.

A Spray Program for the Control of Insects and Diseases of the Pecan

Dormant Spray—applied after the leaves have fallen from the tree but before the buds begin swelling in the spring. Unless otherwise noted, the active ingredients that are listed below are applied in one hundred gallons of water.

1. To control the Obscure Scale, the spray should contain three gallons of either Lubricating or Superior oil in the form of an emulsion.
2. To control Spanish Moss, the spray should contain ten pounds of copper sulfate PLUS ten pounds of calcium arsenate (Bordeaux Mixture).

First Pre-pollination Spray—applied when the buds are bursting and the leaves begin to show one to two inches of green growth. All sprays are listed as active ingredients that are to be suspended in one hundred gallons of water prior to application.

dosage	*to control*
1. 1 pound Cyprex (65% wettable powder)	Scab
OR	
4 pounds copper sulfate PLUS 1 pound calcium arsenate (Bordeaux Mixture)*	Scab
OR	
2 pounds Zineb (65% wettable powder)	Scab
PLUS	
2. 2.5 pounds BHC (12% wettable powder)	Phylloxera
OR	
0.4 pounds Parathion (15% wettable powder)	Leaf Casebearer
OR	
0.6 pounds Malathion (25% wettable powder)	Leaf Casebearer

*Note: The Bordeaux Mixture is not compatible with the insecticides that are listed above.

Second Pre-pollination Spray—applied when the leaves are half grown. The active ingredients listed below should be dispersed in one hundred gallons of water before use as a spray.

dosage	*to control*
1. Bordeaux Mixture 4 pounds copper sulfate PLUS	Scab, Nursery Blight, Downy Spot

dosage	*to control*
1 pound calcium arsenate	
OR	
0.5 pounds dodine (Cyprex)*	Downy Spot
OR	
2 pounds Zineb	Downy Spot
(76% wettable powder)	
PLUS	
1 quart emulsion of summer oil	
PLUS	
2. 2 pounds Parathion	Pecan Leaf Casebearer
(15% wettable powder)	
3 pounds Malathion	Pecan Leaf Casebearer
(25% wettable powder)	

*Note: Since dodine (Cyprex may cause defoliation of the Moore and Van Deman varieties, this ingredient should NOT be used on these varieties.

First Cover Spray—applied when the young nuts first appear. The active ingredients listed below should be dispersed in one hundred gallons of water prior to use as the spray.

dosage	*to control*
1. Bordeaux Mixture	Scab, Brown Leaf Spot,
4 pounds copper sulfate	Nursery Blight
PLUS	
1 pound calcium arsenate	
OR	
0.5 pounds Cyprex	
(65% wettable powder)	
OR	
0.08 pounds Du-Ter	
(50% wettable powder)	
PLUS	
2. 2 pounds Parathion	Pecan Leaf Casebearer,
(15% wettable powder)	Pecan Nut Casebearer, Mites,
OR	Aphids
3 pounds Malathion	
(25% wettable powder)	
OR	
2 pounds EPN	
(25% wettable powder)	
OR	
2 pounds Guthion	
(25% wettable powder)	
OR	

dosage	*to control*
2 pounds Thiodan (50% wettable powder) OR 1.25 pounds carbaryl (Sevin) (80% wettable powder)	

Note: A large population of Casebearers may warrant a second application of the insecticide approximately seven days after the first application. A serious infestation of either Aphids or Mites would lead to the selection of Parathion as the second ingredient in the spray.

Second Cover Spray—applied three to four weeks after the first cover spray. This spray consists basically of the fungicide of the First Cover Spray to control Scab, Powdery Mildew, Nursery Blight, Brown Leaf Spot, Blotch, and Rosette, with the addition of four pounds of zinc sulphate if rosette is present for its control. If Mites, Casebearers, or Aphids are a problem, the use of an insecticide in the Second Cover Spray, such as the previously recommended Malathion or Parathion, is indicated.

Third Cover Spray—applied three weeks after the Second Cover Spray. The same procedure that was outlined for the Second Cover Spray applies for this spray. A repeat application of the insecticide during this spray is recommended if it is necessary. The use of Zineb is recommended if either Downy Spot or Brown Leafspot are prevalent at this time.

Fourth Cover Spray—applied three to four weeks after the Third Cover Spray. As was mentioned before, all dosages are per one hundred gallons of water.

dosage	*to control*
1. Bordeaux Mixture 4 pounds copper sulfate PLUS 1 pound calcium arsenate OR 1 pound Cyprex (65% wettable powder) OR 2 pounds Zineb (65% wettable powder) OR 2 pounds Maneb (70-80% wettable powder) PLUS	Scab
2. 1 pound Karathane OR	Powdery Mildew

dosage	*to control*
2 pounds sulfur (325 mesh) PLUS*	
3. 2 pounds Parathion (15% wettable powder) OR	Black Aphids, other insects
2 pounds Guthion (25% wettable powder) OR	
3 pounds Malathion (25% wettable powder) OR	
2 pounds Endosulfan) (50% wettable powder)	

*Note: These insecticides should be used only as necessary to control insects, e.g. Black Aphids, NOT as a preventative measure.

Fifth and Sixth Cover Sprays—applied at two to three week intervals as needed to control insects and/or diseases. Many of the insecticides must NOT be applied within twenty-one to thirty days prior to the harvest. As always, these chemicals are designed to kill the pest, and the labels should be completely read and understood, and the directions provided by the manufacturer should be carefully followed so that no species of life that are not intended to be controlled are not controlled through either accident or carelessness. Such materials as EPN, Parathion, Demeton, and Guthion are not designed for use in small home plantings. These materials are to be applied by trained operators.

THE PLUM

A Spray Program for the Control of Insects and Diseases of the Plum

Delayed Dormant Spray—applied immediately after the buds begin to open. The spray generally consists of two gallons of Superior Oil per one hundred gallons of water and is used to control Scale Insects and European Red Mites. A Bordeaux Mixture containing twelve pounds of copper sulfate and twelve pounds of calcium arsenate may be added to the dispersion of Superior Oil if Black Knot is a problem.

Pink Spray—applied when the buds show pink. This spray contains one-and-one-half pounds of fifty percent wettable Genite powder per one hundred gallons of water. The Pink spray is used to control European Red Mites.

Petal Fall Spray—applied at the time the petals fall from the tree. This spray, containing two pounds of Captan (50% wettable powder) plus either one-and-one-quarter pounds of Guthion (25% wettable powder) OR three pounds of lead arsenate, is used to control Curculio.

First Cover Spray—applied after the shucks have fallen. A repeat application of the Petal Fall Spray is used to control Curculio and Brown Rot. Either Tedion or Morestan may be used in addition to the conventional Petal Fall Spray if mites are present at this time.

Second Cover Spray—applied ten days after the application of the First Cover Spray. The Second Cover Spray is a repeat application of the First Cover Spray.

Subsequent Cover Sprays—application of these sprays is begun three weeks prior to the harvest and are spaced at seven day intervals. These sprays are used to control Brown Rot. The sprays may be composed of either eight pounds of microfine Sulfur or two pounds of fifty percent wettable Captan per one hundred gallons of water.

Aphid Control Sprays—applied as necessary to control the Aphid population. The spray is composed of two pounds of twenty-five percent wettable Malathion powder per one hundred gallons of water.

Diseases of the Plums

For the discussion of **Brown Rot**, *Monilina fructicola* (Winter) Honey, see the section on Cherry.

Black Knot, a fungus disease caused by *Dibotryon morbosum* (Schweinitz) Theissen and Sydow, infects not only Plum and Prune trees, but also both cultivated and wild cherry trees. The Black Knot infects the woody parts of the tree, especially the twigs. The raised knots vary in length from one-half to twelve inches and reach a circumference of up to two inches. The young knots are rather soft and olive green in the spring, becoming dark and brittle with age.

Plum Leaf Spot, caused by the fungus *Coccomyces prunophorae* Higgins, has a life cycle similar to that of the Cherry Leaf Spot fungus (see the section on Cherries). Plum Leaf Spot may be noted by the appearance of tattered leaves on the tree.

Prune Dwarf, a virus disease, is a serious problem among a limited variety of trees, such as the Damson Plum, and the Albion, Fellenberg, Tragedy, and Lombard Prune Trees. The presence of Prune Dwarf may be noted by the coarse, thick, wrinkled, narrow leaves that are often brittle and more prominently veined than normal leaves. Infested trees should have the diseased limbs pruned off and the removed wood should be burned. Severely infected trees should be destroyed completely, with virus-free trees selected for replanting.

BIBLIOGRAPHY

1. Agricultural Handbook, #331, Suggested Guide for the Use of Insecticides to Control Insects Affecting Crop, Livestock, Households, Stored Products, Forests, and Forest Products, 1967, Agricultural Research Service and Forest Service, U.S.D.A., Wash., D.C., 1967.
2. Anthon, E. W., Journal of Economic Entomology, Volume 53 (1960), p. 1085.
3. Barnes, G., Dowell, G. C., and McDaniel, M. C., Control of Diseases and Insects of Pecans, Leaflet #272, Agricultural Extension Service, University of Arkansas, Division of Agriculture and U.S.D.A. Cooperating, Fayetteville, Ark., March, 1965.
4. Bobb, M. L., Journal of Economic Entomology, Volume 54 (1961), p. 799.
5. Brogdon, J. E. and Lawrence, F. P., Insects and Mites of Florida Citrus, Circular 137-B, Florida Agricultural Extension Service Institute of Food and Agricultural Sciences, University of Florida, Gainsville, Fla., November, 1965.
6. Brogdon, J. E. and Lawrence, F. P., Control of Insects and Diseases of Dooryard Citrus Trees, Circular 139-C, Agricultural Extension Service Institute of Food and Agricultural Services, University of Florida, Gainsville, Fla., 1966.
7. Brogdon, J. E. and Wolfenbarger, D. O., Papaya Insect Control, Circular 136-B, Agricultural Extension Service Institute of Food and Agricultural Sciences, University of Florida, Gainesville, Fla., April, 1967.
8. Chapman, P. J., et al., Journal of Economic Entomology, Volume 55 (1962), p. 737.
9. Chemical and Engineering News, November 6, 1967, p. 45.
10. Commercial Citrus Insect and Mite Control Calendar, Florida Agricultural Extension Service, University of Florida, Gainesville, Fla., July 8, 1966.
11. Compendium of Plant Diseases, Rohm and Haas Company, Philadelphia, Pa., 1959.
12. Controlling Phony Disease of Peaches, Leaflet #515, U.S.D.A., Wash., D.C., 1966.
13. Cox, J. A., Journal of Economic Entomology, Volume 54 (1961) p. 853.
14. deOng, E. R., Chemistry and Uses of Pesticides, 2nd Edition, Reinhold Publishing Corporation, N.Y., 1956.
15. deOng, E. R., Chemical and Natural Control of Pests, Reinhold Publishing Corp., N.Y. 1960.
16. Dunegan, J. C., "Apple Bitter Rot", Leaflet #406, U.S.D.A., Wash., D.C., 1964.
17. Flanders, S. E., Journal of Economic Entomology, Volume 53 (1960), p. 757.
18. Garrett, W. N., Fire Blight of Apples and Pears, Leaflet #10, Cooperative Extension Service, University of Georgia College of Agriculture, Athens, Ga., August, 1964.
19. Glass, E. H., Journal of Economic Entomology, Volume 54 (1961), p. 915.
20. Harris, E. D., Jr. and McGlohon, N. E., Pecan Insects and Diseases and Their Control, Bulletin #644, Cooperative Extension Service, University of Georgia College of Agriculture, Athens, Ga., 1962.
21. Harris, E. D., Jr. and McGlohon, N. E., Georgia Peach Disease and Insect Control Guide, Bulletin #630, Cooperative Extension Service, University of Georgia College of Agriculture, Athens, Ga., January, 1967.
22. Hickey, K. D. and Wooley, P. H., Insects and Diseases of Stone Fruit Trees, Extension Bulletin #1113, New York State College of Agriculture, Cornell, N.Y., June, 1963.
23. Hoyt, S. C., Journal of Economic Entomology, Volume 54 (1961), p. 1127.

24. Jeppson, L. R., et al., Journal of Economic Entomology, Volume 55 (1962), p. 17.
25. Johnson, R. B., Journal of Economic Entomology, Volume 54 (1961), p. 977.
26. Kelthane for Superior Mite Control, Bulletin #AG-222a, Rohm and Haas, Philadelphia, Pa., November, 1966.
27. Koehler, C. S., et al., Extension Bulletin, New York State College of Agriculture, Cornell, N.Y., 1962.
28. Maxwell, C. W., Journal of Economic Entomology, Volume 54 (1961), p. 526.
29. McGlohon, N. E., Bramble Diseases (Blackberry and Dewberry), Leaflet #41, Cooperative Extension Service, University of Georgia College of Agriculture, Athens, Ga., October, 1965.
30. McGlohon, N. E., Bunch Grape Diseases, Leaflet #43, Cooperative Extension Service, University of Georgia College of Agriculture, Athens, Ga., October, 1965.
31. McGlohon, N. E., Strawberry Diseases, Leaflet #62, Cooperative Extension Service, University of Georgia College of Agriculture, Athens, Ga., December, 1966.
32. McGlohon, N. E. and Harris, E. D., Jr., Georgia Apple Disease and Insect Control Guide, Revised, Bulletin #643, Cooperative Extension Service, University of Georgia College of Agriculture, Athens, Ga., January, 1966.
33. McGrew, J. R., et al., Farmers Bulletin #1893, U.S. Department of Agriculture, Wash., D.C., 1961.
34. Mills, W. D. and LaPlante, A. A., Extension Bulletin #711, New York State College of Agriculture, Cornell, N.Y., 1954.
35. Mitchell, A. E., Dowdy, A. and Klos, E., Pest Control Program for Home Grown Fruit, Extension Folder F-17, Michigan State University Cooperative Extension Service, East Lansing, Mich., April, 1966.
36. Peach Spray Program for Control of Insects and Diseases, Stauffer Chemical Company, New York.
37. Perennial Canker of Peaches, Bulletin #806, University of Illinois, Urbana, Ill.
38. Phillips, A. M., Large, J. R., and Cole, J. R., Insects and Diseases of the Pecan in Florida, Bulletin #619-A, Agricultural Experiment Stations, University of Florida, Gainesville, Fla., June, 1964.
39. Phillips, J. H. and Smith, E. H., Journal of Economic Entomology, Volume 56 (1963), p. 175.
40. Polivka, J. B., Journal of Economic Entomology, Volume 53 (1960), p. 320.
41. Powell, D. and Meyer, R. H., Pest Control Guide for Blueberries, Brambles, Grapes, Gooseberries, Currants, Fruit Leaflet #4, University of Illinois Extension Service in Agriculture and Home Economics in Cooperation with Illinois Natural History Survey, Urbana, Ill., June, 1965.
Powell, D. and Meyer, R. H., and Owen, F., Pest Control in Commercial Fruit Plantings, Circular 936, University of Illinois College of Agriculture Cooperative Extension Service in Cooperation with Illinois State Natural History Survey, Urbana, Illinois, May, 1966.
42. Roberts, J. E., Barnes, G., and McDanial, M. C., Spraying Plums and Cherries, Leaflet #265, Agricultural Extension Service, University of Arkansas Division of Agriculture and U.S. Department of Agriculture Cooperating, April, 1966.
43. Sanford, K. H., Journal of Economic Entomology, Volume 55 (1962), pp. 659, 724.
44. Shannon, E. L., Control Fire Blight in Pear and Apple Trees and Ornamentals:, Bulletin #400 H-103, Plant Science Guide, Cooperative Extension Service, New Mexico State University, University Park, N.M.

45. Smith, E. H., Journal of Economic Entomology, Volume 54 (1961), p. 668.
46. Smith, E. H., Journal of Economic Entomology, Volume 56 (1963), p. 463.
47. Smith, E. H. and Phillips, J. H. H., Journal of Economic Entomology, Volume 54 (1961), p. 1165.
48. Taschenberg, E. F., et al., Journal of Economic Entomology, Volume 53 (1960), p. 856.
49. Turnquist, O. C., Lofgren, J. A., and Johnson, H. G., Home Fruit Spray Guide, Extension Pamphlet #184, University of Minnesota Agricultural Extension Service, University of Minnesota, St. Paul, Minn., 1967.
50. Walker, J. C., Plant Pathology, 2nd Edition, McGraw Hill Book Co., Inc., New York, 1957.
51. Westcott, C., Plant Disease Handbook, D. Van Nostrand Co., Inc., New York, 1950
52. Wolfenbarger, D. O., Controlling Mango Insect Pests, Circular 147-A, University of Florida Agricultural Extension Service, Gainesville, Fla., May, 1963.
53. Wolfenbarger, D. O., Insect Pests of the Avocado and Their Control, Bulletin 605-A, University of Florida Agricultural Extension Service, Gainesville, Florida, May, 1963.

VIII

Insects and Diseases in the Garden and Truckfarm

TREATMENT OF THE SEED

Although the seed of the vegetable may be treated to control various insect pests, a second purpose for the treatment of seed is the control of diseases. If the seed was not treated by the producer, the user should consider the treatment of the seed to thwart seed-borne diseases and to control the damage that is caused by maggots to the seedlings. A common, small-scale treatment consists of the placement of a pinch of forty percent Chlordane wettable powder and a pinch of a commercial combination of seventy-five percent active Thiram and seventy-five percent of Captan in the seed packet. Coat the seeds by shaking the packet carefully.

If more than a few seeds must be treated, more precise methods should be employed. An "eradicative" treatment can be employed to destroy seed-borne disease organisms and confer a protection against soil-borne diseases to both the seed and the seedling. The two "eradicative" treatments that are available are the use of hot water or the use of mercuric chloride. The treatment with mercuric chloride consists of soaking the seed of vine crops for ten minutes in a 1:1000 mercuric chloride:water mixture. This ratio is equivalent to one ounce of mercuric chloride per seven-and-one-half gallons of water. Pepper seeds should be soaked for five minutes in a 1:2000 solution of mercuric chloride in water. After the seeds are soaked for the recommended time, they are drained and rinsed in cold, running water for fifteen minutes. The seeds are spread in a thin layer to dry to complete the treatment. Sweet potato roots can also be treated with mercuric chloride immediately before they are planted. Since the mercuric chloride and its mixtures are very corrosive, the treatment process must not be conducted in metal containers.

The treatment of the seeds with hot water will destroy most seed-borne diseases, but to be effective the prescribed temperatures and times must be followed exactly. Too low a temperature or exposure will fail to destroy the diseases while either extended temperature or exposure may damage the seed. The first step in the treatment consists of soaking the seeds in warm water before they are placed in the hot water treatment bath. The hot water treatment bath must be at least five times the volume of the seed that is to be treated to prevent the cool seed from chilling the bath too severely. The times of exposure and temperatures of the water bath recommended are shown in Table VIII-1. (Shurtleff and Linn)

TABLE VIII-1

*TEMP**	*TIME*	*SEED*
118	30 minutes	lettuce, celery
122	15 minutes	cress, mustard, radish
122	20 minutes	broccoli, carrot, cauliflower, chinese cabbage, collard, cucumber, rale, kohlrabi, rape, rutabaga, turnip
122	25 minutes	brussels sprouts, cabbage, eggplant, spinach tomato
125	30 minutes	pepper

*temperature in degrees Fahrenheit at with the bath must be maintained.

After the seed is treated, it should be *gently* plunged into cold water before being spread out to dry.

Various chemicals are available for the treatment of the seeds to protect against various pests. For protection against seed decay, seed-surface borne smuts, and damping off, the use of one of the following is generally effective when used according to the directions of the manufacturer: Chloranil, Thiram 50 or 75, or Captan 75. Various types of seed rots and seedling blights may be controlled by the proper application of Dichlone. Both Sweet-potato roots and Irish potato tubers are treated with some forms of Semesan, a fungicide that contains a mercuric compound.

INSECT CONTROL IN THE GARDEN

Control of Insects in Asparagus

The control of the **Asparagus Beetle** has been affected by the application of either one percent rotenone dust or one-and-one-quarter percent active malathion either in dust or spray form as prepared from the emulsifiable concentrate. These agents should not be applied during the cutting season.

Control of Insects of the Bean

Aphids can be controlled by the application of either of the following two pesticides per acre of crop: one-and-one-quarter pounds of active malathion in the form of either a dust or emulsifiable concentrate, OR one quart of the twenty-five percent parathion emulsifiable concentrate.

The application of one-half to one pound of carbaryl (Sevin) dust or wettable powder has controlled infestations of the **Bean Leaf Beetle**. In the home garden only, the application of rotenone at the rate of one-quarter of a pound of active ingredient per acre in the form of either a dust or powder.

The term **Cucumber Beetle** is applied to a number of species of the genus *Diabrotica*, some of which are found in the United States. These insects are similar in size, ranging from five to six millimeters in length. The wing covers are pale green, marked with either twelve black spots or three stripes that extend the length of the body. The thorax of these insects is distinctively yellow. The life history of these similar species begins when their eggs are laid around the base of the plant that is to be their host. The larvae that emerge feed on the roots of grasses and other plants, while the adults feed on a great variety of plants that include fruit trees, vegetables, and flowers. The application of one-half to one pound of active Sevin per acre in the form of a dust has been effective in the control of these **Cucumber Beetles.**

The **Cutworms** are gray or greenish caterpillars, without hair, that reach a maximum length of thirty to forty millimeters. During the daylight hours, these pests are usually curled in a tight roll immediately beneath the surface of the soil. These pests are very destructive in their attacks on cereals, vegetables, and both fruit and shade trees. Dusting the surface of the soil with two pounds of active DDT per acre is effective in the control of the Cutworms, especially if the application is made before the beans are planted.

To control infestations of the **Japanese Beetle**, one of the following pesticides can be applied to the foliage of the plant: one to two pounds of active carbaryl (Sevin) either in the form of a dust or a wettable powder, OR one-and-one-quarter pounds of active malathion dust or emulsion concentrate or wettable powder per acre of beans to be treated.

The **Mexican Bean Beetle** is one member of the family *Coccinellid* (Lady Beetles) that is destructive. These pests possess the same general appearance as their better known and regarded cousins, a grayish brown or yellow color with a few black spots on the covers of the wings. Both the adults and the larvae feed actively, especially on the various types of beans. To control this pest, the application of one of the following insecticides per acre of crop has proven to be effective: one-half to one pound of carbaryl (Sevin) in the form of either a dust or a wettable powder, OR one-half pound of active diazinon as a wettable powder, OR one-quarter of a pound of active rotenone dust, extract, or powder to the foliage of the plant.

The **Root Maggot** can be controlled by treatment of each one hundred pounds of seed with one pint of a slurry that is formed from two thirds of an ounce of seventy-five percent wettable aldrin powder plus three ounces of seventy-five percent wettable captan powder with enough water to make up to the pint required.

Spider Mites have been controlled by the application of 1.2 pounds of carbophenothion dust or three-quarters of a pound of either the emulsifiable concentrate or the wettable powder, OR three to five tenths of a pound of demeton emulsifiable concentrate, OR ten grams of either the emulsifiable concentrate or wettable powder prepared from Dicofol per acre of crop.

Control of Insects of the Beet

The **Beet Webworm** can be controlled by the application of either one ounce of the two-tenths of a percent pyrethrum dust or one-half ounce of the emulsifiable concentrate, OR one-and-one-half pounds of active DDT dust or wettable powder per acre. Tops of beets that have been treated with DDT must not be fed to livestock.

Insects of the Cabbage, Cauliflower and Kale

Aphids can be controlled by the application of either diazinon (six to sixteen ounces of active ingredient) as a dust, OR one-and-one-quarter pounds of malathion from either the emulsifiable concentrate, wettable powder or as a dust, OR one to two pounds of active naled either as a dust or emulsifiable concentrate per acre of the crop. These applications may be repeated weekly as needed, but seven days must elapse between the last application of either malathion or diazinon and the harvest (one day is required in the case of naled).

A number of species of *Caterpillars* feed upon these and related plants during the period of the growth and maturity of the plant. The green Caterpillars are without hair with the exception of the **Salt-Marsh Caterpillar**, which is brown and quite hairy. The application of twenty-five billion spores per gram preparations of the commercial culture of *Bacillus thuringiensis* applied as a wettable powder at a dosage of two pounds per acre has been effective in the control of most Caterpillars with the exception of the Cabbage Looper. Alternatively, either one-and-one-quarter pounds of active malathion (for the home garden only) in the form of the emulsifiable concentrate, OR one to two pounds of active naled applied either as a dust or in the form of the emulsifiable concentrate can be used in the control of these pests in each acre of infested cropland. One day must elapse between the last application of the naled and the harvest, while seven days must be allowed in the case of the malathion.

The black **Harlequin Bug** has a bright red or yellow spot on the wing covers and the thorax. The eight to eleven millimeter long adults hibernate during the winter to emerge with the first warm days of the spring. This pest feeds on the cabbage and various other related plants. One-and-one-half pounds of active naled in the form of the emulsifiable concentrate OR 1.2 pounds of active DDT in the form of either the dust, emulsifiable concentrate, wettable powder, or granules may be applied per acre of crop for control of this pest. At least one day between the last application of naled and the harvest must be allowed, while the DDT may be applied only before the heads have begun to form.

The **Root Maggot** may be controlled by the application of one-and-one-half pounds of active chlordane as a dust, wettable powder, or emulsifiable concentrate per acre of crop land. The chlordane is applied at the base of the plants when the leaves appear and the application is repeated at the time of thinning.

Insects of the Cantaloupe

Aphids can be controlled by the application of either three-quarters to one pound of active diazinon dust, OR one pound of active endosulfan dust, OR one-half pound of active endosulfan emulsifiable concentrate per acre of cropland.

For a brief discussion of the **Cucumber Beetles**, which also infest the cantaloupe, see the section dealing with Bean Insects. For the control of these pests in fields of Cantaloupe, the application of 1.7 pounds of active either Malathion or methoxychlor per acre has been used. While the Malathion may be supplied in the form of either a dust or an emulsifiable concentrate, the methoxychlor is found in the form of a dust or a wettable powder.

The **Pickleworm** can be controlled by the application of one pound of active carbaryl (Sevin) as either a dust or a wettable powder per acre of crop. The pesticide is used on the foliage when the worms appear in the blooms of the plant.

Spider Mites are controlled by the application of one-quarter to one-half pound of active parathion dust or emulsifiable concentrate or one pound of active ingredient tetradifon dust. If a tetradifon emulsifiable concentrate or wettable powder are used, only one-half pound of the concentrate active ingredients are necessary. While seven days are required between the last application of a parathion treatment, only one day is required in the case of the tetradifon.

Insects of Celery

The **Leaf Tier** is the primary pest of the Celery. The reddish-brown moths possess gray hind wings. In the adult stage, these pests are approximately twenty millimeters long. While these insects are feeding, the foliage of the plant is bunched together by the insects. This pest may be controlled by the application of either twenty to thirty pounds of two-tenths of a percent pyrethrum dust or one-and-one-quarter pounds of active DDT either as a dust, or from the emulsion concentrate or wettable powder per acre of Celery. Generally two applications of the pesticide are made at an interval of one-half hour. DDT must not be applied after the bunch of Celery begins to form, while the pyrethrum must not be used within at least one day of the harvest.

Insects of the Cucumber

Aphids may be controlled by the application of one of the pesticides shown in Table VIII-2.

TABLE VIII-2

AGENT	*FORM*	*CONCENTRATION**
diazinon	dust	0.8 - 1.0
	emulsifiable concentrate or wettable powder	0.5
endosulfan	dust, emulsifiable concentrate or wettable powder	0.5 - 1.0
Malathion**	emulsifiable concentrate	1.7
mevinphos	dust, emulsifiable concentrate or wettable powder	0.5

*pounds of active ingredient per acre
**for use on home gardens only

The following minimum number of days should be allowed to elapse between the last application of the insecticide and the beginning of the harvest: diazon (seven days), endosulfan (one day), Malathion (one day), mevinphos (one day).

Cucumber Beetles are controlled by application of one-half to one pound of active endosulfan dust, emulsion concentrate, or wettable powder, OR one to two pounds of active methoxychlor dust, or wettable powder, OR one-half pound of parathion dust, emulsifiable concentrate or wettable powder. These dosages are sufficient for the treatment of one acre of crop. The minimum number of days that must elapse between the last application of the insecticide and the beginning of the harvest are: endosulfan (one day), methoxychlor (one day), and parathion (eighteen days).

For control of the **Pickleworm**, see the section dealing with Cantaloupe.

The application of twenty to thirty-five pounds of a five percent malathion dust or one-quarter to one-half pound of active parathion in the form of either a dust, emulsifiable concentrate, or wettable powder will control the infestation of an acre of cropland by the **Spider Mites**. Whereas one day should elapse between the last application of malathion and the beginning of the harvest, fifteen days are needed if parathion is used.

Insects of the Eggplant

The **Eggplant Lacebug** is rather flat and possesses lace-like wings. The light brown or grayish adult is one-sixth of an inch long and the spiny, rather yellow nymphs are one-tenth of an inch long. Both stages of this pest feed in groups on the underside of the leaves of the Eggplant, causing the plant to turn yellow. Infested plants often die from the attack of these pests. These insects are most frequently found in the southern, rather than northern United States. The Lacebug may be controlled by the application of a spray that consists of two teaspoons of a fifty-seven percent malathion emulsifiable concentrate per gallon of water. The malathion spray must not be applied within three days of the harvest.

The **Colorado Potato Beetle** is the primary insect pest of the Eggplant throughout most of the United States of America. The description of this pest and its control are discussed in the section dealing with Potatoes.

Insects of Lettuce

To control **Aphids**, the application of one of the following insecticides has proven effective: six to eight ounces of active diazinon dust, emulsifiable concentrate, or wettable powder, OR one-half pound of active parathion dust, emulsifiable concentrate, or wettable powder OR one to two pounds of active malathion dust, emulsifiable concentrate, or wettable powder per acre of crop. The minimum days between the last application of the insecticide and the beginning of the harvest are as follows: diazinon (ten days), malathion (seven days on head lettuce and fourteen days when applied to leaf lettuce), parathion (fifteen days for head lettuce when the maximum temperatures are above 70°F and twenty-one days on leaf lettuce).

The **Caterpillars, Cabbage Looper**, and **Imported Cabbageworm** can be controlled by the application of 1.2 pounds of active malathion (for use in the home garden only), one-half pound of active mevinphos, one-and-one-half pounds of active naled, OR one-half pound of parathion. Each of the dosages are stated in the amounts of the active ingredient that is recommended to be applied per acre of crop. Each of the insecticides may be applied either as a spray, prepared from either the emulsifiable concentrate or the wettable powder, or as a dust, with the exception of the naled that is not available in the form of the wettable powder. The following minimum number of days must be allowed to elapse between the last application of the insecticide and the beginning of the harvest: malathion (seven days when applied to head lettuce, fourteen days when applied to leaf lettuce), mevinphos (two days), naled (one day), parathion (the same interval as was recommended in the preceeding section on Aphids).

The disease that is known as Aster Yellows on lettuce is caused by a virus that is carried by the **Six-Spotted Leafhopper.** To control this pest, the application of one of the following amounts of active ingredient should be applied per acre of crop: 1.2 pounds of malathion, 1.2 pounds of methoxychlor, or one-and-one quarter to two-and-one half pounds of DDT. The malathion and DDT are available in the form of a dust, wettable powder, or emulsifiable concentrate, while the methoxychlor is available as a wettable powder or dust. DDT must not be used on head lettuce after the heads have begun to form, or to leaf lettuce after the appearance of leaves that are to be eaten or marketed. The minimum number of days that must elapse between last application of the chemical and the harvest of the crop are: malathion (fourteen days on leaf lettuce, seven days when applied to head lettuce), methoxychlor (fourteen days on leaf lettuce). In the home garden, a four percent malathion dust may be applied and repeated in one week for the control of Leafhoppers. This dust must not be applied within ten days of the harvest of head lettuce and fourteen days of the harvest of leaf lettuce.

Wireworms are the larval stage of the Click Beetles. These yellowish-brown pests are shiny and reach one inch in length at maturity. Wireworms are especially injurious to the crops in the irrigated regions of the western United States where they attack the fleshy root stalk of the lettuce plant. In the northeastern United States, the application of four to eight pounds of active chlordane in the form of granules, emulsifiable concentrate, or wettable powder had been found effective. The chlordane is broadcast on the soil and thoroughly worked into the top six to nine inches of the soil. This treatment must not be applied after the crop has been planted and the application can be made once every three years at a maximum. In the irrigated regions of the western United States, DDT is used at the rate of ten pounds of active ingredients per acre. The DDT is applied in the same fashion as the chlordane above. Again, one application should give control for at least three years. DDT must be applied at least six weeks before the planting is begun. Fumigation of the field with ethylene dibromide yields immediate control of the Wireworms, but is more expensive than the use of DDT and lasts but one season.

Insects of the Onion

The **Onion Maggot** may be controlled by the application of one of the following insecticides and the control agents may be applied directly into the seed furrow after the seed has been dropped (using separate hoppers for the pesticide and the seed): carbophenothion granules (0.6 to 0.8 ounces per one thousand feet of row in mineral soils or 1.1 ounce per thousand feet of row in muck or peat soils), OR one ounce of Ethion granules per one thousand feet of row. Alternatively, one-and-one-half pounds of malathion, either from the dust, emulsifiable concentrate, or wettable powder, may be applied to the foliage when the flies appear. This dosage is per acre of cropland and the treatment may be repeated in four days. The malathion must not be applied within three days of the harvest of the crop.

Onion Thrips are tiny, winged insects that pierce the tissue of the onion and suck its juices when the Thrips is in the larval or the adult stage. The Onion Thrips may be controlled on both dry and green onions by the application of either 1.2 pounds of active diazinon in the form of a dust, OR two pounds of active malathion dust. For control of the Thrips on dry onions only, three to four pounds of active DDT dust or five pounds of active toxaphene dust may be applied per acre of cropland when the plants are in bloom. The DDT and toxaphene must not be used on plants that will be used for spring or green onions. The following minimum number of days from the last application of the pesticide until the beginning of the harvest must be ovserved: diazinon (ten days) and malathion (three days).

Garden Insects of the Pea

The **Pea Aphid** is one of the larger aphids, with the winged female ranging from one-eighth to three-sixteenths of an inch long. This pale green insect has antennae that are slightly longer than the body. (Muka, 1959) The application of either one to 1.2 pounds of malathion active ingredient in the form of a dust, emulsifiable, or wettable powder, OR one-half pound of active parathion dust, emulsifiable concentrate, or wettable powder is effective in the control of the Pea Aphid. These concentrations are per acre of peas. Parathion must not be used in the home garden. The minimum number of days that must elapse between the last application of the insecticide and the first day of the harvest are three days in the case of the use of malathion and fifteen days if parathion was used.

The brownish **Pea Weevil** is a small, stout beetle whose back is mottled with black, white, and dark brown coloring. The short wing covers of this pest do not quite cover the tip of the abdomen. The eggs are laid on the outside of the young pod of the pea. The larva that hatch from the egg cut through the wall and enter a pea, where they develop. (Muka, 1959) In order to control this pest, either of the following insecticides has been commonly used: one to 1.2 pounds of malathion active ingredient from either a dust or emulsifiable concentrate OR one to 1.2 pounds of active DDT in the form of the dust, emulsifiable concentrate, or wettable powder per acre of cropland. If DDT is used, seven days must elapse between the last application of the insecticide and the beginning of the harvest, whereas if the malathion is used, eight days are required. The application of DDT to edible podded peas should be avoided after the bloom appears and must not be applied to those peas that will be fed to either dairy animals or livestock that is being finished for market.

Insects of the Potato

A few general comments concerning the insects that infest potatoes in general, and Aphids in particular, are appropriate. "Field tests with insecticide treatments on Irish potatoes at planting (time) in south Texas have shown that aphid, psyllid and white-fly control was effective for 77 days and evident for 96 days after treatment. Flea beetle injury was controlled." Tuber yields were increased. "Phorate at 1.5 to 3, DiSyston at 1.5 to 4, demeton at 1.5 to 3 and dimethoate at 3 pounds of active ingredient per acre were most efficient". Potato Leafroll-Psyllid Yellows infected plants and Potato Crinkle Plants were significantly reduced. Tuber yields were increased. (Harding, 1962)

Infestation by fungus caused diseases of the Aphids on potatoes in northeastern Maine is reported by Shands, et al., (1962). The Potato Aphid was the species that was most affected. No effect on the infestation was noted from the application of pesticides.

Tests with a number of insecticides as a control of Aphids on potatoes in New Brunswick, Canada, are reported by Pond (1962). Malathion was very effective against all species of the Aphid if the temperature was about 70°F at the time of application. Cool, damp weather almost completely nullified the toxic action of the insecticide. Diazinon, endosulfan, lindane, phorate, and dimethoate were effective in the trials over two or more years. Malathion, diazinon, and endosulfan are currently recommended for the control of Aphids on potatoes in New Brunswick.

In the northeastern United States, the application of ten ounces of active malathion or three to five ounces of parathion per acre of potatoes has been used for the control of **Aphids**. One to two pounds of naled are used for the control of these pests in the home garden in the northwestern states, while one-and-one-half pounds of naled per acre of home garden is used in the southeastern areas. The malathion may be found in the form of either a wettable powder or an emulsifiable concentrate, whereas the naled is available as a dust or emulsifiable concentrate, and the parathion is found in all three forms: dust, emulsifiable concentrate and wettable powder. The parathion should not be used in the home garden, and at least five days must elapse from the last application of the parathion until the product may be harvested.

The widely distributed **Blister Beetles** frequently attack the foliage of the potato. These seven to twenty-eight millimeter long pests vary in color from gray through green to black. One-and-one-half to two pounds of DDT, either in the form of the dust or wettable powder, may be used to control the Blister Beetles if the tubers from the plants are not fed to dairy animals.

The **Colorado Potato Beetle** is a native of the Rocky Mountain regions of the United States. This oval, brown pest has spread over a large part of the North American continent. Ten black stripes run the length of the wing covers of this six to twelve millimeter long insect. Two broods of the Colorado Potato Beetle are produced each year from the orange eggs that are laid in clusters of eight to ten eggs on the under side of the leaf of the potato. The application of carbaryl (Sevin) to the foliage of the plant when the eggs hatch is effective in the control of this pest. In the northeastern United States eight-tenths of a pound of active ingredient per acre is recommended, while in the western areas one to one-and-one-half pounds per acre and in the southeast one to two pounds per acre are commonly used. The application of the insecticide from the wettable powder form may be repeated in ten to fourteen days if necessary.

The slender, pale green **Leafhoppers** have thickened forewings and range from two-and-one half to four millimeters in length. The young, very pale green

Leafhoppers are quite active as they run about on the under-side of the leaf. Both stages of development of the pest pierce the leaf and suck the juices of the plant. Severe infestations of Leafhoppers may cause the leaves of the plant to dry up and die. In addition, Leafhoppers carry a number of plant diseases. The application of one-half to one pound of malathion, 2.2 pounds of methoxychlor, or one to one-and-one-half pounds of DDT is generally sufficient to control these pests. The dosages of the insecticides are given in pounds of active ingredient per acre of cropland. The DDT may be found as either a dust or wettable powder, the malathion is available in the form of an emulsifiable concentrate, whereas the methoxychlor is available in the form of a wettable powder.

Thė **Potato Tuberworm** becomes a small, silvery-gray moth in the adult stage. The white eggs of this pest are laid on the leaves and stems of the plant and on exposed tubers both in the field and in the warehouse. In the field, the larvae mine the leaves and stems, continuing their feeding on the tuber in the warehouse. This pest may be controlled by the application of one pound of active DDT in the form of the emulsibiable concentrate when the worms begin webbing the leaves together. Alternatively, one-half to one pound of active endosulfan per acre may be applied from either the dust or the emulsifiable concentrate.

In the western United States, **Wireworms** may be controlled in the same manner that the Wireworms of the Lettuce that infest irrigated lands were controlled. In the east, four to eight pounds of active chlordane in the form of granules, wettable powder, or emulsifiable concentrate should be broadcast per acre. The insecticide should be worked thoroughly into the soil to a depth of four to six inches. One application of the insecticide should provide control of the Wireworms for at least three years. The chlordane must not be applied to the field after the crop has been planted.

Insects of the Pumpkin and Squash

The **Squash Bug** is a large and serious pest of pumpkin, squash, and related varieties of plants. The eight to ten millimeter long adults are grayish brown and possess thick wing covers. The pale green very young insects have antennae and pink legs. The shiny brown eggs are laid in clusters on the under side of the leaves and on the stems. All stages of the development of this pest pierce the leaves of the plant to suck the juices. In the small garden, hand removal of the adults and the eggs may be a satisfactory means of control of this pest. The adults and the more mature young are very resistant to control by the applications of chemicals, but a four percent nicotine sulfate dust will help to check the young. In the field, the application of one pound of active carbaryl dust or wettable powder or one-half pound of active parathion dust, emulsifiable concentrate, or wettable powder per acre may be effective. The carbaryl (Sevin) may be applied once or twice per week and both insecticides are applied at the base of the plants. The parathion should not be used in the home garden. If carbaryl is used, one day must elapse between the last application and the beginning of the harvest, but ten days must be allowed if parathion is applied.

The large, white **Squash Vine Borer** burrows into the stems of both the squash and the pumpkin. In the garden, the burrows may be located by the presence of sawdust-like waste what is expelled from the burrow. The larva may be killed by puncturing with a wire or slitting the stem and removing it. After removal of the larva, soil should be drawn over the cut stem to reduce further injury. (Pratt, 1959) In the field one pound of active carbaryl (Sevin) may be applied from either the dust or wettable powder form per acre of cropland. At least one day between the last application of the carbaryl and the beginning of the harvest must be allowed.

Insects of Sweet Corn

The **Corn Earworm** may be controlled by the application of one to one-and-one-half pounds of active carbaryl in the form of a wettable powder that is dispersed in twenty-five gallons of water. This slurry is applied to the larvae in the silk and ears at one to two day intervals beginning on the day that the silks appear. Wet the silks thoroughly. If the infestation is heavy, five to eight applications of the slurry may be necessary. Alternatively, two pounds of DDT emulsifiable concentrate dispersed in twenty-five gallons of water can be applied on the larvae that are in the whorls of the leaf (Budworm Damage). The spray should be directed so that it will run into the whorls of the plant before tasseling and silking. If DDT is used, the forage, including the ensilage, must not be fed to dairy animals or animals that are being finished for slaughter.

The **Corn Sap Beetle** may be controlled by the application of one pound of active malathion in the form of the emulsion concentrate per acre of the home garden only. The first application should be made six days after the appearance of the silks, and a repeat application made after ten days.

The female moth of the **European Corn Borer** is yellowish-brown with dark lines across its wings. The slightly smaller male is darker in color. The wing spread of the European Corn Borer is about one inch, the same length as the mature dirty white larvae. The larvae overwinter in the cornstalks where they pupate to emerge as adult moths with the first warm weather of the spring. The larvae bore into all parts of the cornstalk and ear, weakening the stalk and reducing the yield. (Muka, 1959) The European Corn Borer can be controlled by the application of either one-and-one-half pounds of active carbaryl (Sevin) in the form of the wettable powder or granules, or two pounds of DDT dust, one-and-one-half pounds of DDT emulsifiable concentrate, or one pound of DDT granules per acre of cornland. The insecticide should be used when the borers begin to hatch and the treatment should be reapplied at five day intervals as necessary. The DDT should not be applied to corn that is used for silage or ensilage, but the carbaryl-treated corn can be used if one day elapses between the last application and harvest, or seven days if the corn is to be used for fodder.

Insects of the Sweet Potato

The adult **Sweet potato Weevil** is blue-black with red legs. This one-fourth of an inch long pest lays its eggs in shallow punctures in the stem of the plant, in its roots, or on the stored sweet potatoes. As the larvae feed, they burrow into the roots and stems of the plant. The adults feed on both the roots and the stems of the vine. This insect damages the crop both by its feeding and the bitter flavor that it imparts to the sweet potato. Weevil-free seed, preferable from an uninfested area, is the first and foremost means of control of the Sweet potato Weevil. Seed potatoes should be stored separately. The seed should be stored one layer deep and covered with a ten percent dust of DDT. One pound of dust should be sufficient for each six to eight bushels of seed. The storage room should be emptied, cleaned, and coated thoroughly with a ten percent DDT dust before the new crop is planted. Two or more applications of a two percent dieldrin dust may be applied as follows: first, when the plants in the seed beds begin to show color; second, when all the plants are up; the surface of the soil can be dusted around all of the plants. After the seed bed is emptied, it should be dusted thoroughly. The sweet potatoes should be planted in new, clean fields. Two percent dieldrin dust should be applied around the plants, one coating for the deep-rooted varieties and two light dustings for the shallow-rooted varieties of the sweet potato. The dust can be applied along the row in strips six to eight inches wide. The dieldrin must not be applied within twenty-one days of the harvest and DDT treated potatoes must not be fed to dairy animals or those that are being finished for the market. (Staff, U.S.D.A., Research Division, 1960)

Insects of the Tomato

A few general notes concerning the control of insects in the tomato are germane. A series of tests were conducted in southern California during 1961 and 1962 to determine the control of insects on poled tomatoes and the effect of the control techniques on the populations of beneficial insects. "High dosages of the *Bacillus thuringiensis* Berliner showed promise for the control of the Cabbage Looper and the Beet Armyworm, as well as the prevention of fruit damage by the Tomato Fruitworm." However, this treatment did not adequately control the Potato Aphid, the Western Flower Thrips, or the Dipterous Leafminers. The bacillus was applied at the relatively high rate of twenty-five to thirty pounds of agent per acre-foot (poled tomatoes) and appeared to cause a reduction in the damage that is caused by the Fruitworm that is comparable to the control achieved with an application of two thirds of a pound of active DDT per acre-foot. The application of carbaryl, in the form of a dust and DDT, diazinon, naled, and Zectran in the form of sprays provided good control of Leafminers, Cabbage Loopers, and Beet Armyworms. The proporation of Leafminers that were parasitized . . . was reduced by the use of DDT, naled, carbaryl, Zectran, and Bacillus thuringiensis (Shorey and Hall, 1963).

The application of one of the following insecticides is effective in the control of **Aphids** on the tomato: three-quarters of a pound of active diazinon dust, or one-quarter to one-half pounds of active diazinon in the form of the emulsifiable concentrate or the wettable powder, OR 1.2 pounds of active malathion in the form of either the dust, emulsifiable concentrate, or wettable powder, OR one-half pound of active mevinphos dust, wettable powder, or emulsifiable concentrate per acre of the crop. One day must be allowed to elapse, at least, between the last application of any of these chemicals and the beginning of the harvest.

The tiny *Drosophila*, or **Vinegar Fly**, is light brown or yellow in color. This pest breeds in fermenting fruit and vegetables and is of increasing importance as a pest of ripened tomatoes. The rough handling of the tomatoes in marketing tends to form small cracks in the fruit in which the fly lays its eggs. The attack of this insect may occur in both the field and the packing house. Five percent diazinon-walnut granules prepared at three-quarters of a pound of active ingredient in the form of the wettable powder per acre of crop offers longer residual suppression of *Drosophila* populations than do diazinon sprays. The granules fall through the foliage of the tomatoes and the weeds after which the toxicant volatilizes from the sufrace of the granule and a portion redeposits on the lower surfaces of the plant. The adult flies, found on these lower surfaces, are thereby controlled. (Riley, et al., 1964) Dust applications of one-tenth percent pyrethrins plus piperonyl butoxide at a dosage of one-half to one pound per ton of fruit is also recommended. If the insecticide, diazinon is used, one day must elapse between the last application of the chemical and the commencement of the harvest.

The **Hornworms** were described in the section dealing with Tobacco. To control these pests in fields of tomatoes, an application of either two to four pounds of active carbaryl (Sevin) or two to three pounds of TDE per acre may be used. Whereas the carbaryl may be supplied in the form of a dust or a wettable powder, TDE is available in the form of the emulsifiable concentrate or wettable powder. One day should be allowed to elapse between the last applications of these materials and the beginning of the harvest of the crop.

The **Tomato Fruitworm** is the same insect that was described in the section dealing with Corn (the Corn Earworm) and the section that dealt with Cotton (the Cotton Bollworm). This pest attacks the tomato by cutting through the side of the fruit and feeding upon its contents. The application of two pounds of active carbaryl (Sevin) dust or wettable powder, OR three pounds of active DDT dust, emulsifiable concentrate, or wettable powder, OR five pounds of active toxaphene dust, emulsifiable concentrate, or wettable powder per acre is sufficient to control this pest. The following minimum number of days must be allowed to elapse between the last application of the insecticide and the beginning of the harvest: carbaryl (one day), DDT (five days), and toxaphene (three days).

The control of the **Tomato Pinworm** is described under the discussion of the Tomato Fruitworm.

The **Tomato Psyllid** is a sucking insect that is similar in size to the Aphids. The young stage (nymph) of this pest lies flat upon the surface of the leaf, attached by its sucking beak in the same fashion as the Scale insects. While feeding, this green or orange-yellow insect injects a substance into the plant that is toxic and causes the leaves to turn yellow and curl. The condition of the plant with yellow, curled leaves is known as "**psyllid yellows**". The application of twenty-five to thirty-five pounds of a dust that contains five percent DDT will generally control the Tomato Psyllid. This dust should be applied to the foliage of the plant when the plant is six inches high. The dust may be reapplied in ten to twenty days if necessary.

The **Tomato Russet Mite** is a serious pest of tomatoes that are grown under conditions similar to those that exist in California. This Mite sucks the juices from the leaves and stem of the plant, causing attacked leaves to assume a brown or russet color as they dry and fall from the plant. the application of one-half pound of parathion in the form of a dust, or a spray that is prepared from either the wettable powder or the emulsifiable concentrate to the foliage when the fruit begins to set is an effective means of control. Alternatively, thirty pounds per acre of a twenty to fifty percent sulfur dust may be used. The minimum amount of time that must be allowed to pass between the last application of the insecticide and the beginning of the harvest is one day if sulfur was used, but ten days are required if the parathion is used. All sulfur must be carefully washed or wiped from the fruit at the time of the harvest.

Insects of the Turnip

Aphids that attack the turnip can be controlled by the application of four-tenths of a pound to one pound of active diazinon dust or four to five-tenths of a pound of diazinon in the form of the emulsifiable concentrate or the wettable powder, OR 1.2 pounds of malathion active ingredient, OR one-quarter to one-half pound of mevinphos per acre of crop. The malathion and mevinphos may be applied from the dust, emulsifiable concentrate, or wettable powder. The mevinphos should not be used in the home garden and should only be applied by a trained operator. The following number of days should be allowed to elapse between the last application of the insecticide and the commencement of the harvest, at a minimum: diazinon (ten days), mevinphos (three days), and malathion (eight days or, if the tops are not to be used for food or feed, three days).

The tiny, oval shaped **Flea Beetles** are black or brown and, as the name indicates, capable of jumping. The adults feed on the leaves of the plant while the larvae feed on its roots. The application of one of the following four insecticides generally controls infestations of Flea Beetles: one to one-and-one-half pounds of DDT active ingredient, OR 1.7 pounds of methoxychlor active ingredient in the form of either a dust or wettable powder, OR one-and-one-half pounds of active naled emulsifiable concentrate, OR three-tenths of a pound of rotenone dust or spray per acre of crop. The DDT may be applied as a spray that is formed from the emulsifiable concentrate or wettable powder or may be in the form of the dust. DDT must not be applied to the leaves of the plant if they are to be either eaten, marketed, or fed to livestock. the following minimum days must elapse between the last application of the insecticide and the beginning of the harvest: naled (one day), rotenone (one day), and methoxychlor (fourteen days or one day if the tops are not to be used for either food or feed).

Root Maggots may be controlled by the application of one pound of diazinon granules per acre in the seed furrow at the time of planting or one-half of a pound per acre in the form of the wettable powder or emulsifiable concentrate that is mixed with water and sprayed around the stems of the plants.

DISEASE CONTROL IN THE GARDEN

Diseases of the Asparagus

Asparagus Rust, a fungus disease, is characterized by elongated, red-orange, powdery blisters on the foliage and stems of the plants. The tops of plants that are infected die prematurely and the yield of the crop is reduced in the following year. The refuse of disease plants is a site for the overwintering of the disease. Seasons that obtain a high degree of rainfall are especially hospitable for this disease. The use of clean seed from a resistant variety of the plant, such as Martha Washington or Mary Washington, is the first step in the control of this disease. Any plant stalks that are diseased should be cut and burned. If the growth of the fern should become infected, one of the following fungicides may be applied at seven to ten day intervals per acre of crop: one-and-one-half pounds of Dithane M-45 wettable powder or two pounds of zinde dust or wettable powder.

Diseases of the Beet

The **Cercospora Leaf Spot** is fully described in the section on Field Crops that deals with Sugar Beets. For the control of this disease in garden beets, treated seed that has been obtained from a reputable source is the first step. Of course, the use of a disease-resistant variety of the beet is desired whenever possible. Beets should be rotated on a three year basis with non-susceptible crops and all crop litter should be cleanly plowed under each year. When the Leaf Spot is a serious problem, every ten days the application of a spray that consists of two to three pounds of eighty percent Maneb wettable powder per acre should be considered. This spray is applied beginning when the symptoms of the disease are first noticed. For additional control, irrigation should be accomplished early in the day to allow the plants to dry before the cool of the evening.

Diseases of Garden Beans

Lima Beans, Green Beans, Wax Beans, and many similar plants are attacked by *Xanthomonas phaseoli*, the pathogen that causes **Common Bacterial Blight**. The first symptoms of an infection by this disease are the translucent lesions that appear on the underside of the leaf. These spots become larger, coalesce, and develop dry, brownish-red centers as the disease progresses. The haloed centers of the lesions accompany the scorched, withered leaves that may fall from the plant. The pods of the bean develop water soaked lesions that coalesce, become sunken and dry, and have reddish-brown edges. The lesions that appear on the stem of the plant are similar to those that appear on the pods. The plant may break at the site of cankers on the stem.

Green Beans, Lima Beans, and Wax Beans are also attacked by the bacteria *Pseudomonas phaseolicola*. This bacteria causes the disease known as **Halo Blight** on these plants. The symptoms of Halo Blight are similar to those of Common Bacterial Blight with the exception that the lesions of the former is wider and the infected leaves eventually turn yellow. At temperatures over 75°F, the haloes of Halo Blight nearly disappear and the lesions become dry and brownish. A plant that is infected with Halo Blight may wilt and die or may be stunted, with the upper leaves wrinkled. A greenish-white bacterial ooze exudes from the pods that are infected by this disease.

Pseudomonas syringae is responsible for the **Brown-Spot Blight** of Lima Beans. This disease is also known as the **Bacterial Spot** and the disease may be found on such plants as roses, lilacs, ash, pears, hibiscus, clovers, etc. The upper leaf surfaces of the lima bean that are infected by this disease contain small, pale-green lesions that have reddish-brown margins. These spots often split and fall from the leaf. The lower surface of infected leaves develop reddish-brown veins, while brown spots that have water-soaked haloes appear on the infected pods. The stems of a diseased plant often develop sunken, reddish-brown lesions.

Various practices can be followed that will aid in the control of the Bacterial Blights. Disease-free seed of a resistant variety of the bean should be planted whenever possible. The beans should be rotated in the garden plots with non-susceptible plants, while the debris from the plants should be destroyed each year. Since the diseases are easily spread when the plants are wet, they should not be handled or worked near when wet.

Stem Anthracnose of Lima Beans is a result of an infection by the fungus *Colletotrichum truncatum*. This disease often is erroneously called **Stem Rust**. Stem Anthracnose causes blight of the pods, a deep red discoloration that spreads over the entire surface of the pod. As this disease progresses, these spots fade to a tan or gray color as they develop dark fruiting bodies. Usually the leaves of an infected plant are not affected, but the stems may be afflicted with red streaks. Since the debris of the plant is a great source of infection, it must be destroyed or plowed deeply under the surface of the soil. Beans and similar crops should be placed on a three year rotation plan with non-susceptible plants. Lima Beans should not be planted in fields that adjoin those in which alfalfa, cowpeas or soybeans are planted. The use of a fungicide is suggested if the attack of Stem Anthracnose is a problem. Beans that are used for commercial processing, and therefore picked all at one time, may not need as many applications of the fungicide as those that are produced for home kitchen use to affect the desired level of control. The application of one of the following sprays or dusts every seven to ten days from the time of blossom until the time of harvest will control most outbreaks of Stem Anthracnose: twenty-five pounds of six percent maneb or zineb dust, OR one-and-one-half pounds of sixty-five percent zineb wettable powder, OR one-and-one-half pounds of eighty percent maneb wettable powder. The wettable powders must be dispersed in sufficient water to cover the plants thoroughly and all concentrations were given in quantities of the active ingredient per acre of crop to be treated. Maneb must not be applied within four days of the harvest and zineb must not be applied within seven days of the harvest.

Diseases of the Cantaloupe

Anthracnose, a fungus diseas often extensively damages crops of melons. The first symptoms of an infection by *Collectotrichum lagenarium*, the fungus that is responsible for this disease, are the reddish-brown, round spots that appear on the older leaves of the plant. Occasionally the centers of these spots drop from the leaf leaving a leaf with a lacy appearance. Frequently the crown of the plant becomes defoliated by the action of this fungus. The infected fruit is marked by sunken, water-soaked, round spots that become dark colored as the disease progresses. The spores of the fungus manifest themselves in a pink tinged exudate. This disease may be controlled by the application of either maneb or zineb at the rate of two pounds of wettable powder or twenty to thirty pounds of dust per acre of cropland.

During the middle of the growing season, premature defoliation occurs on melons that are attacked by *Alternaria cucumerina*, the fungus that causes **Alternaria Leaf Spot**. Small, watersoaked spots on the upper surface of the leaf develop into concentric rings and the leaves often curl down as the disease progresses. Healthy, vigorous plants generally are not affected by attacks of the Alternaria Leaf Spot, but if the disease should become prevalent, the same control is effective in the control of this disease as recommended for the control of Anthracnose.

Downy Mildew, a result of an infection by the fungus *Pseudoperonospora cubersis*, is capable of causing much damage to crops of cantaloupes. The yellow areas of infection first appear on the upper surface of the leaf as a gray mold appears on the lower surface. Infected leaves become yellow before turning brown and dying. Cool, damp weather favors the development of this disease. The infection of Downy Mildew can be controlled by one of the following sprays that are applied at seven to ten day intervals during dry weather: two pounds of fifty percent copper wettable powder, OR two pounds of either zineb or maneb wettable powder, OR twenty to thirty pounds of copper, zineb, or maneb dust per acre. During wet weather, the application of these fungicides should be made every five to seven days.

The fungus *Erysiphe cichoracearum* causes **Powdery Mildew**, a potentially destructive disease. This disease is characterized by the grayish-white mold that appears on the leaves and young stems of the plant. As the disease progresses, the leaves become yellow, then brown and shriveled. Although the fruit is not directly attacked by Powdery Mildew, it is often malformed if the damage to the leaves is extensive. Powdery Mildew is distinguished by the growth of the mold on the upper surface of the leaf. This disease is capable of overwintering on the debris of the plant and is able to grow during hot (80°F) weather. Powdery Mildew can be controlled by applications of Karathane according to the directions of the manufacturer.

The fungus *Mycosphaerella citrullina* attacks the stems and leaves of cantaloupes to cause the disease **Gummy Stem Blight**. This disease causes cankers to appear on the vine and the death of an apparently healthy vine. The cankers that are caused by this disease often yield a gummy exudate while brownish spots may appear on the older leaves of the plant.

Cucumbers, Cantaloupes, Honeydew Melons, West Indian Gherkins, Summer Squash, and Pumkins are some of the cucurbits that are attacked by the fungus that causes **Scab**, *Cladosporium cucumerinum*. The symptoms of Scab include the small, water-soaked spots on the leaves. These spots turn white, then gray and often die and drop from the leaf. Small sunken, gray, gummy spots develop on the fruit. These spots enlarge, become very sunken, and develop velvety spores that line the cavity. Knoblike lumps may appear on varieties of cucurbits that are resistant to this disease, such as the Squash. The first step in the control of Scab is the use of disease-free or treated seed in well drained soil. The crop should be rotated on a three year basis with non-susceptible crops. A consistent program for the application of a fungicide with effective insect and weed control programs will aid the control of Scab. Of course, disease resistant varieties of plants should be used whenever possible.

The fungus that causes the **Fusarium Wilt** of muskmelons, *Fusarium oxysporum f. melonis*, may be either seed-borne or existent indefinately in the soil or on plant refuse. This pathogen enters the melons through the wounds and quickly plugs the vascular tissue of the plant. The result of the attack of this disease is a wilted plant. The pathogen attacks melons at all stages of their growth; infected seedlings may die before they emerge from the soil; seedlings that emerge may "damp-off" after they emerge. The plant is also vulnerable to attack by this fungus at any stage of later development as well. An infected plant is marked by the streaks of dead tissue that extend along its runners. The best control method that is currently available consists of the use of disease resistant varieties of the plant.

The bacterium *Pseudomonas lachrymans* attacks many cucurbits, or vine-type plants, including melons, cucumbers, squashes, pumpkins, and gourds. Symptoms of an attack by this agent include small, water-soaked lesions that become grayish-tan and die, that appear on the upper surface of the leaf. The spots that appear on the under surface of the leaf are shiny and gummy and become dry, tearing away from the healthy tissue of the leaf to give the leaf a tattered appearance. Ripening fruit that is infected suffers small, water soaked spots that dry and become white and cracked. Infected stems may develop patches of whitish-crust that are formed from the dried bacterial exudate. This disease may be controlled by rotation of the cucurbits with non-susceptible crops on a three to four year cycle. Vines that are wet should not be disturbed to avoid spreading the disease. Since insects may be vectors of this disease, they should be controlled. Areas that are troubled by infections of this disease may be sprayed with one tablespoon of maneb or zineb per gallon of water. Finally, all crop litter should be destroyed each year.

Erwinia tracheiphila, the causative agent of **Bacterial Wilt** of the Muskmelon and the Cucumber, attacks several other vine crops, including pumpkins, gherkins, and squashes. The first symptoms of this disease are the wilting of the leaves. The wilted leaves turn brown, shrivel, and die. This disease is spread through the entire plant which soon dies. A sticky sap may be found inside the stem of an infected plant. Squashes and pumpkins are not quite as susceptible to this disease as other plants and sometimes exhibit abnormal growth patterns, such as dwarfing, rather than dying from the infection. Spotted and Stripped Cucumber Beetles are vectors for this Bacterial Wilt. When an adult beetle feeds on the infected plant, his mouth parts become contaminated. When the insect punctures the leaf of a healthy plant, the disease is introduced into the plant. The digestive tracts of these insects are the sites in which this disease over-winters. As a result, the control of these insects is necessary to control infestations of Bacterial Wilt. To control the insects, the application of carbaryl (Sevin) should be begun as soon as the plant emerges. The directions of the manufacturer should be followed concerning the dosage and spacing of applications of the carbaryl.

Diseases of the Cabbage

Once the ground has become infected with *Plasmodiophora brassicae*, the fungus that is responsible for **Clubroot** of cabbage and similar plants, the fungus is able to survive for years. The fine root hairs of the plant are the site of entrance for the infection by the fungus. Infected roots have big, club-shaped swellings. The yellowish, infected plants are not vigorous and may wilt during dry days. Good, marketable heads are not produced by infected plants. Disease resistant plants should be planted to control this disease. If Clubroot is expected to be a problem, a wettable powder formulation of PCNB(Terraclor) should be added to the transplant water following the directions of the manufacturer. The emulsion of Terraclor should not be used to control the Clubroot.

Black Rot, Stump Rot, or **Bacterial Blight**, caused by the bacterium *Xanthomonas campestris*, is one of the most destructive diseases that infect cabbage and related crops. This disease is particularly destructive during moist seasons. Any injuries to the tissue of the leaf becomes the site of entry of this disease, as are the water pores that are on the margin of the leaf. The leaves of diseased plants are marked by triangularly shaped areas of infection that are first yellow, then brown, and finally dried. The veins of infected leaves are black or dark brown and the lower leaves of an infected plant may wilt and die. A black ring extends around the stem of an infected plant if the stem is viewed in cross-section. Finally, the infected plant may grow in a lopsided fashion. Control methods include the treatment of the seed and rotation of the crop as well as proper drainage of the soil and effective insect control. Plant litter must be destroyed and the garden should not be worked when it is wet. The fertility of the soil should be maintained and the plants should not be crowded to enable them to resist the attack of the bacterium.

The soil-borne fungus *Rhizoctonia* is distributed throughout most of the world and various strains of the fungus are responsible for a host of diseases that infect a variety of plants. Specifically, *Rhizoctonia solani (Pellicularia filamentosa)* causes **Wirestem** and **Damping-Off** of the seedlings of cabbage and related plants, **Root Rot** of the Horseradish, Rutabaga, Radish, and Turnip, and **Head Rot** and **Bottom Rot** of the Cabbage, both while the plant is growing and later in storage. A portion of the stem on either side of the soil line of seedlings that are infected with Wirestem shrivels and becomes both dark and tough. The diseased plants are unproductive if they manage to survive. Damped-Off plants may die before they emerge or they may develop brown, water-soaked areas on their stems above the line of the soil before they fall and die. Vegetables that are infected with Root Rot decay both in the garden and later in storage. This rot is characterized by sunken, dark brown, spongy spots and a brownish or tan mold that features rich brown, irregular fruiting bodies. Bottom Rot often occurs during the middle of the growing season when the lower leaves wilt, become dark, and finally decay. This infection may occur when the leaves are in contact with wet, infected soil. Head Rot is characterized by a dark decay of the base of the leaves and head combined with a wilting and blackening of the outer leaves. The decayed tissue of the plant may display a cream or brown colored mold. Dark brown, sunken areas develop inside the head. An infection of the plant by *Rhizoctonia solani*, if not deleterious enough in itself, may provide the means of entrance for the bacterium that produce the **Foul Storage Rots**.

To control Rhizoctonia diseases, treated seed should be planted in fertile, well-drained soil. Any seedlings that show discoloration of the stem should be disposed. If Rhizoctonia disease develops in a garden, cabbage and related plants should not be planted for at least three years. After heavy rains, the soil should be cultivated as soon as possible, but the leaves must not be covered and the soil must not be banked around the plant. Badly infected crops should not be placed into storage, especially with crops that are not infected.

The soil-borne fungus, *Fusarium oxysporum f. conglutinans*, that causes **Fusarium Yellows** of the cabbage, also attacks related crops such as brussels sprouts, cauliflower, radishes, broccoli, and turnip, yielding similar symptoms in each of the infected plants. The first symptom is the appearance of dull or yellowish-green coloration on the leaves of the young plants. The lower, infected leaves curl length-wise, and often one-half of the leaf is distinctly more yellow than the other half. The yellowed leaves often turn brown and die as the disease progresses. Infected plants that manage to survive until maturity fail to head normally. The inside of the infected plant is ringed with a dark brown discoloration that is similar to that of Black Rot. If the soil becomes contaminated with this fungus, resistant varieties of the crop must be grown.

Diseases of the Carrot

The aster-yellow virus, *Chlorogenus callistephi*, is responsible for the disease **Carrot Yellows.** This virus is capable of infecting over three hundred types of plants, some of which are related to carrots. The first symptom of the infection of carrots by this virus is dwarfing of the plant as well as yellowing of the young leaves. The outer leaves that are older and normal size may have their edges tinged with red or purple. The taproot of an infected plant is often stunted, deformed, and possessed of tufts of hair roots. The inner cluster of leaves is yellow and bunched. The infested carrots are bitter and have a greenishy-yellow, tough core. This virus is capable of living within the body of the Six-Spotted Leafhopper. The insects carry the disease from plant to plant. If the temperature rises above 100°F for two to three weeks, the virus becomes inactive both in the insects and the plants. As a result, this virus enjoys greater distribution in the temperate areas of the world. Control of the virus is difficult. Effective control of weeds and insects, especially the Six Spotted Leafhopper, is important. Infected carrots should be destroyed as they appear and are diagnosed.

Diseases of the Cucumbers

Anthracnose affects cucumbers as well as other plants. The detailed description of this disease is presented in the section dealing with Cantaloupe. Cucumbers that are infected with Anthracnose have leaves that bear circular, reddish-brown spots and stems that are marked with tan cankers. The infected fruit is marred by round, sunken spots that have pinkish-tan centers which later become dark as the disease progresses. Cucumbers and Melons should not be grown in the same plot more frequently than every three years. The application of a spray or dust that contains either ziram or captan applied according to the directions of the manufacturer is effective in the control of this disease.

For a description of the symptoms and control of **Bacterial Wilt**, see the section "Cantaloupe".

The symptoms of **Downy Mildew** on the Cucumber are similar to the infestation of the cantaloupe. For control of this disease, the application of one of the following fungicides is recommended if the directions of the manufacturer are followed: a fixed copper fungicide, or an organic fungicide, such as captan, maneb, zineb, or ziram (but not ferbam). Alternatively, and generally additionally, the mildew-resistant varieties of cucumber include Ashley, Fletcher, Palomar, Palmetto, Santee, Smoothie, and Stono.

Stunted plants that have galls or knots on their roots are likely to be the victims of Nematodes. These pests are most numerous in areas that are south of 40° latitude. Sometimes the use of a nematocide is practical to control these pests. The local agricultural extension agent should be contacted for his recommendations concerning the control technique that is most effective in any particular locality.

Mosaic, a virus, manifests itself in the curled and mottled leaves of the stunted plants in addition to the spotted and misshapen fruit. Various perennial weeds harbor this virus and Aphids are vectors of the virus. Control of both the weeds and the Plant Lice (Aphids) is important to prevent damage to the crop from this disease. Varieties of cucumbers that are resistant to Mosaic include the following pickling types: Ohio MR-17, Ohio MR-25, Spartan Dawn F Hybrid, Wisconsin SMR-18, and Yorkshire Pickling. Disease resistant slicing varieties of cucumber include: Burpee Hybrid, Niagra, Ohio MR-200, Sensation Hybrid, Surecrop, and Table Queen.

Diseases of the Eggplant

The **Fruit Rot** of the Eggplant is caused by a fungus. This disease results in shrunken, brown areas on the stem near the soil line. Infected leaves have gray or

Brown spots, while the infected fruit has large, circular, brownish spots that contain blisters. This widespread disease can be avoided by growing one of the varieties of the Eggplant that are resistant to Fruit Rot, such as Florida Market or Florida Beauty.

Wilt, a fungus disease, causes stunting of the plant in addition to wilting of the leaves. The soil-borne fungus often causes the death of the plant. Crop rotation is important in the control of this disease. Eggplant should not be planted where potatoes or tomatoes have been grown within three years.

Diseases of the Onion

Several species of the fungus *Botrytis* are responsible for a destructive, **Post-Harvest Rot** of the Onion. The first symptoms of the Post-Harvest Rot are the soft, sunken, dried spots on the neck area of the onion. When the onion is cross-cut, these spots are brownish and water-soaked, appearing as if they had been cooked. A grayish-white fungus often grows between the scales and a gray mold appears on the infected area. Dark fruiting bodies grow on the bulb and may form a crust. Finally, the infected bulbs become mummified. This fungus lives in the soil and on plant debris. White onions are more susceptible than the red and yellow varieties of the fruit. Moist, cool weather at the time of the harvest favors the development of this disease. Control methods begin with regular applications of either Dyrene, maneb, or zineb to control Blast, Downy Mildew, and Purple Blotch. Since the *Botrytis* fungus can only enter the bulb through a wound, careful handling of the fruit is important. The bulb should be allowed to mature and the top to die naturally. The tops should be cut rather closely and the onions cured on shallow trays in forced air for forty-eight to seventy-two hours at 90°-120°F. The onions should be stored in a dry, cool area (32°-34°F) that is well ventilated. In addition, all plant litter in the field should be destroyed each year.

Smut, caused by the fungus *Urocystis cepulae* Frost, appears as spore-filled, black blisters on the leaves of the plant. Since this disease becomes inactive at soil temperatures above 80°F, Smut is a problem if the climate is cool. If the soil becomes infested by this fungus, the fungus may be controlled by the application of a solution of one teaspoon of formaldehyde per quart of water to the seeds before they are covered.

Diseases of Peas

Several soil-borne fungi are responsible for the rotting of the roots of the Pea plant, including *Aphanomycis euteiches, Fusarium solani f. pisi,* and *Rhizoctonia solani (Pellicularia filamentosa)*. The Pea plant is most likely to be damaged by **Root Rot** when it is placed in poorly drained soil. Plants that are affected by this disease often occur in patches that enlarge as the season progresses. The control of these fungi is difficult if they become well established in the soil. A crop rotation plant that includes peas every four to six years is a helpful control measure, especially if the plot features fertile, well-drained soil. Of course, the plant litter should be destroyed and varieties of peas that are resistant to Root Rot should be planted whenever possible.

Wilt and **Near-Wilt** of peas, both fungus diseases, result from infection by *Fusarium oxysporium f. pisi Race 1* and *Fusarium oxysporum f. pisi Race 2*, respectively. The lower leaves of plants that are infected with this disease are yellowed, stunted, and curl downward. The water-conducting tissue inside the stem of the diseased plant is often yellow or dirty orange. The fungi enter the plant through the roots and travel up the stems, interfering with the water conducting tissue and causing the typical symptoms of wilt before killing the plant. If the soil becomes infested with this disease, varieties of pea that are resistant to this disease must be raised or the soil must be diverted to some other crop since the disease will persist in the soil for extended periods of time. Otherwise, general clean cultural practices, such as the use of treated seed and the destruction of the plant refuse, are helpful in the prevention of an infestation.

Large, water-soaked spots on the pods of the pea indicate the presence of **Bacterial Blight**, a disease that is caused by *Pseudomonas pisi* Sackett. Dark, irregular spots that yield a shiny, off-white exudate appear on leaves that are infected with Bacterial Blight. This disease is seed-borne and may overwinter on plant refuse. The only measure that is effective in the control of this disease is the use of disease-free seed.

Diseases of the Turnip

The fungus *Peronospora parasitica* causes **Downy Mildew** of the greens of the turnip and overwinters on the roots and old tops of the turnip. Infected leaves exhibit spots of grayish mold on their underside, while yellowish areas develop on the tops of the leaves. These spots often become large and resemble inserts of pale brown paper. Severely infested leaves may die before the mold is apparent. Spores of the fungus drift from infected areas to adjacent healthy plants, especially during cool, damp weather. New areas of mold can develop in three to four days during favorable conditions, such as in the spring. Measures for the control of Downy Mildew include the planting of seed that is resistant to the disease and adherence to a plan of crop rotation. Two pounds of seventy-five percent wettable zineb powder per acre can be applied every three to five days during periods when the outbreak of Downy Mildew is serious. The directions of the manufacturer must be followed and the zineb cannot be used within seven days of the harvest.

Pale oval spots with yellow or brown borders characterize the **Cercospora Leafspot**, a result of an infection of the turnip by *Cercospora brassicicola*. A heavy infestation of this fungus can cause the leaves of the plant to die.

Cercospora Leafspot overwinters on turnips and related plants. This disease is most likely to be a problem in the fall crop of greens. Control of the Cercospora Leafspot may be effected in the same manner as control of the Downy Mildew with the exception that the fungicide may need to be applied more frequently.

Diseases of Sweet Corn

Narrow, yellowish streaks, that become brown, on the leaves of the Sweet Corn are an indication of an infection by the bacterium *Xanthomonas (Bacterium) stewartii*. These bacteria are often introduced to healthy plants by the feeding of the adult Corn Flea Beetle. The body of this insect is the overwintering site of the bacterium. The Twelve-Spotted Cucumber Beetle and the White Grub may spread the disease while they feed on the corn. The kernels and inner ear husks may be covered with a yellow, slippery substance, or the kernels may exhibit dark-edged, gray spots if the plant is infected by this disease. Cavities may develop inside the lower portion of the stalk of a severely infected plant. Disease-resistant varieties of sweet corn should be planted if this disease is a problem, but even the resistant varieties are susceptible to the disease when they are seedlings. The varieties usually achieve their resistance by the time they reach the third leaf stage. Therefore, it is important to apply an insecticide on the day before the corn emerges or just as it is "cracking" the ground. The application of carbaryl (Sevin) according to the directions of the manufacturer every three to five days is generally effective in the control of the insect pests that spread this disease. Treatment of the seed is not effective in preventing this disease, therefore disease-free seed must be obtained from a reputable supplier as a part of the disease control program.

Diseases of the Potato

Common Scab, caused by the fungus *Streptomyces scabies*, is borne by the soil and may be carried on the tubers of the plant. Scabby, rough, irregular areas on the tubers are symptoms of the Common Scab, which is widely distributed throughout the United States. For the control of this disease, Scab resistant varieties of the potato should be planted in uninfested fields and a sensible plant of crop rotation should be followed. The use of woodashes, fresh manure, or lime should be avoided if the soil is infected with this fungus. The soil should be acidified or treated with a fungicide such as pentachloronitrobenzene. Resistant varieties of potato that can be planted include: Cayuga, Early Gem, Ontario, Menominee, Cherokee, and Seneca.

Alternaria solani, the soil-borne fungus that is responsible for **Early Blight**, causes small, dark brown spots to appear on the leaves while reducing the yield of the potatoes. The planting of clean tubers and the application of an organic fungicide, such as captan, ferbam, maneb, zineb, or ziram, in addition to the destruction of the refuse of the plant are effective in the control of this disease.

Fusarium Wilt and **Dry Tuber Rot** are caused by various species of the *Fusarium* fungus. Yellowing of the leaves, darkening of the vascular system of the plant, and the general wilting of the plant are characteristics of Fusarium Wilt. The Dry Rot of the Tubers is one of the most destructive diseases of potatoes that are stored. The affected tubers shrink and their decayed areas display white or bluish protuberances. Since varieties of the potato that are resistant to these diseases have not been developed, control is sought through planting disease-free seed that has been treated in soil that is not infested with the fungus. Other control techniques include avoiding injury to the tubers during harvest and storage and the adherence to an effective crop rotation program.

Phytophthora infestans, the fungus that causes the **Late Blight** of potatoes, produces irregular, dark, dead spots on both the leaves and the stems of the plant. This disease, which is carried by the tubers, causes the death of the plant early in the season, especially during a cool, wet year. Tubers that become infected from the above-ground portion of the plant may rot during subsequent storage.

Mosaic, a virus, causes stunted plants that have curled and mottled leaves. The eventual result of this disease is reduced yields of the crop. Since the Aphids are the vectors of the virus, they must be controlled to control Mosaic. Any infected plants must be destroyed and cleaned, certified seed should be planted to the exclusion of other types of seed. The following varieties of potato are resistant to some of the Mosaic infections: Chippewa, Cherokee, Pungo, Katahdin, Kennebec, Sebago, Saco, Norland, and Irish Cobbler.

Diseases of Sweet Potatoes

The presence of **Black Rot** may be observed by the blackened area at the base of the stem of the young plant as well as the stunted growth of the plant and its yellowed leaves. Infected potatoes may develop small, black spots that enlarge until the entire potato is covered with the Black Rot that has a rather greenish, metallic sheen.

A black or grayish-brown discoloration is produced on the surface of the skin of the roots of the sweet potato by **Scurf.** This disease is unusual since it does not penetrate the skin of the potato and can be scraped from the fruit.

Plants wilt and develop yellowish and puckered foliage when they are infected with **Wilt.** The older leaves of the plant fall while the younger leaves turn yellow and numerous short stalks develop from the center of the plant.

Each of these fungus diseases are controllable by similar means. The planting of disease-free, treated seed from vine cuttings of healthy stock in well-drained soil is the first step in the control program. Sweet potatoes should never be fertilized with manure from the stable. The sweet potato should be rotated with other crops and must not be planted in the same soil more frequently than every three to four years, ten years if Wilt is present. A new site should be selected each year for the seedling beds and, ten days prior to bedding, the soil should be treated with one pound of methyl bromide per one hundred square feet of bed. If permanent seedling beds are used, the top twelve to fourteen inches of the soil should be removed and discarded each year. The frames, beds, and surrounding area should be drenched with either one part of formaldehyde per twenty parts of water or eight pounds of copper sulfate per fifty gallons of water. The following day, the beds should be filled with fresh soil and the methyl bromide treatment should be conducted. Transplant containers, storage houses, and handling containers should be disinfected before they are used. During the harvest, the sweet potatoes should be handled carefully so as not to bruise or break their skin and therefore provide an entrance for the disease organisms.

Diseases of the Tomato

Damping-Off, caused by a number of fungi, may occur either before or after emergence of the seedling. The seeds that rot in the soil may be the victim of Pre-Emergence Damping-Off, while Post Emergence Damping-Off manifests itself when the tiny seedling bends over at a water-soaked site on its lower stem. Overcrowded and damp environments favor this disease. The disinfestation of the soil, treatment of the seed, and/or application of a water dispersion of captan, zineb, or ziram every five to ten days according to the directions from the manufacturer are useful in the control of Damping-Off.

Bacterial Wilt, that results from an infection by *Pseudomonas solanacearum*, causes infected plants to wilt suddenly and die. The internal portion of a stem that is infected is discolored, water-soaked, and exudes a slimy material when it is pressed. Infected plants should be removed and destroyed. Areas in which tomatoes are grown should be rotated for four to five years with crops other than tomatoes, potatoes, eggplants, or peppers.

Corynebacterium michiganense, the agent that causes **Bacterial Canker**, causes drooping, wilting, and dying of the leaves of the plant while the *petioles* (fruit stalks) remain green and firmly attached to the stem. One entire stalk may die while the remainder of the plant is green for some time. Stems that are infested exhibit yellowish streaks and cracks on their surface with a pale-colored discoloration of the internal areas. The fruits that are infected have rough brown spots with white borders. Old plant refuse should be destroyed and treated and/or disease-free seed should be used as a basis for any program to control Bacterial Canker. A three year plan of crop rotation should also be considered.

Yellows, or **Fusarium Wilt**, is characterized by the yellowing, wilting, and dying of the plant. A leaf or stem, turned yellow by the infection of the soil-borne fungus *Fusarium oxysporum f. lycopersici*, will drop from the plant from the slightest nudge. The inner part of the lower portion of the stem contains a brown discoloration if the plant is infected by this fungus. The small side roots of the diseased plant suffer from Black Rot and the growth of a pinkish mold on the wounds of the plant may be noted during humid weather. Crop rotation and the use of disease resistant varieties of plant are the basis for the control of the Fusarium Wilt. Varieties of the Tomato that are resistant to this disease include: Campbell KC-146, Floradel, Harvester, Heinz 1350, Heinz 1370, Homestead, Manalucie, Roma, Superman, and Supreme. A consultation with a seed supplier or his catalogue will reveal other disease-resistant varieties.

Verticillium Wilt is caused by the soil-borne fungus *Verticillium albo-atrum.* This Wilt possesses symptoms that are so similar to those of Fusarium Wilt that the two diseases can only be distinguished from each other by laboratory analysis. Varieties of Tomato that are resistant to Verticillium Wilt include Heinz 1350, Harvester, Roma VF, and Superman.

Small water-soaked spots on the leaves of the tomato that become dark as the disease progresses are an indication of the presence of **Bacterial Spot**, caused by *Xanthomonas vesicatoria*. The black pustules are enclosed in water-soaked areas that appear on the fruits. These areas become sunken before being scabbed. When the stems of the flower are infected with Bacterial Spot, the blossoms will drop. The treatment of the seed is the most important single measure in the control of Bacterial Spot.

Gray Leaf Spot, a potentially destructive fungus disease, results from the infection of the plant by *Stemphylium solani*. The spots that are characteristic of this disease assume a glazed, shiny appearance, while the affected stems develop elongated, brown spots. The treatment of the seed, crop rotation, and the application of a fungicide are important measures in the control of this disease. Recommended fungicides include the application of zineb or maneb according to the directions that are supplied by the manufacturer. Disease-resistant varieties of tomatoes can be planted, including Floradel, Floralou, Manaluci, Monapal, and Ohio WR Red Jubilee.

Small, dark bordered gray spots that have tiny, black centers are characteristic of **Septoria Leaf Spot**. During periods of wet, cool weather, this disease progresses rapidly and the plants that are attacked may be severely defoliated. The casual fungus, *Septoria lycopersici*, attacks most parts of the plant and, occasionally, the fruit. Measures to control the Septoria Leaf Spot include the use of clean cultural practices, effective control of the weeds, and the application of fungicides such as Daconil 2787, Dithane M-45, maneb, Polyram, and zineb.

The fungus that causes **Early Blight**, *Alternaria solani*, can produce brown sunken spots on the stems and brown spots on the leaves of the infected plant. These spots resemble "Bull's Eyes". The stem end of the fruit is the most frequent site of the brown, moldy spot that is composed of concentric rings. The treatment of the seed, effective control of the weeds, good cultivation practices, and the application of one of the fungicides that were recommended for the control of Septoria Leaf Spot form the basis for an effective control program.

Tomatoes, as well as potatoes, are susceptible to attack by the fungus *Phytophthora infestans* that results in the disease known as **Late Blight**. Dark brown, water-soaked spots, that have light green haloes, are formed on the upper surface of the leaves while a gray mold is formed on the underside of the spots to indicated the presence of Late Blight. A plant that is severely infected appears to have been damaged by frost. Blackish-green, wrinkled, rotted areas first appear at the stem end of infected fruit. The application of either Dithane m-45, maneb, polyram, zineb, or a fixed-copper fungicide is effective in the control of the Late Blight. The effectiveness of the fungicide is enhanced if it is applied before the disease is noticed.

Poorly drained soils that have a low fertility are the likely locations for the development of **Ripe Rot**, or **Anthracnose**. The causative fungus, *Colletotrichum phomoides*, is capable of causing extensive damage to crops of tomatoes. This disease appears as small, round, water-soaked, sunken spots on the nearly ripe fruit. As this disease progresses, the middle of the affected area on the fruit darkens and may produce a pink mold during damp weather. Effective measures for the control of Ripe Rot include staking the plant, the use of crop rotation, the timely picking of the crop, and the application of one of the following fungicides according to the directions of the manufacturer: captan, Daconil 2787, Dithan M-45, maneb, polyram, zineb, and ziram.

Buckeye Rot also appears on plants that grow in areas that are poorly drained. This disease is caused by the fungus *Phytophthora parasitica*. The blossom end of the fruit is the site for the water-soaked spots that form where the soil contacts the fruit. Occassionally, concentric rings appear on the diseased area of the fruit as well as a pale gray mold. Measures that may be taken to control the incidence of this disease include the establishment of effective drainage of the soil, the staking of the tomato plant, and the application of one of the following fungicides according to the directions of the manufacturer: Dithane M-45, polyram, zineb, or a fixed-copper fungicide.

Virus Diseases of the Tomato

The infection of the Tomato by **Aucuba** or **Yellow Mosaic** virus causes a bright yellow mottling on the leaves. The upper portions of the infected plants may be so yellowed that they appear bleached. The foliage of the diseased plant may be stunted, distorted, and abnormally curled. The fruits that are infected become unmarketable due to their mottled appearance, that is dark and light green before maturity and red and yellow when the fruit is ripe.

The **Common Tomato Mosaic** causes puckering of the texture of the leaves and mottling of their color. The young leaves may unfold unevenly while the plants may be stunted and have a yellowish appearance. The fruits are unaffected by the Common Tomato Mosaic with the exception that their size and number is reduced.

Plants that are infected with the **Cucumber Mosaic** are stunted, abnormally bushy, and yellowish. The most distinctive symptom of this disease is the "shoestring" discoloration of the leaves. The fruits from infected plants may be small, mottled, and lumpy.

A late infection by some strains of the common **Tobacco Mosaic** virus may cause the vascular tissue in the fruit near the stem to become "corky" and brown. If the infection is severe, most of the internal tissue may be discolored and the discoloration even may be partially visible through the skin of the fruit. This virus is highly contagious and may be spread by insects, humans that handle the plants, and the use of tobacco products that contain the virus in the vicinity of the plants.

The same general control procedures for each of the above viral diseases are appropriate. Virus-free seedlings should be used or disease-free seed should be treated. Tomatoes should not be planted near peppers, potatoes, tobacco, or vine crops. Effective weed control should be maintained since many weeds are reservoirs of the disease. Effective insect control should be practiced since Aphids, Cucumber Beetles, and other insects often act as vectors of the viral diseases. Infected plants should be destroyed. At the time that the tomatoes are planted, they may be sprayed with powdered milk at the rate of one pound of milk per gallon of spray that is applied to twenty square yards of plant bed. This treatment inhibits the virus. Hands that are dipped frequently in milk when handling the plants is another excellent control method.

Diseases of the Pepper

Round, small, water-soaked spots on the fruit are the first symptoms of **Anthracnose** on the Pepper. These spots are caused by the fungi *Colletotrichum piperatum (Gloeosporium piperatum* or *C. nigrum)* and *C. capsici*. The dirty green spots are composed of concentric rings that contain dark, raised specks filled with sticky, pink spores.

After the peppers have suffered Sunscald or other injuries, the dark brown lesion that forms and is covered with a dark mold may indicate the presence of **Alternaria Rot**.

The control of both Anthracnose and Alternaria Rot includes a program of good cultivation practices: the use of treated seed in clean, well-drained soil. An effective spray program may include the application of either seventy-six percent ziram OR fifty percent captan at a rate of two pounds per gallon. The spray is applied every five to seven days after the seedlings have emerged. The transplanted seedlings may be sprayed according to the directions of the manufacturer with either seventy-five percent zineb OR eighty percent maneb every seven to ten days. The frequency of application may be increased to every five to seven days during rainy periods. The fruit should be handled carefully to avoid damage during the harvest and the fruit should be kept cool and dry in storage.

Diseases of Lettuce

The virus *Marmor lactucae* is the cause of the **Common Lettuce Mosaic.** Plants that are infected when they are young are severely stunted. Light green and yellow mottling on leaves that have irregular lobes are evidence of this infection. The Common Lettuce Mosaic is especially apparent during cloudy, cool weather. Head lettuce that is infected with this disease often will not head properly. Since many common weeds are susceptible to the Common Lettuce Mosaic, their control is important in the control of the disease. Aphids are vectors of the virus, making the control of these insects important in the control of the disease. Diseased plants and the litter from all plants should be destroyed in combination with an effective plan for the rotation of the crops to prevent the disease from being harbored in the soil. Paris Island is the strain of lettuce that is most resistant to the Common Lettuce Mosaic, although no variety of lettuce that is currently available is completely resistant to this disease.

BIBLIOGRAPHY

1. ---, "Vegetable Production", Cornell University Experimental Station, Ithaca, N.Y., (1962).
2. Linn, M.B., Shurtleff, M.C., and Vandemark, I.S., Fusarium Yellows of Cabbage and Related Crops, Report on Plant Diseases No. 901, Cooperative Extension Service, University of Illinois, Urbana, Ill.
3. Linn, M.B. and Shurtleff, M.C., Rhizoctonia Diseases of Cabbage and Related Crops, Report on Plant Diseases No. 902, Cooperative Extension Service, University of Illinois, Urbana, Ill.
4. Linn, M.B., Shurtleff, M.C., and Petty, H.B., Carrot Yellows, Report on Plant Diseases No. 903, Cooperative Extension Service, University of Illinois, Urbana, Ill.
5. Linn, M.B. and Luckmann, W.H., Tomato Diseases and Insect Pests—Identification and Control, University of Illinois College of Agriculture Cooperative Extension Service in cooperation with the Illinois Natural History Survey, Circular 912, Urbana, Ill., (June, 1967).
6. Linn, M.B., Shurtleff, M.C., and Vandemark, J.S., Fusarium Wilt of Watermelon and Muskmelon, Report on Plant Diseases No. 904, Cooperative Extension Service, University of Illinois, Urbana, Ill.
7. Linn, M.B., Shurtleff, M.C., and Petty, H.B., Bacterial Wilt of Cucumber and Muskmelon, Report on Plant Diseases No. 905, Cooperative Extension Service, University of Illinois, Urbana, Ill.
8. Linn, M.B., Shurtleff, M.C., Petty, H.B., and Vandemark, J.S., Stewart's Disease or Bacterial Wilt of Sweet Corn, Report on Plant Diseases No. 907, Cooperative Extension Service, University of Illinois, Urbana, Ill.
9. Linn, M.B., and Shurtleff, M.C., Root Rots of Pea, Report on Plant Diseases No. 911, Cooperative Extension Service, University of Illinois, Urbana, Ill.

10. Linn, M.B., Shrutleff, M.C., and Vandemark, J.S., Wilt Diseases of Pea, Report on Plant Diseases No. 912, Cooperative Extension Service, University of Illinois, Urbana, Ill.
11. McGlohon, N.E., Stem Anthracnose of Lima Beans, Leaflet No. 44, Cooperative Extension Service, University of Georgia, Athens, Ga., (December, 1965).
12. McGlohon, N.E., Leafspot Diseases of Turnip Greens, Leaflet No. 48, Cooperative Extension Service, University of Georgia, Athens, Ga., (January, 1966).
13. McGlohon, N.E., Cantaloupe Diseases, Leaflet No. 67, Cooperative Extension Service, University of Georgia, Athens, Ga., (January, 1967).
14. Michelbacher, et al., (1959).
15. ---, 1967 Chemical Control of Insects and Diseases on Commercial Vegetables, Extension Bulletin No. 312, Farm Science Series, Cooperative Extension Service, Michigan State University, E. Lansing, Mich., (March, 1967).
16. Reed, L.B. and Webb, R.E., Insects and Diseases of Vegetables in the Home Garden, Home and Garden Bulletin No. 46, U.S.D.A., Washington, D.C., (March, 1967).
17. ---, Potatoes U.S.A., Publication AG-229, Rohm and Haas Corp., Philadelphia, Pa., (February, 1966).
18. Shurtleff, M.C. and Linn, M.B., Vegetable Seed Treatment, Report on Plant Diseases No. 915, Cooperative Extension Service, University of Illinois, Urbana, Ill.
19. Shurtleff, M.C. and Linn, M.B., Damping-off and Seedling Blights of Vegetables, Report on Plant Diseases No. 916, Cooperative Extension Service, University of Illinois, Urbana, Ill.
20. Shurtleff, M.C. Thornberry, H.H., and Linn, M.B., Virus Diseases of Tomato in Illinois, Report on Plant Diseases No. 917, Cooperative Extension Service, University of Illinois, Urbana, Ill.
21. Shurtleff, M.C. and Linn, M.B., Angular Leaf Spot of Cucurbits, Report on Plant Diseases No. 919, Cooperative Extension Service, University of Illinois, Urbana, Ill.
22. Shurtleff, M.C. and Linn, M.B., Bacterial Blights of Garden Beans, Report on Plant Diseases No. 921, Cooperative Extension Service, University of Illinois, Urbana, Ill.
23. Shurtleff, M.C. and Linn, M.B., Clubroot of Cabbage and Other Crucifers, Report on Plant Diseases No. 923, Cooperative Extension Service, University of Illinois, Urbana, Ill.
24. Shurtleff, M.C. and Linn, M.B., Black Rot of Cabbage and Other Crucifers, Report on Plant Diseases No. 924, Cooperative Extension Service, University of Illinois, Urbana, Ill.
25. Shurtleff, M.C. and Linn, M.B., Scab of Cucumber and Other Vine Crops, Report on Plant Diseases No. 928, Cooperative Extension Service, University of Illinois, Urbana, Ill.
26. Shurtleff, M.C. and Linn, M.B., Botrytis Neck Rot of Onion, Report on Plant Diseases No. 930, Cooperative Extension Service, University of Illinois, Urbana, Ill.
27. Shurtleff, M.C. and Linn, M.B., Anthracnose and Alternaria Fruit Rots of Pepper, Report on Plant Diseases No. 940, Cooperative Extension Service, University of Illinois, Urbana, Ill.
28. Shurtleff, M.C., Linn, M.B., and Petty, H.B., Jr., Lettuce Mosaic, Report on Plant Diseases No. 946, Cooperative Extension Service, University of Illinois, Urbana, Ill.

29. Shurtleff, M.C. and Linn, M.B., Cercospora Leaf Spot of Beets and Swiss Chard, Report on Plant Diseases No. 951, Cooperative Extension Service, University of Illinois, Urbana, Ill.
30. Taylor, J.L., Janes, R.L., Nicklow, C.W., and Potter, H.S., Home Vegetable Garden, Extension Bulletin 529, Farm Science Series, Cooperative Extension Service, Michigan State University, E. Lansing, Mich., (May, 1966).
31. ---, Handbook No. 120, U.S.D.A., Washington, D.C., (1962).
32. ---, Vegetable Pest Control, Extension Guide, University of California, (1962).
33. ---, Diseases of Sweet Potatoes, Leaflet No. 47, Cooperative Extension Service, University of Georgia, Athens, Ga., (January, 1966).

IX

Pests of Ornamentals, Lawns, Flowers and Shrubs

The protection of flowering plants, shrubs, and even trees against the attack from insects requires continual alertness and, frequently, the use of pesticides. Early recognition of leaves that are infested with Aphids, Leaf Miners, or hungry Caterpillars, accompanied by the disposal of the infested leaves, may prevent serious attacks of these pests. If the invaders are well established and no natural enemies are controlling the pests, the application of sprays or dusts that contain chemical control agents may be necessary.

To a varying extent, all pesticides, and especially insecticides, are hazardous to handle. The label on the package was provided for the information and safety of the user; its contents should be read carefully, and understood before the instructions are followed. The selection of a particular insecticide for example, should be made with regard to the tolerances of the plants. Dormant trees and shrubs may tolerate double the concentration of a refined spray oil or a lime-sulfur solution that would be used in the growing season. The delicate opening buds and blooms of the daffodil and similar types of plants are very easily damaged, therefore the control sprays should be applied before this stage is reached. If you are uncertain as to the effect of a particular pesticide, the application of a small amount followed by waiting for a few days will enable you to assess the immediate, but not long term, effect of the pesticide.

Materials that are applied in the form of a dust are usually purchased in the prepared form and require no preparation by the user. Emulsifiable concentrates must be mixed into the water before the spray is applied. Such mixing should be done very carefully to avoid the contact of the concentrated mixture with the skin. If some of the pesticide contacts the skin or the clothing, contaminated clothing should be removed and the skin washed and flushed with running water as soon as possible. Chlordane, even in the powdered form, may be absorbed through the skin in sufficient quantities to cause concern. After insecticides have been handled and applied, the hands and face must be washed carefully, especially before eating. Care must be exercised to avoid inhaling either drifting dusts or the mists of a spray. The wetting action of a spray on the foliage may be improved by the addition of one-half teaspoon of a mild detergent per gallon of spray. Unused insecticides must be stored out of the reach of children and persons who are not familiar with them, and caution should be observed about their use. The stability and effectiveness of chemical compounds, such as the insecticides, is improved by selecting a storage site that is also cool, but not necessarily freezing. All containers and equipment that are used for the application or storage of pesticides should be cleaned after use and empty containers that held the concentrate should be disposed of with care.

The application of sprays in the garden can usually be handled more satisfactorily with the use of compressed air sprayers than with atomizers. The aerosol spray can is helpful in the house and garden, especially for use on house plants, but the label must be consulted to avoid damage to plants that have delicate foliage.

The more toxic insecticides, such as EPN, heptachlor, and methyl parathion, should be used only by operators who are trained to observe all of the warnings that have been issued by the manufacturer to ensure both the safety and the effectiveness of the product.

The following list of insecticides contains those that are most commonly used on ornamentals and lawns. These materials can be handled with safety, if the foregoing recommendations and the instructions of the manufacturer are followed.

Chlordane: Used in the control of Ants where they are associated with Aphids and Scale insects. This chemical is also used in the control of Earwigs and Earthworms.

DDT: This chemical has a long "residual life" that makes it especially useful in the control of biting or chewing insects, such as caterpillars and beetles. This long residual life is the greatest sin of the material since it may harm species other than the one whose control is desired, for many years into the future. The DDT gives protection from Leafhoppers and Thrips, but the long residual action may be dangerous to predators, such as the Lady Beetles, if they are present in the garden. Because of the long residual action of this compound, many state and federal governments are banning or restricting the use of DDT. This chemical is included in this book since the benefits that may be gained by its use outweigh the costs, including to the ecology, in many areas of the world.

Malathion: This chemical is used against Aphids (Plant Lice), Caterpillars, Mealy Bugs, Scale insects, Thirps, Spider Mites, and Whiteflies.

Nicotine: This naturally derived compound is an effective control agent for use against Aphids, most Thrips, Leafhoppers, and the younger stages of

Rose Slugs and Leaf Miners. Nicotine is largely lost from the foliage of the plant withing three days of its application and hence is less dangerous to beneficial insects than the more persistant materials that are listed above.

Pyrethrum: This naturally occurring material and the extracts of the Pyrethrins are used in the control of Aphids.

Refined Petroleum Oil (White Oil): These oils are available both alone and in combination with nicotine, pyrethrum, rotenone, and a number of organic insecticides. This oil may be used alone in the dormant season against Scale insects and the overwintering eggs of the Spider Mites. Combinations of the oil with other materials contain labels that list the effectiveness which is claimed for the mixture.

Rotenone: This chemical is used as a five percent active ingredient emulsifiable concentrate or as a wettable powder. Rotenone is effective against certain Aphids and the young stage of the Whitefly.

THE CONTROL OF INSECTS AND MITES

The African Violet

The **Cyclamen Mite** is the most serious pest of the African Violet. Leaves of the plant that are infested with the Mite are twisted and brittle. The buds and flowers of infested plants are deformed and the latter are streaked with a dark color. Injured leaves, buds, and flowers may be blackened by the attack of this pest. New leaves that are attacked are very hairy and soon die. (Johnson and Smith, 1960) The injured parts of the plant should be removed before the plant is sprayed or dipped at weekly intervals with one teaspoon of the twenty-five percent Kelthane emulsifiable concentrate or one level tablespoon of the twenty-five percent Kelthane wettable powder per gallon of water. To improve the wetting power of the treatment, one teaspoon of a mild household detergent may be added to each gallon of water.

Azaleas

The Azalea may be infested by both the **Spider Mite** and the **Privet Mite**. Apply a Kelthane spray in the manner described for the control of the Mites in the African Violet.

Bulbs

The Bulbs of Narcissus, Daffodil, and Amaryllis, as well as those of many other similar plants, are attacked by the **Bulb Fly**, the **Lesser Bulb Fly**, and the **Bulb Mite**. The eggs of these pests are laid near the crown of the host plant, near the surface of the soil. The usual method that is used to control these pests is the fumigation of the Bulb in a tight fumatorium. Three pounds of methyl bromide per one thousand cubic feet for four hours is the usual fumigation specifications. A temperature between 70° and 80°F, and a humidity of seventy to eighty-five percent provides the highest degree of effectiveness. (de Ong, 1960)

The **Gladiolus Thrips** is one of the most difficult insects of this group to control. Weekly applications of DDT as recommended by the manufacturer is the treatment commonly used. The bulbs are generally dusted with DDT before they are stored for additional control. The Gladiolus is also subject to the attack, in the western areas of the United States, by a Leafhopper that is the carrier of the virus that causes the disease that is known as **Aster Yellows.** (Smith, 1961)

Columbine

The **Bristly Rose Slug** is a black, wasp-like insect that is six to eight millimeters long. The mature larva is pale green, bristly, and approximately fifteen millimeters long. This pest feeds on the underside of the leaf, cutting irregular shaped holes in the foliage. This pest may be controlled by the application of either DDT, TDE, or methoxychlor as directed by the manufacturer. If **Mites** are also a concern, the inclusion of Kelthane or sulfur in the spray is often helpful. (Micheloacher, et al., 1959)

The **Columbine Leaf Miner** is a pest of this plant. The larvae of this fly feeds between the surfaces of the leaf, making long serpentine trails between the surfaces that are plainly visible from the upper side of the leaf. The lower leaves of the plant are attacked first. If these leaves are removed at the first indication of attack and are disposed of, the infestation may be checked. Sprays at one week intervals with nicotine sulfate (See Aphid, under Chrysanthemum) will kill the younger stages of the larvae.

Carnations

The grower of carnations under conditions that exist in southern California, is subject to severe losses of the crop from injury to the bloom. The damage to the bloom chiefly is due to the attack of the **Thrips** and the **Green Peach Aphid.** Experiments in the control of these insects with a systemic insecticide have yielded favorable results. The insecticides included dimethoate, phorate, Di-Syston, and Phosphamidon, applied at the rate of five pounds of the active ingredient per acre. Granular formulations, soil drenches, and foliage sprays were applied in April, May, and June during the experimental procedure. "In one experiment in which there was a very heavy infestation, flowers cut from the dimethoate plots one month after treatment averaged 75.3 percent saleable (blooms) compared with 16.1, 14.7, 5.7, and 0.7 for phorate, Di-Syston, phosphamidon, and the check (control), respectively. Dimethoate was also effected as a soil drench and as a foliage spray." (Jefferson, et al., 1964)

Chrysanthemum

The **Aphid**, a pest of the chrysanthemum may be controlled by the application of pyrethrum or malathion as directed by the manufacturer. Alternatively, one tablespoon of forty percent nicotine sulfate per gallon of water may be used. Condron, et al., 1962, found that the vapors of Dibrom in the fumigation of a greenhouse gave good control of the Aphids on the chrysanthemums. The Dibrom liquid that contained eight pounds of Dibrom per gallon was applied to the cold heating pipes at the rate of one ounce per ten thousand cubic feet. The steam was turned on immediately after the application of the liquid to the exterior of the pipes.

Thrips may be controlled by the application of either DDT or malathion as directed by the manufacturer on the label of the insecticide.

Spider Mites may be controlled by the application of Kelthane as suggested in the control of Mites (African violet). The greenhouse fumigation that is described under Aphids above was also effective in the control of the Mites.

Gardenia

To control the **Aphid** in the gardenia, the same methods are used for the control of this pest in the chrysanthemum.

The **Citrus Mealy Bug** is one of the most destructive species of pest to many kinds of plants. The three millimeter long, yellowish body of this pest is covered with a white, powdery wax. Short filaments of the wax occur along the side, the hind ones being about one-fourth the length of the body. The eggs of this insect are laid in loose, cottony masses of three hundred to four hundred. This Mealy Bug is found on many ornamentals in addition to being a pest of the Gardenia. (Pritchard, 1949) Plants that are infested with the Mealy Bug may be dipped in, or sprayed with, two teaspoons of the fifty to fifty-seven percent malathion emulsifiable concentrate dispersed in one gallon of water. (Johnson and Smith, 1960)

Cedar

For the control of the **Bagworm**, **Spider Mite**, or **Webworm** that infest the Cedar, use the recommendations for the same insect in the section dealing with the Juniper.

Holly

The **Holly Leaf Miner** attacks the leaves of both the American and the English holly. The adult insect is a grayish-black fly about one-tenth of an inch long. This pest passes the winter inside the infested leaves on the trees. With the opening of the new leaves in the spring, the adults emerge and deposit their eggs in punctures that are made in the foliage. The larvae that hatch from these eggs feed between the surfaces of the leaf, making long serpentine mines that are plainly visible from the upper surface of the leaf. The leaves that are infested by the Holly Leaf Miner have an unhealthy appearance, being somewhat faded in their color, (Weigel and Baumhoffer, 1948). The application of a systemic insecticide, dimethoate, in the form of the five percent active granules can be used to control the Holly Leaf Miner. The granules should be raked into the soil in an area 9.4 square feet per plant, with a dosage of four ounces per plant. As the granules are dissolved by the rain, the solution is absorbed by the plant, killing the larvae that reside in the leaves, (Schread, 1960).

Ornamental Juniper

The Juniper, a common ornamental tree in California, is widely infested by a **Twig Girdler,** *Periploca nigra* which is the larvae of a moth. Heavy infestation results in severe damage to the plant. Presently, the insect seems to be important, in an economic sense, only in the western United States, but it has been found in the Atlantic coastal regions of the United States. "Injury to juniper twigs is caused by the larvae which tunnel beneath the bark in the region of the cambium. Some penetration into wood also occurs". The larvae tend to move in a circular direction while they feed, thus girdling the twig. A number of larvae may be found on a single twig. Although the damage that is caused by this pest may be seen during any season of the year, infestations are more severe during the hotter times of the year. Pupation of the larvae occurs in the gallery that is formed by the larvae during their feeding. These cream colored larvae have a brown head capsule, and reach six millimeters in length at maturity. The Twig Girdler is attacked by a tiny parasitic wasp, one of the braconids. "The relatively high number of parasites found suggests that the species may be an important agent of biological control", (Koehler and Tauber, 1964).

Bagworms may be controlled by the application of either three tablespoons of fifty percent carbaryl (Sevin) wettable powder or three tablespoons of the arsenate of lead per gallon of water. This spray is applied in June.

A variety of insecticides are available for use in the control of the **Spider Mite**. The following concentrations of recommended insecticides may be dispersed in one gallon of water and applied as a spray: one tablespoon of fifteen percent Aramite wettable powder, two teaspoons of twenty-five percent chlorobenzilate emulsifiable concentrate, two teaspoons of twenty-five percent Diazinon emulsifiable concentrate, two teaspoons of twenty-five percent Dimite emulsifiable concentrate, two teaspoons of the eighteen percent Kelthane emulsifiable concentrate, OR one tablespoon of Tedion twenty-five percent active wettable powder.

The application of either two tablespoons of twenty-five percent malathion wettable powder OR two teaspoons of twenty-five percent Diazinon emulsifiable concentrate per gallon of water may be used to control infestations of **Webworms.** The concentrates should be dispersed in the gallon of water to form the spray that is applied at high pressure to the infested plants.

The **Pine Tip Moth** may be treated with an application of either two tablespoons of fifty percent **DDT** wettable powder or two tablespoons of fifty percent carbaryl (Sevin) wettable powder per gallon of water if its control is desired. The spray that is formed by dispersing the concentrated insecticide in the gallon of water may be applied in the first week of May and the first week of June. This spray should be applied to the tips of the shrub.

Many ornamental plants are plagued by **Leaf Miners**, a general term that is applied to the larvae of a number of different moths and flies. The adult insects deposit their eggs on the underside of the leaves. The emergent larvae feed between the two surfaces of the leaf, resulting in the formation of serpentine trails or flat mines between the surfaces of the leaf. The application of either dimethoate (Cygon) or Zectran according to the directions that are supplied by the manufacturer are effective in the control of these pests. Dimethoate (Cygon) should not be used on any plant that is not specifically recommended for treatment on the label that is prepared by the manufacturer of the chemical.

Iris

The larvae of the **Iris Borer**, *Macronoctua oneusta* Grote, enter the leaves of the Iris in the spring, working their way to the root where they hollow the rhizomes. The destruction of the plant litter at the end of the growing season is the first step in the control of the Iris Borer. When the leaves of the plant are five to six inches high, weekly application of one of the following insecticides may be commenced: three tablespoons of a fifty percent DDT wettable powder per gallon of water, OR a five percent DDT dust, OR two tablespoons of a two-and-two-thirds pounds of dimethoate (Cygon) per gallon emulsifiable concentrate per gallon of water to form a spray.

Rose

Since **Ants** often place **Aphids** on the Roses, the ants must be controlled as well as the aphids to effect the control of the latter. The ants may be controlled by the broadcast application of either one pound of ten percent chlordane or five percent dieldrin granules. The granules should be applied per one thousand square feet of the rose garden and the area immediately surrounding it. The treated area should be watered and allowed to dry before children are permitted to enter the treated ground. The **Aphids** and **Thrips** may be controlled by the application of a spray that contains either two teaspoons of 23.4 percent dimethoate (Cygon) emulsifiable concentrate or one tablespoon of the fifty-seven percent malathion emulsifiable concentrate per each gallon of water.

To control **Mites**, the Rose plants should be sprayed with either two teaspoons of the 23.4 percent dimethoate (Cygon) emulsifiable concentrate, or one-and-one-half tablespoons of the 18.5 percent Kelthane wettable powder dispersed in each gallon of water.

Proper cultivation in the Rose garden include the destruction of the old blossoms and the effective control of weeds in the garden.

Nematodes are microscopic worms that feed upon the roots of the plant, forming small galls on the tiny feeder roots. Stock that is infected with Nematodes should not be planted and a preventative measure is the fumigation of the soil with a substance such as methyl bromide before the plants are placed in the soil. The methyl bromide must be applied and handled with a great deal of care and with strict adherence to the directions supplied by the manufacturer to avoid misuse of this poisonous chemical.

Figure IX-1. Soil fumigation for control of nematodes, insects, weeds and soil diseases. (Morton Chemical Company.)

The attack of **Spider Mites** results in yellow foliage and/or defoliation of the plant. This attack leaves the plant more susceptible to winter killing and the disease that is known as Black Spot, (Hennenberry, et al., 1961). The use of clean cultivation is essential in the protection of the rose since the Mites overwinter on hardy plants such as **Horseweed** (*Erigeron canadensis* L.) and **Chickweed** (*Stellaria media* (L.)) in the Atlantic coast regions of the United States, (Hennenbery and Taylor, 1961). In areas such as California, the Mites overwinter on **Malva** and the **Wild Morning Glory** or **Bindweed** in addition to overwintering in the underground roots. In the early spring, the overwintered Mites attack the rose plants, extending the infestation throughout the growing season. (de Ong, 1960) The use of pesticides in the control of Aphids and plants diseases in a number of instances has resulted in an increased population of Spider Mites. The inclusion of an effective Acaricide, such as aramite, Kelthane, or Tedion, in the spray program may often be necessary to avoid this side effect, (Hennenberry and Taylor, 1962).

A General Program to Control Diseases of the Rose Plant

All diseased growth should be remove from the plant and destroyed, while the plant should be pruned well below the area that appears to be diseased. Just before the buds open in the spring, a dormant spray containing one part of liquid lime-sulfur per ten parts of water should be carefully applied to the rose. When the new leaves start to unfold, weekly applications of a fungicide should be commenced to control **Mildew** and **Black Spot**. The fungicide may be a spray composed of one tablespoon of the seventy-five percent Phaltan (folpet) wettable powder per gallon of water OR it may be a dust containing seven-and-one-half percent of the active ingredient. If **Black Spot** is particularly prevalent, the application of a spray consisting of two tablespoons of an eighty percent maneb wettable powder per gallon of water is recommended. However, if **Mildew** is the primary concern, one-and-one-half teaspoons of Karathane wettable powder or Acti-dione PM, applied according to the directions of the manufacturer, per gallon of water is called for.

Note: The application of pesticides when temperatures exceed 85°F is not recommended since injury to the plant may result. If the application of a pesticide is required during hot weather, one half the recommended dosage should be used.

Several fungi are responsible for the diseased condition of the rose that is commonly referred to as **Dieback**. The pathogens that are responsible for this disease gain entrance to the plant through either injuries to the plant or a general weakened condition of the plant. This infection is manifested in the dying of the branches of the plant from the tip back to the main stem.

Black Spot, caused by an infection by *Diplocarpon rosae*, is characterized by the appearance of black spots on the foliage of the plant as well as black or dark purple spots on the stems. The severe infection of the plant with this disease can result in the defoliation of the plant.

Crown Gall is a bacterial disease that is caused by an infection of an injury to the plant with *Agrobacterium tumefaciens*. The roots and crowns of the infected plants develop galls before these stricken plants die. When rose plants are purchased, the graft union should be inspected to ensure that it is not infected. When roses are transplanted, care should be exercised to avoid any injury to the roots that might provide entrance for this disease. If this soil-borne disease develops on any plant, the plant must be destroyed and the soil fumigated with a substance such as methyl bromide before any new roses are planted in the infected ground. The methyl bromide must be used with extreme caution and following the directions of the manufacturer since this chemical is **very poisonous.**

A grayish-powdery mold growth on the rose plants indicates the presence of **Powdery Mildew** that is caused by the fungus *Sphaerotheca pannosa* var. *rosae*. This disease also causes the new growth of the plant to be stunted and twisted.

Rose canes that are infected with **Stem Canker** are stunted or killed. A number of fungi are able to cause Cankers on the stems of the rose plant.

Lilac

The application of an insecticide spray in mid-May, and repeated three times at three week intervals, can be used to control infestations of the **Lilac Borer.** Among the sprays that have proven to be effective are the use of three tablespoons of twenty-five percent DDT emulsifiable concentrate or three tablespoons of eighteen percent dieldrin emulsifiable concentrate per gallon of water.

Powdery Mildew, caused by *Microsphaera alni*, can be controlled by the application of two pounds of wettable sulfur per one hundred gallons of water, OR one-half pound of 22.5 percent dinocap per hundred gallons of water OR cycloheximide applied according to the directions provided by the manufacturer. The sulfur-containing spray may be applied weekly as necessary to control this disease.

The presence of **Phytophthora Blight**, caused by *Phytophthora cactorum*, or **Bacterial Blight**, caused by *Pseudomonas syringae* can be controlled by the adherence to the following procedure. The dense growth should be pruned from the plant along with any infected growth, which should be destroyed. Two or three times every ten days, beginning when the leaves begin to unfold, the plant should be sprayed with an 8-8-100 Bordeaux mixture. A cultural procedure to help control this disease consists of avoiding the placement of rhododendrons in the same area as the lilacs.

LAWN PEST CONTROL

The following recommendations for the control of insects of turf are given for the use of commercial appliers of the compounds. Of course, the dosage that is recommended on the label is provided by the manufacturer to insure the safety and effectiveness of the product and should always be followed.

The application of one of the following insecticides may be used in a program for the control of the **Cinch Bug**: Aspon, Baygon, Diazinon, Dursban, Ethion, Trithion, U-C 13, or Zytron.

Lawn Caterpillars can be controlled by judicious application of DDT, diazinon, Sevin, or toxaphene.

The use of various types of bait, Aldrin, chlordane, heptachlor, or Kepone will control infestations of **Mole Crickets**.

Diazinon is used for the control of the **Bermuda Grass Mites**.

Figure IX-2. Lilac-French Hybrid Bacteria Blight. (Illinois Natural History Survey.)

White Grubs may be controlled by the application of Aldrin, chlordane, or heptachlor.

The addition of Summer Oil to either malathion or parathion may be used to control the **Grass Scales**. The parathion must not be used in residential areas.

Ants in the lawn may be controlled by the application of either Aldrin or chlordane.

Baygon, Di-Syston, or Thimet are recommended for the effective control of **Billbugs**, but the Di-Syston and the Thimet must not be used in residential areas.

One of the most destructive and common disease of turfgrass, **Brown Patch** results from the infection of the lawn with *Rhizoctonia solani*. This fungus, which will attack most turfgrasses, is most prolific during warm, moist weather and following the application of heavy concentrations of nitrogen. The appearance of this disease is aptly described by its common name, Brown Patch. Affected areas of the turf may turn brown or merely take on a pale color. The leaves of the plants in an infected area may be the only loss, but if the disease is severe, the entire plant may die. Measures for the control of Brown Patch include the removal of the clippings of the grass from the lawn, the avoidance of the application of excessive amounts of nitrogen during fertilization, infrequent watering of the turf and application of the water early in the day, and the application of a fungicide. Typical fungicides include mercury, PCNB, or thiram as active ingredients. One of the following commercial products may be used according to the directions that are supplied by the manufacturer: Ortho Lawn Disease Control, Ortho Lawn and Turf Spray, Panogen Turf Spray, and Tersan-OM.

Small tannish spots, approximately the size of a silver dollar, that are scattered about the lawn suggest the infection by the fungus *Sclerotinia homeocarpas*. This disease, commonly called **Dollar Spot**, can be controlled by correcting any nitrogen deficiency and the application of one of the following per one thousand square feet of lawn (all active ingredient mixtures are dispersed in ten gallons of water to form the spray): four ounces (one-and-one-third cups) Acti-dione-thiram, OR eight to twelve ounces (two-and-one-half to three-and-one-third cups) Ortho Lawn Disease Control, OR six ounces (two cups) Ortho Lawn and Turf Fungicide, OR two ounces (one-half cup) PSMA (phenyl mercury), OR six ounces (one-and-one-half cups) Terson-OM, OR one-and-one-half ounces Panogen Turf Fungicide, OR Scutl, according to the directions of the manufacturer.

The fungus disease that is called **Melting-Out** attacks the crown, leaf sheath, and the leaf of the grass plant. The infected areas are afflicted with spots that vary from dark brown to pale gray. The infection of the leaf sheath results in its browning and demise. Lawns that are extensively infected with *Helminthosporium spp.* become thin and exhibit a brown undercast. This disease is capable of killing an entire lawn. The same fungicides that are recommended for the control of Dollar Spot may be used for the control of Melting Out.

During warm, damp weather, the leaves of the grass may become covered with the dark, bluish-gray colored spores of the **Slime Mold**. The application of five ounces (one-and-one-half cups) of fifty percent Captan wettable powder, or maneb, or zineb per ten gallons of water is sufficient to control Slime Mold on one thousand square feet of turf.

Rust, caused by *Puccinia spp.*, is manifested on the leaves of the plant as small yellow to brownish-red pustules. The lawn may be treated with zineb, sulfur, or cycloheximide in combination with thiram, following the directions of the manufacturer, for the control of this disease.

During rainy weather, St. Augustine grass sometimes becomes infected with **Gray Leaf Spot,** *Piricularia grisea*. This disease is recognized by the tan or brown spots that have purple or brown margins. The treatment of one thousand square feet of lawn with four ounces of thiram, captan, or Dyrene generally is adequate to control this disease.

Small, purplish or brownish spots with tan centers may indicate the infection of the lawn with *Helminthosporium spp.*. A heavy infestation of this disease will thin the turf. This disease may be controlled by the application of thiram, organic mercury, or the combination of cycloheximide and thiram, according to the directions supplied by the manufacturer.

Most types of grasses are susceptible to the attack by **Fairy Ring,** *Lepiota sp., Marasmius sp.*, etc. This growth in a circular area of the lawn is initially stimulated and, within a few weeks, the area may decline or even die. Mushrooms or Toadstools may appear at the margin of the ring. Fairy Ring is difficult to control. Five to ten pounds of Captan fifty percent wettable powder may be worked into one thousand square feet of the soil to afford some measure of control.

Poorly drained areas are susceptible to **Cottony Blight** during warm, damp weather. *Pythium aphanidermatum*, the causative agent, kills the grass in well-defined areas. Initially, the stricken grass is greasy in appearance and blackened, often sporting the growth of a cottony fungus. Measures to control Cottony Blight include the application of either Memmi, Dexon, or Panogen Turf Fungicide, according to the directions that are supplied by the manufacturer.

When the snow is melting in the spring, various patches of grass may be covered with a dark or light pink growth of a mold, especially in shaded areas. **Pink Snow Mold**, or **Fusarial Patch**, results from the infection of the grass by *Fusarium nivale (Calonectria). Typhula itoana* causes the dark mold growth of **Snow Scald**. Lawns should not be fertilized after September 15 in the northern areas as part of a plan to control these diseases. Before winter, the application of Thimer, Calocure, Calo-Clor, or Tersan-OM according to the direction of the manufacturer is effective in the control of these diseases. The treatment of the lawn may be repeated during a mid-winter thaw. A final control procedure is keeping the lawn well drained.

BIBLIOGRAPHY

1. ---, Insect Control Recommendations for Ornamental Plants and Lawns, E.C. 67-1502, University of Nebraska, Extension Service.
2. Coleman, V.R., Iris Borer, Leaflet No. 64, Cooperative Extension Service, University of Georgia, Athens, Ga., (January, 1967).
3. Coleman, V.R., Control of Insects and Mites on Roses, Cooperative Extension Service, University of Georgia, Leaflet No. 65, Athens, Ga., (January 1967).
4. Coleman, V.R., Leaf Miners on Ornamental Plants, Leaflet No. 66, Cooperative Extension Service, University of Georgia, Athens, Ga., (January, 1967).
5. Freeman, T.E. and Mullin, R.S., Turfgrass Diseases and their Control, Circular 221-B, Agricultural Extension Service, Institute of Food and Agricultural Sciences, University of Florida, Gainesville, Florida, (February, 1966).
6. Garrett, W.N., Brown Patch of Lawn and Turn Grasses, Leaflet No. 16, Cooperative Extension Service, University of Georgia, Athens, Ga., (December, 1964).
7. Garrett, W.N., Common Rose Diseases, Leaflet No. 68, Cooperative Extension Service, University of Georgia, Athens, Ga., (February, 1967).

8. Weidhaas, J.A. Jr., Sinclair, W.A., and Pridham, A.M.S., Cornell Recommendations for Trees and Shrubs Commercial Production and Maintainance, Cooperative Extension Service, New York State College of Agriculture at Cornell University, Ithaca, N.Y., (April, 1966).
9. Weihing, J.L. and Shurtleff, M.C., Lawn Diseases in the Midwest, North Central Region Extension Publication No. 12, Extension Service, Kansas State University, Manhattan, Kans., (June, 1961).
10. Williams, T.G. Jr. and Smith, G.E., Georgia Lawns . . . Establishment and Maintainance, Bulletin 642, Cooperative Extension Service, University of Georgia, Athens, Ga., (January, 1966).

X

Household Insects and Rodents

Protections against insects and rodents in and about the home depends upon the persistent use of sanitary procedures supplemented by the judicious use of insecticides and repellents as necessary. Without the adherence to sanitary practices in barns that house livestock, flies are produced faster than chemicals can check their production. With the prompt disposal of vegetable wastes, moist manures, and the improved handling of garbage both at the site of creation and storage and at the site of disposal, such as the city dump, dysentery and other intestinal diseases can be checked.

Malaria, once a menace throughout the world, is now under control in many districts. For example,

> "In Greece and Italy as late as the early 1940's, the number of cases of malaria exceeded a half million and the number of deaths ran into the thousands annually. The death rate dropped from about 4,000 in 1944 to 7 in 1951. The concerted use of DDT for spraying homes and to some extent for larvicidal purposes practically eliminated the incidence of malaria."
>
> "While all the credit cannot be attributed to DDT alone, it is well to recall that prior to the advent of DDT, malaria ran rampant despite the use of water-level control and of predaceous fish, oil and Paris Green for larval control." (Decker, 1962)

Typhus fever is another insect-borne disease that has yielded to DDT and cleanliness. In California, Bubonic Plague is fought by attacks on flea-bearing rats and ground squirrels.

Pesticides are, of necessity, toxic substances, otherwise they would not be effective. However, as a group they are one of the most, if not the most, carefully supervised of any group of chemicals and drugs, both in their manufacture and their use. Decker (1962) stated that in 1956, the accidental deaths in the United States resulted from:

Miscellaneous causes	55.38%
Motor vehicles	41.77%
Poisonous Substances	2.78%
Pesticides	0.15%

"Data compiled by the U.S. Public Health Service reveal that in 1956, of the 152 accidental deaths attirbuted to pesticides, 94 (62 percent) involved children under 10 years of age and 78 of these were children under 4 years of age . . . The careless storage of household pesticides in pantries and cupboards represented the principal cause of serious illness and deaths due to food contamination."

As with any tool that has been developed by ancient or modern science, chemical control of pests is neither good nor bad in and of itself. The costs of the use of chemical means to control pests, including the costs such as altering the ecology of the surroundings, must be constantly weighed against the benefits that are expected from the control of the pests. The careless and indiscriminate use of some pesticides when their use is not warranted has a secondary effect in the development of pesticide resistant pests, resulting in more difficult control when such control is essential. The adoption of either motto that pesticides are all good or pesticides are all bad is a too simplistic view.

Flies

The **House Fly** and the **Lesser House Fly** breed in waste vegetable material, moist manure, lawn cuttings, compost piles, and exposed garbage. The **Vinegar Fly** breeds in fermenting and decaying fruits, vegetables, and their porducts. The first step in the elimination of these pests is the proper disposal of suitable breeding material. Lawn clippings should be left where cut or spread in a thin layer to dry. Garbage should be kept in tightly closed garbage cans. Accumulations of poultry manure should be treated with a five percent solution of suspension of DDT or malathion as directed by the manufacturer on the label. Barn walls and porch ceilings where flies rest may be sprayed with a two-and-one-half to five percent spray of methoxychlor applied at a rate of one to two gallons of spray per one thousand square feet of surface. Alternatively, one of the following may be applied around areas that are frequented by the flies: a two percent spray (or four to five percent dust) composed of premium grade malathion, OR a one percent naled spray, OR a one to two percent ronnel spray. These sprays should be kept away from children and they should not be allowed to contaminate water, food, or dishes and other food utensils. The following are available as aerosol sprays and may be (judiciously) sprayed into the air: 0.5 percent dichlorvos, two percent malathion, 0.1 to 0.25 percent pyrethrins plus one to two percent piperonyl butoxide, or 0.5 ronnel. Since the house fly is greatly reduced in numbers during the winter by both unfavorable weather and disease, every effort should be made to destroy the few emerging adults in the spring. The descendants from a single pair of flies will amount to many thousands during the season if no effort is made to check their development. The airborne, or aerosol sprays are most effective if they are applied within a closed room, with less effectiveness if sprayed in open, but still air, and only marginal effectiveness applied in moving air.

Mosquitoes

The first step in the control of **Mosquitoes** is the location and elimination of their breeding places. If the destruction of the breeding sites is not practical, the use of larvicides should be considered. The larvae and pupae are usually found in shallow, quiet, and often stagnant water, but are very seldom found in swift running water. Waste irrigation water, tin cans, and watering troughs are frequent sources of temporary breeding places for the Mosquito.

Adult Mosquitoes may be controlled within *dwellings* by a spray that is composed of one of the following and applied as necessary: four to five percent DDT, one-half percent Dichlorvos, two percent malathion, five percent methoxychlor, one percent naled, OR 0.025 percent pyrethrins plus two-tenths of a percent piperonyl butoxide or sulfoxide. Any of the above sprays may be applied either from the emulsion concentrate or disolved in oil. Deposits of methoxychlor retain their effectiveness for days and even weeks.

To control the Mosquito larvae on *non-agricultural lands*, one of the ways shown in Table IX-1 may be applied to standing water that contains the mosquito larvae (the weights indicate the pounds of active ingredient to be applied per acre of treated area).

TABLE IX-1

AGENT	*CONCENTRATION*			
	Emulsifiable Concentrate	*Granules*	*Wettable Powder*	*Oil Solution*
Abate	0.016-0.046	0.05-0.20		
DDT	0.05-0.40		0.05-0.40	0.05-0.40
EPN*			0.0375-0.0750	
Fenthion*	0.05	0.05-0.10	0.05	0.05
Lindane	0.1-0.15		0.1-0.15	0.1-0.15
malathion	0.5			0.5
parathion*	0.063-0.100			0.063-0.100
Paris Green		0.6-0.15		

*To be applied only by a trained operator

The DDT may be used at the rate of 0.5 pounds per acre from either the wettable powder or emulsifiable concentrate form in the breeding areas before the larvae hatch.

The adult mosquitoes may be controlled in the non-agricultural areas by either spraying or fogging with one of the following pesticides: 0.25 to 0.5

pounds of carbaryl (Sevin) wettable powder, OR 0.2 pounds of DDT, OR 0.05 to 0.1 pounds of dichlorvos, OR 0.1 pound of Fenthion*, OR 0.1 to 0.2 pounds of lindane, OR 0.1 to 0.5 pounds malathion, OR 0.1 to 0.25 pounds of naled, OR approximately 0.1 pound of pyrethrins plus 0.08 pounds of piperonyl butoxide or sulfoxide. All concentrations of the active ingredient are given in pounds per acre and the pesticide may be formulated from either the emulsifiable concentrate that is suspended in water or the oil-soluble concentrate that is mixed with a suitable solvent, with the exception of the Sevin which is available as a wettable powder to be dispersed in water.

Mosquitoes in Rangelands and Pastures may be controlled by the following treatment programs:

Larvae: the application of 0.5 pounds of active malathion per acre or the use of 0.063 to 0.1 pounds of active parathion per acre.

Adult: the application of 0.25 to 0.5 pounds of active carbaryl (Sevin), OR 0.1 to 0.5 pounds of active malathion, OR 0.1 to 0.25 pounds of active naled, OR approximately 0.01 pounds of pyrethrins plus 0.08 pounds of piperonyl butoxide active ingredients per acre of land to be treated.

Mosquitoes inhabiting Rice Fields may be controlled in the same manner as those that inhabit Rangelands and Pastures while the mosquito is in the larval stage. The adult mosquito can be controlled through the application of either carbaryl (Sevin), malathion, or naled according to the directions given for the use of these materials in the Rangeland and Pasture section above.

A number of chemicals, when applied to either the surface of the skin or the clothing, act as repellents for mosquitoes and biting flies. These repellents may keep the pest at a distance, discourage the pest from lighting on the surface that is treated with the repellent, or discourage the pest from biting. Although the user desires a repellent that accomplishes the former task, most repellents are only effective in preventing the actural injury and not the annoyance. A common formula for a repellent consists of a mixture of dimethyl phthalate (forty percent), dimethyl carbate (thirty percent), and ethyl hexanediol (thirty percent). These chemicals may be used alone as well as combined in the preceeding formulation. Diethyltoluamide, as a fifty percent solution in ethanol, was found by Gilbert (1957) to be more effective as a repellent than the formulation stated above. Later Gertler, et al. (1962) tested a large number of N-alkyl toluamides and found that the *ortho, meta,* and *para* isomers of N,N-diethyltoluamide were the most effective of the group as mosquito repellents, "the *meta* and *para* analogs persisting for thirty-nine and seventy-four days".

Sorptive dusts, such as Dri-Die, have been found to be equally effective in the control of susceptible and DDT-resistant mosquitoes (Micks, 1960). Such dusts

*To be applied by a trained operator.

are inert, achieving their pesticidal activity by removing the protective outer layer of wax from the insect, thus hastening the death of the pest by desiccation. These dusts consist of a combination of sodium silicate and sulfuric acid that forms a hydrogel that may be dried and ground into a fine powder. The addition of water-soluble flourides increases the effectiveness of the mixture, especially when the relative humidity is high (Ebeling and Wagner, 1959).

Control of Ants

The **Argentine Ant** is the prevailing form of ant in the coastal United States, with the exception of the more northern states. This ant is prolific and usually displaces the native species. Unless it is checked, the Argentine Ant becomes a serious pest. This ant nests in the soil with one or more egg-laying queens and frequently with a number of supplementary colonies. During the winter, this ant frequently nests in a hidden nook in a dwelling. Almost every kind of food is eaten by this pest, but preference is shown for sweets and meats. In the garden, their trails lead to plants infested with Aphids, Mealybugs, and Scale insects.

Satisfactory chemical agents for the control of this pest include either a two-and-one-half percent solution or three percent dust of chlordane, OR a one-half percent solution of dieldrin, OR the combination of Strobane plus pyrethrins, as directed on the label supplied by the manufacturer. Any one of these chemicals may be used sparingly in the house as a repellent wherever the ants find entrance to the dwelling. The solutions may be applied with a small paint brush along the baseboards, around windows and plumbing, and wherever the trail of the ant is visible. The best control of the ant is obtained if, at the same time the repellent is used within the dwelling, a bait, such as a fraction of one percent of arsenic, is placed outside the dwelling along the trails. To control various native species of ants, higher concentrations of arsenic may be used in the bait. In addition, a five percent concentration of chlordane in a spray or dust may be used around the nests of the ants and their principal feeding areas in the garden.

Control of Roaches (Cockroaches)

Roaches are flattened, active insects that are usually nocturnal and range in color from light brown to almost black. Because roaches contaminate foodstuffs and impart a sickening odor to their environment, they are serious house pests. Since these pests hide in very small cracks and are not active during the day, their presence may not be noticed until they have become abundant. This small size and consequently difficult determination of their presence is especially true of the common **German Cockroach**, that is only about one-half of an inch long compared with the other species that range from one to one-and-one-half inches in length.

Chlordane and malathion are frequently used as agents to control roaches, but many roaches have developed a resistance to both insecticides. Roach powders that contain sodium fluroide or pyrethrins are of value, but must be used very carefully since the former material is very toxic. All contact of these chemicals with foods must be avoided during the application of such powders. Presently the most effective control agents are the sorptive dusts. These dusts may be used freely along the baseboards, moldings, and under and around furniture and fuse boxes (as described in the section dealing with Mosquitoes). Where kitchen utensils are stored, the dust coating should be covered with a sheet of paper to avoid contamination of the utensils. The coating of dust should be left for several days while the level of the infestation is reduced. In heavily infested buildings, the dust should be applied at the rate of three-quarters to one pound per one thousand square feet of floor and shelf areas. Before applying the dust, all kitchen equipment, drawers, pans, dishes, and all foods must be removed from the area. This precaution permits more thorough coverage and avoids the possibility of contamination. Needless to say, applications of these materials in areas in which pets or children might come in contact with the bait must be well supervised at the very least. It is important to remove all traces of foods, even small particles, and any litter, such as beneath the lining paper of the drawers to maintain control of the roaches by the elimination of their source of food. (Meichelbacher, et al., 1961)

Control of Grain and Food Beetles

A number of small beetles attack grain and its products, dried fruits, nuts, and seeds. A very common pest is the **Confused Flour Beetle**, a one-seventh of an inch long, reddish brown beetle. The slender, flat **Saw-Toothed Grain Beetle** is about one-eighth of an inch long, having six saw-toothed projection on each side just back of the head. The **Drugstore Beetle** is somewhat cylindrical, about one-tenth of an inch long, and brown.

To control these pests, destroy the infested portions of the food and place the remainder in a container that is as nearly airtight as possible, such as a glass jar with a tight cover or a new garbage can. The enclosed foodstuff can then be fumigated or exposed to an extreme of either heat or cold.

Since the insects become dormant below temperatures of 40° to 50°F, stocks of food and fur that are maintained at such temperatures are effectively protected against damage from these insects. The exposure of the food at 40°F for three days will kill many insects, while two to three weeks in a deep freeze unit will kill the more resistant forms of the insect. When large amounts of goods are chilled, allowance must be made for the time that is necessary for the cold to penetrate the bulk of the product so that all of the insects are fully exposed to the cold.

Exposure to temperatures of 125° to 130°F for two or more hours will kill all stages of these insects in stored products. At temperatures of 160° to 180°F, shorter times of exposure are equally effective. Finely ground meals and flour require either higher temperatures or longer exposures to allow for the penetration of the heat to the bulk of the product. Small quantities of product may be heated in the oven if care is taken to avoid scorching the product. The material to be treated should be spread in shallow pans and left for two to three hours at 125° to 130°F. Feed grains and dried fruit may be sterilized by immersion in boiling water for one to two minutes, allowing time for the heat to penetrate the food completely if large quantities are being treated. Seed grain should not be exposed to such high temperatures since the germination of the seed will be affected. The treated material should be spread out to dry following immersion in the water bath.

To control these pests through the use of chemicals, a pyrethrum spray that contains an activator, such as piperonyl butoxide or sulfoxide, can be applied to the empty storage bins and drawers. This spray will kill the insects that are present at the time of application, but will not leave a residue that is toxic. This application can be made with either a hand sprayer or a household aerosol. Sprays or dusts that contain chlordane, dieldrin, or DDT must not be used since these leave residues that are toxic.

Sorptive dusts, such as were described in the section on Mosquitoes, may also be used on the floors and sides of storage containers. These dusts are left for twenty-four hours before they are removed. A sheet of paper should be placed on the bottom of drawers in which food is stored for the minimum protection of the product.

Fumigation is necessary only if large amounts of material require treatment. A fumigant that is used commonly in the treatment of grain is a mixture of three parts by volume of ethylene dichloride per one part by volume of carbon tetrachloride. This mixture of liquids is non-flammable and sold under a number of trade names. Carbon tetrachloride is used alone, but is somewhat less effective than the previous combination. The container that is used for fumigation should be filled with the material to be treated and a shallow pan should be placed on the top of the material. The fumigant should be pored into the pan, the lid closed and sealed, such as with tape. A dosage of about one-quarter ounce per cubic foot, or one-quarter cup for a medium sized garbage can is commonly used. All of the material that is treated should be removed and aired thoroughly in shallow piles for at least twenty-four hours. Fumigation is most effective at temperatures above 70°F. As with any chemical, fumigants and all insecticides should be stored in a cool place that is out of the reach of the most active children. (Michelbacher, et al., 1961).

Control of Clothes Moths

The two species of **Clothes Moths** are so similar in appearance and feeding habits that they are discussed together. Although the larvae alone are responsible for the destruction, the adult moth that flutters about gives the warning of the presence of this pest. The light brown moths are about one-quarter of an inch long when their wings are folded. These moths hide during the day in dark corners and behind furniture, emerging at twilight to fly about. The larvae partially conceal themselves by the web which they spin; this distinguishes them from the larvae of the Carpet Beetles.

Woolen clothing, especially when spotted with food or grease and seldom worn, is most frequently infested by this pest. The natural food of the Clothes Moth is fur, hair, wool, and bits of dried animal matter that are often found in the cracks of the floors.

Cold storage of valuable furs, as described under Grain and Food Beetles, is the surest protection against this pest during the summer months. Clothing that is seldom worn should be taken out of the closet and hung in bright sunshine occasionally. The clothing should be brushed thoroughly, especially along the seams and folds. Dry cleaning will destroy all stages of the insects, but immediate reinfestation is possible. Woolens may be stored safely in trunks or tightly closed boxes if crystals of para-dichlorobenzene or naphthalene are scattered through the folds, using about one-half pound of crystals to a large trunk. For a tight closet, the applicaiton of the above crystals to the floor and shelving at the rate of one pound to each twenty cubic feet of space is an effective means of control. The closet should be closed as much as possible for at least one week for the most effective action.

Pyrethrum sprays that contain activators, if applied very thoroughly as a mist spray, are effective in the control of the adult moth, but may not reach all of the hidden larvae. Following the application of such a spray, the closet should be closed for at least two hours for maximum control.

Chemically treated garments are available from time to time in some markets. The insecticides that are most commonly used in their treatment are sodium fluosilicate or DDT. Such treated garments are protected from infestation for several months or until they are dry cleaned. Clothing may be treated in the home by the application of a five percent solution of DDT or a saturated solution of sodium fluosilicate.

Rugs and carpets should be treated with a long-lasting chemical, such as DDT. Wall-to-wall carpets should be turned back at each edge and either sprayed or dusted with DDT. The insecticide must be forced under the base board as well as being applied to the carpet. Cloth based rugs and carpets should be dusted on the under side with DDT that will be drawn up into the fiber by the normal action of a vacuum cleaner.

Control of Carpet Beetles

Carpet Beetles include two different forms. The mottled black and white form is about one-eighth of an inch long while the black variety of Carpet Beetle is about three-sixteenths of an inch long. Both adults and larvae move freely and go from room to room in the house. Adults are attracted by light and may be found in the vicinity of the windows. No webbing is spun by the larvae, but their presence may be detected by the molted, quite hairy larval skins. The larvae feed in sheltered locations along the edges of carpets, behind heavy furniture, and in the cracks in the floor. The larvae attack nylon and dacron fabrics in addition to woolen fiber.

To control these pests, remove all lint from the furnace vents, about heavy furniture and the cracks in the floors. Out-of-doors, old bird nests and other dried animal matter should be destroyed. Woolen clothing, drapes, and other articles that may be infested should be brushed and hung out-of-doors on sunny days.

Pyrethrum sprays, (see Clothes Moths) applied to areas in which the insects are present, will kill those insects that are wetted, but the spray has no lasting effect. Alternatively, five percent DDT sprays or dusts can be applied in the same manner as for the control of the Clothes Moth. Finally, malathion, used at the dosage directed by the manufacturer on the label, may be used in the place of either aforementioned sprays.

Control of Bedbugs

The oval, flattened, **Common Bedbug** is brown, reddish after feeding, and about three-sixteenths of an inch long. This pest overcomes its wingless state by running rapidly. Its presence may be detected by a careful examination of the bedstead, mattress, and springs for the dark stains of excrement and the characteristic odor of this pest. The younger stages in the development of this pest have the same general appearance as the adult.

For temporary control of this pest, the various hiding places of the bedbug around the bed can be brushed with kerosene. Better results will occur from the application of two-tenths of a percent pyrethrins plus one percent of an activator, such as piperonyl butoxide or sulfoxide, in a kerosene base. For heavy infestations, this spray should be applied to upholstered furniture as well as cracks in the floor and walls. The spray can be repeated as necessary. Light applications of coarse sprays of five percent DDT dissolved in kerosene can be applied to beds and will yield lasting protection; a five percent DDT dust may be used as a substitute for the spray. Alternatively, one of the following may be sprayed around the areas that are infested with bedbugs: one tenth percent lindane, one percent premium grade malathion, one percent ronnel or one-half percent dichlorvos. After treating the springs, bedframes, and tufts and seams of the mattress, all areas should be allowed to dry for at least four hours before they are used.

Sprays that contain kerosene are flammable and should not be used around open flames.

The sorptive dusts that were described in the section dealing with the control of mosquitoes are also used for controlling infestations of bedbugs.

Control of Fleas

The fleas that attack dogs and cats and those forms that attack man directly are annoying pests in their feeding, but should be viewed with concern since some forms of fleas are carriers of plague, typhus, and other diseases.

Ten percent DDT dust may be applied to mature dogs for the control of their fleas, but should not be applied to either cats or puppies. One to two teaspoons of dust per animal is usually sufficient. Sorptive dusts (See Mosquitoes) may be used on animals at a dosage of one-third to two-thirds of an ounce per animal. The dust should be thoroughly worked into the fur or hair of the animal.

Flea powders usually contain one to two percent pyrethrins plus the activators, and/or one to two percent rotenone and either thiocyanate compounds or malathion. Since they may paralyze but not kill all of the insects, the survivors should be combed out of the coat of the animal and destroyed. The directions on the package of commercial products should be followed to ensure maximum safety and effectiveness.

If the sleeping quarters of the pet should become infested with fleas, the floor and bed can be dusted with either a sorptive dust or flake napthalene. The treatment should be left for twenty-four hours before sweeping up the material. If flake naphthalene is used, the room should be closed tightly for the entire time. If the human dwelling becomes contaminated by fleas, the rugs, floors, baseboards, and the bottom twelve inches of the wall can be treated with either of the following: a two percent spray or four to five percent malathion dust, OR a five percent spray or ten percent methoxychlor dust, OR a one percent ronnel spray.

Control of Lice

The presence of Lice on the body and clothing is not only a source of irritation, but also carries the threat of disease. Personal cleanliness and cleanliness of the clothing is the first step in the prevention of an infestation of lice.

The **Head Louse**, found in the hair of the head, is a flat insect, about one-eighth of an inch long and bluish-gray in color. To control the Head Louse, one teaspoonful of a ten percent DDT powder can be rubbed into the hair. A second treatment may be necessary in ten days to destroy the lice that hatch during this time. If resistance to DDT has developed or the use of DDT is to be avoided, the application of a three percent rotenone dust or wash can be used. The material should be washed off the head on the morning following the application. Since the rotenone is an irritant to the eyes, care should be taken to avoid getting the rotenone in the eyes. The wash is prepared by combing the rotenone powder with soap in one pint of warm water. The application of this insecticide should be repeated in about eight days.

The control of **Body Lice** requires treatment of both the person and the clothing. Ten percent DDT powder can be applied to both the hair and very thoroughly to the clothes, especially the seams. Before the treatment, but not for twenty-four hours afterward, the body should be thoroughly bathed. An alternative method of control consists of a light application of three percent rotenone powder over the body at intervals for three weeks. Infested bedding and clothing can be dusted with a ten percent DDT powder. Leather, felt, and rubber goods should be heated for at least twenty minutes at 140°F. Cotton goods can be soaked in water at 150°F for five or more minutes. When a large amount of goods are treated, time should be allowed for the heat to penetrate the bulk of the material so that all insects receive the recommended treatment.

Pubic (Crab) Lice should be treated with a ten percent DDT powder as described above, with special care being given to the armpits and pubic regions. (Michelbacher, et al., 1961)

Cole and Clare (1962) found, in a laboratory investigation that Sevin was effective for several days in the control of Body Lice, and was especially effective when activated with either peperonyl butoxide or sulfoxide. The former combination was most effective against the adults and the latter combination was particularly effective against the eggs.

Control of Wasps

A great number of species of Wasps are represented in the United States, but only one, the **Yellow Jacket**, is discussed. The following remarks also apply to the life and control of the Hornet. The fertilized queen of the Yellow Jackets is the only member of the colony who survives during the winter. In the spring, the queen selects the site for the nest and begins constructing the paper-like substance that is used in making the cells for the new brood. As the workers emerge from the first cells, they assume the work of enlarging the nest and supplying the food. The larvae are fed captured flies and small caterpillars. The variegated pattern of black and yellow characterizes the approximately one-half inch long Yellow Jacket. The black and white Hornet is slightly larger than the Yellow Jacket.

The application of a ten percent DDT dust, five percent chlordane dust, or either dieldrin or lindane dusts that are applied as directed by the manufacturer will control both the Yellow Jacket and the Hornet. Sprays that are formed by the suspension or solution of the above ingredients at about one half the dust-concentration may also be used. Since the solutions may be flammable, they should be kept away from open flames.

The control of these insects should be accomplished at night when most of the insects are in the nest to reduce the danger of getting stung and to treat the largest number of the pests. A garden duster with extension tubes is a convenient means of applying the dusts to as many of the entrances as can be reached. Underground nests can be treated by pouring a few ounces of carbon tetrachloride into the entrance before closing it with a small piece of either sod or cotton. Carbon disulfide may be substituted for the carbon tetrachloride, but since this chemical is both flammable and explosive when it is mixed with air, it must be handled with care. The nests may also be treated directly with a spray that contains one-half percent of either heptachlor or dichlorvos.

Control of Termites

Both the Eastern and Western forms of **Subterranean Termites** are serious pests, especially of pine and fir lumber. The heartwood of redwood and certain cedars is quite resistant to the attack of these insects, but the sapwood has considerably less resistance. Under favorable conditions of soil, moisture, light, and temperature, colonies of termites reaching immense size are developed. These colonies include the blind, wingless, dirty white workers which resemble ants, the large-headed soldiers, the egg-laying queen, and, at certain times of the year, the reproductive forms of the termite. The latter are winged and, under Western conditions, emerge following the first heavy rain in the fall to disperse, looking for a favorable place to establish another colony. Under eastern conditions, the mating flight is more apt to occur in the spring or early summer.

The Subterranean Termite lives in dark, moist environments and, when these are not present naturally, artificial means are used to secure them. Tightly sealed feeding tubes are built over and around stones and other exposed areas, such as concrete foundations, to bridge the distance between the soil home and the adjoining timber foodstuffs. The relative humidity in the tubes and the nest is maintained at above ninety percent. The first requisite for the control of these termites begins during the construction when all timbers are isolated from contact with the soil by the use of concrete or brickwork. Coarse sand of ten to sixteen mesh has been found to provide a satisfactory barrier in fills beneath proches (de Ong, 1960). Concrete slab-on-ground construction is quite susceptible to attack by these pests and treatment of such infestations is very difficult. Before laying such a slab, low places should be filled with coarse gravel. Plans for the construction should include an ample number of ventilators in the basement or crawl space to furnish a certain amount of light and insure the movement of currents of air.

Following the construction of the building, the next important step in the control of termites is the chemical treatment of the soil. The following insecticides, when properly applied, have been found to give protection for a number of years: Aldrin (one-half percent in fuel oil or as an oil-in-water emulsion), one percent chlordane, eight-tenths of a percent of the gamma isomer of benzene hexachloride, one percent heptachlor, eight percent DDT, OR ten percent sodium arsenite in water. For buildings without basements, one of the above materials should be applied along the exterior foundation walls, in shallow trenches at the rate of two gallons for each ten linear feet of trench. Some of the solution should be poured in the bottom of the trench and covered with a few inches of soil. More of the solution should be added to the layer of soil, mixed thoroughly, and tamped down. This procedure should be continued until the trench is filled. For buildings with a basement, a deeper trench should be made and twice the dosage that is stated above should be used. A similar trench is made along the interior side of the wall and treated as above. Infestations of termites in "slab-on-the-ground" construction can be treated by drilling holes at one foot intervals in the slab and forcing a few gallons of one of the above solutions under the slab.

Plants can be protected from these insecticides by lining the sides of the trench next to the plants with heavy building paper. Sodium arsenite should not be used along outer walls where plants are growing or desired, (USDA Leaflet 324, 1952).

A comparative test of the toxic action of several insecticides against termites, under laboratory conditions in Florida, was conducted. For a testing period of fifteen years in soil, the chemicals that were rated in toxicity as "medium" for dilutions of one to two hundred included DDT, TDE, methoxychlor, BHC, and pentachlorophenol. Only chlordane and toxaphene showed a still higher range of toxicity at dilutions for one to two thousand. For a testing period of twelve years, Aldrin showed a high degree of toxicity at the dilution of one to fifty thousand, dieldrin at one to one hundred thousand, and heptachlor at one to twenty-five thousand, (Hetrick, 1962).

A desirable precaution to avoid poisoning the soil is the application by brush of one of the above insecticides to all structural timbers placed near the soil as a repellent to attack by the termites. Joints between two timbers should receive special attention. Timbers that are impregnated with creosote or pentachlorophenol are protected against both insect and fungus attack for a period of years.

Dry Wood Termites are similar to the Subterranean species, but differ in both their feeding habits and the location of their attack. Infestations of the Dry Wood Termite are quite common in the southern coastal United States as well as in California, but are spreading to other regions of the United States. The presence of this pest may be indicated by small piles of light brown fecal pellets that fall from timbers that are infested. The mating flight of this pest usually occurs in October. Infestation in the home may occur in the structural timbers, rafters, joists, and window frames. The attack in the open may be on bridge timbers, crossarms of utility poles, and posts. Insecticides are applied through holes that are bored in the infested timbers or the chemicals may be brushed on the timbers. (St. George, et al., 1960)

Control of Borers

The different species of **Powder Post**, or **Lyctus Beetles** vary in color from reddish brown to deep black. These pests are long and slender, ranging from one-tenth to one-eighth of an inch long. Lyctus Beetles lay their eggs in the exposed pores in the sapwood of seasoned oak, ash, hickory, walnut, maple, and other hardwoods. The emergent beetles may reinfest the wood in which they developed as larvae or they may infest other wood. These pests commonly attack hardwood flooring, tool handles, and unfinished lumber, (Staff, Division of Forest Insect Research, 1954).

Unfinished lumber may be protected from the attack of the Lyctus Beetle by its immersion for one-half minute in a five percent DDT emulsion. Since the dip does not penetrate the wood, a second treatment is necessary after the wood is planed. Finished articles, such as flooring and tool handles, should be dipped in a refined kerosene solution of five percent DDT for three minutes. Where the infestation of the hardwood is common, including the flooring, furniture, and picture frames, fumigation by a licensed pest control operator is the most practical method of control. The operator is responsible for the proper closure of the house, evacuation of all residents, and other measures that are necessary for both the safety and effectiveness of the treatment.

The **Old House Borer** is a beetle that causes severe damage in the eastern United States. The larval stage bores through, and feeds on, wood. The mining of the larvae of this pest weakens the structural timbers and other wooden parts of the house. This beetle, although commonly found in older houses, is, to an increasing extent, reported in new construction. The Old House Borer may be introduced into a building by the lumber that is used in repairs. Once an infestation is established, it may continue as long as the food and moisture conditions are favorable. The insect seems to prefer pine and spruce wood, but also attacks hemlock and fir with preference shown for the sapwood, rather than the heartwood. Decayed wood is avoided, (McIntyre, et al., 1961).

The adult, brownish black Old House Borer is one-half to three-fourths of an inch long. The slender antenna is almost as long as the body of the insect. The flesh-colored larvae vary in length from the one-twelfth of an inch long newly hatched stage to the one-and-one-quarter inch long fully mature larvae. The larval stage lasts five to seven years in the northern United States, but a shorter period is needed in the southern areas. The mature beetle emerges in the early summer through oval holes of approximately one-quarter of an inch diameter.

Natural control of the insect may occur in new construction by the lowering of the moisture content of the wood as the house is heated. This is especially true in regions of light rainfall. Mature larvae and pupae are not as easily affected by climatic conditions as are the young larvae. The usual method of treatment is the injection of chemicals under pressure at intervals into boreholes that are made in the infected wood. Severe infestations that include the structural timbers and places that are difficult to treat should be referred to an experienced pest control operator who possesses the proper equipment and knowledge for their control.

The location and degree of infestation can sometimes be determined by hearing the rasping or scratching sound that is made by the larvae as they tunnel through the wood. Following this sound and probing with a sharp tool may locate the tunnel. If the larvae are near the surface of the wood, blistering of the wood may indicate the location of the tunnel. Once the infested area is located, holes should be bored into the wood in the area of the tunnel for the injection of the insecticide. The bored holes should be about one-half inch in diameter, one inch deep, and spaced within a few inches of each other in the infested area. The holes should be filled with the solution of the insecticide and the application repeated until the wood is saturated. The following insecticides have been used with success following the above procedure: one-half percent lindane or five percent DDT, dissolved in deodorized kerosene. When the wood is also decayed, a combination of an insecticide and a fungicide is used, such as five percent pentachlorophenol dissolved in light petroleum oil and kerosene with one-half percent lindane. The above chemicals are poisonous and should be kept away from children. Rubber gloves and apron should be worn during the mixing and application steps of the procedure. Any solution that touches the body or clothes should be thoroughly flushed with running water immediately. (McIntyre, et al., 1961)

The **New House Borer** is occasionally found in new subdivisions on the Pacific coast of the United States in the subflooring. These larvae are similar to the subflooring borers known as the **Horntail**, that is a larvae of the wasp. The frequency of the complaints about these pests are a partial result of the active housing industry and the practice of using third-grade lumber as the subflooring. The third-grade lumber may be cut from weak, dying, or fallen trees, many of which may be infested with borers.

The adult is a narrow, black beetle from three-quarters to one-and-one-quarter of an inch long. The eggs of the beetle are laid on bark, but those of the Horntail have been found on seasoned redwood lumber in the forest areas. The life cycle of the beetle from the egg to the adult insect lasts from one to one-and-one-half years. The larvae bore through the solid wood, feeding on the starch and cellulose of the tree. The tunnel is packed tight behind the tunneling larvae with the particles of the wood. The growth of this pest is dependent on the variety of wood and the amount of moisture that is present. The drying of the subflooring of the heated house may kill larvae that are less than six months old, but older stages of the pest are generally unaffected. The mature larvae cut their way to the surface of the wood, preparatory to pupating. The larvae then close the entrance with particles of wood and return into the wood to rest before the adult insect emerges.

The damage caused by the New House Borer is principally the emergence holes that are cut in the upper flooring. If these holes are abundant, replacement of the damaged flooring may be desirable. If only a few emergence holes are found, they may be filled with wax and recoated to achieve a satisfactory surface. Since the structural timbers of the dwelling are rarely attacked, generally little concern is expressed for the weakening of the frame of the house.

"Heat is an effective method for controlling roundheaded borers in lumber and kiln-drying by the usual commercial schedules will kill the insects. For one and two-inch Douglasfir lumber, either green or air-dry, a heat sterilizing treatment in the dry kiln at a temperature of 140°F and a relative humidity of 75 to 80 percent for a period of 6 to 8 hours is recommended." (Eaton and Lyon, 1955)

Since the control of the borers in the house is difficult, emphasis should be placed on the careful culling of the lumber that shows signs of recent insect infestation or decay. The tunnels of wood-boring insects are plainly visible where the saw has cut through the log.

Infested subflooring may be soaked for twelve, and preferrably forty-eight, hours in a solution that is prepared by dissolving five percent pentachlorophenol is a low grade of petroleum oil. Heavy oils must be diluted with twenty percent kerosene to increase the penetration of the solution. This solution acts both as a control for both types of borers, and a control for wood decay (Dry Rot). The toxic action of the solution is increased by the addition of one-half percent of the insecticide lindane. The exposure of newly laid subflooring to the sun in clear weather aids in drying the lumber and killing the younger stage of the borer. Parasitic wasps also attack and kill the borers in their tunnels. Adult beetles that emerge in the basement and find no unseasoned lumber in which to deposit their eggs will die within a few days, effecting their own control.

Control of Rodents

The three kinds of rats, and one form of mouse have been introduced from Europe to the United States, and are now occasional visitors to many homes. These pests include the **Norway Rat**, the **Roof Rat**, the **Black Rat**, and the **House Mouse.** The scaley tails of these pests and their coarse hair serve to distinguish these rodents from the **Native Woodrat.**

The Norway Rat is the largest one of the group of adults, measuring about sixteen inches for its total length. The brown body fur is colored by scattered black hairs. The under parts of this pest are pale gray. The Norway Rat has frequently displaced the Black Rat.

The Roof, or Gray Rat has a total length of about fifteen inches. The slender nose of this rat is sharp and the sides are gray or gray-brown while the belly is white.

The Black Rat is almost solidly black, but otherwise resemble the Roof Rat both in size and structure. Most commonly these rats are found along the waterfronts.

The House Mouse has a total length of about six inches. The upper surface of the body is brown and the mouse has pale under parts, (Storer, 1952).

The most important measures for the control of these rodents are sanitation and the exclusion of these pests. Supplementary measures include both trapping and the use of poisoned baits. Floors and basement walls should be made of concrete with ventilator and other openings closed with metal linings. Tight-fitting doors and window screens are also useful and unscreened doors and windows should be closed at night.

Rubbish piles should not be allowed to accumulate, especially near buildings as they may be used by the rats for nesting. All garbage should be stored in metal cans that have tight fitting lids that are always in place. No foods of any kind should be left exposed to attack by the rodents.

Many effective types of traps, both for rats and mice, are available in many markets. Where rodents are abundant, a number of traps should be used at one time, and carefully placed according to the location of the "runs" of the rodents. Rolled oats or other cereals are effective as baits for the traps and sufficient variety and changes in the bait increases its effectiveness. Prebaiting, without the use of the traps, with a variety of foods is effective in learning the best food to be used as the bait.

Poison baits that are intended for use around the home should be restricted to the anticoagulant types as they are the safest to handle and require little experience for effective use. These baits are available under a number of proprietary trade names and the directions that are supplied by the manufacturer should be carefully followed.

On the Pacific coast of the United States, ground squirrels have been found occasionally to carry organisms that are fatal when they infect human beings. For this reason, control measures against the ground squirrels are in common use. Poisoning campaigns are instituted under the supervision of the county agricultural commission, the state Department of Agriculture, and officials of the Federal government. During these campaigns, baits containing sodium fluoacetate, strychnine, and/or thallium are prepared and distributed, (Dana, 1962).

Control of Scorpions

Approximately fifty species of **Scorpions** are known in the United States, and these are generally found in the southwestern parts of the country. Although Scorpions don't sting man as often as their reputation would indicate, the bite of a Scorpion may be likened to that of a wasp and should be treated by a physician. To control these pests in a building that has been abandoned by humans, the contents should be removed before ducks and chickens are introduced, to kill the Scorpions. One of the following chemical methods of control is also recommended: a two percent solution of chlordane in oil, six ounces of forty-four percent chlordane emulsifiable concentrate per gallon of water, OR three ounces of an eighteen-and-one-half percent dieldrin emulsifiable concentrate per gallon of water. One of these sprays may be sprayed into the hiding places of the Scorpion or a spray containing five percent DDT may be applied around the cracks and crevices of the building. As usual, the chemicals should be applied with caution so that food is not contaminated. The label directions should be followed, and the chemicals kept away from children.

BIBLIOGRAPHY

1. Agricultural Handbood No. 331, p. 183.
2. Chemical Week, p. 44, February 11, 1970.
3. Coleman, V.R., Scorpions and their Control, University of Georgia Cooperative Extension Service, Leaflet No. 38, Athens, Ga., (September, 1965).

XI

Pests of Barns and Livestock

The control of flies on livestock and in and around barns continues to be a difficult porblem. These pests include the House fly and the biting Flies, such as the Horn Fly, Stable Fly and Horse Fly, as well as the Face Fly of the northeastern areas of the United States. Because the house fly has developed resistance to many of the chemicals that are used in its control, increased attention is being given to the use of basic sanitation in the control of this pest. The careless disposal of manure about a barn will cause the flies to develop more rapidly than they can be checked by the use of chemical control agents. Trapping, baiting, fly attractants, and the spraying of areas outside of the barns are now being stressed. Many insecticides, when applied to the body of the dairy cow, or as a residual spray that is applied within the barn, are absorbed through the skin and either stored indefinitely in the animal or are excreted in the milk. For this reason, the use of such chemicals in these areas is forbidden by law in the interest of public health. These restrictions limit the application of the chemicals to the bodies of dairy cows, allowing certain mist sprays of those chemicals that are considered harmless to humans, such as the household type of pyrethrins (a derivative of a plant), rotenone (also a plant derivative), Lethane, Thanite, or to very limited treatments with malathion or methoxychlor dusts. A wider choice of materials is allowed for use on and around beef cattle, provided that twenty-eight to sixty days, varying with the compounds used, are allowed to elapse between the last application of the insecticide and the slaughter of the animal.

The successful use of irradiation for the sterilization of great numbers of male Screwworm Flies that were released to mate with female flies was widely practiced in Florida and the southeastern United States. Chemical control agents are now being evaluated for the control of this pest in an attempt to lower the cost of the control program.

METHODS OF APPLICATION OF INSECTICIDES

The use of large, water-tight vats for dipping sheep and even cattle and horses largely has been discarded in favor of spraying and dusting programs. Power sprayers or dusters apply the insecticide to both sides of the animals as they are driven through a chute. In the dairy barn, the hand sprayer is used either for the application of the pyrethrin spray directly to the animal or by the more expensive method of "fogging" in which the closed room is filled with a fine mist of the spray either before or after the cows are admitted to the barn. Applications of these sprays, often containing one of the pyrethrins plus a synergist, are made at least once and sometimes twice, as necessary, (DeFoliart and Eschle, 1961).

PESTS OF LIVESTOCK

The **House Fly**, although not a biting insect, tends to assemble around barns and other buildings, especially in damp, rainy weather. This pest breeds in the moist, fermenting manure and other filth; even in wet feed mash.

The **Stable Fly** is similar in appearance to the House Fly, but has a little heavier body and a prominent proboscis. Much of the time of this pest is spent about cattle and horses, but it also frequents barns. The Stable Fly breeds in moist, grassy waste, lawn clippings, and in moist manure.

The **Horn Fly**, also a biting insect, spends much of its time on cattle, clustering about the base of the horn, the neck, and the flanks. This pest is smaller and more slender than the House Fly, but similar in color. The Horn Fly breeds in fresh cattle manure and is less affected by common sanitation measures than the preceding insects.

The **Horse Fly (Deer Fly)** varies in size depending on the specific species. This pest is the worst biting or bloodsucking fly that infests cattle and horses. Protection of the animals against the attack from this fly is difficult since it is present only when it is feeding. The Horse Fly breeds in swampy ground, hence sanitation about the barn is of no avail in the control of this pest. The larvae of the fly are round, white, and sometimes marked with black as well as being pointed at both ends. The semi-aquatic larvae live in the mud of ditches and marshes, feeding on the larvae of other insects and other small animals.

The control of the larval stage of the House Fly, Stable Fly, and, to a certain extent, the Horn Fly depends on the proper handling of the manure and other grassy waste or the treatment of these wastes with chemicals. Manure may be stored in fly-tight containers or hauled from the barns five hundred to six hundred feet to be either composted or allowed to dry. A moist medium is necessary for the development of the larvae of the fly. Control of the larval stage of the house fly has been achieved by the application of either Diazinon or ronnel to the manure. 1.86 pounds of active Diazinon in the form of the wettable powder was mixed with thirty-six pounds of attapulgite (clay) and two pounds of Celite. This mixture was sprinkled on the manure at the rate of one hundred and fifty milligrams per square foot. (Wilson and Gahan, 1957) One-and-one-quarter percent ronnel wettable powder was applied at the rate of 107 milliliters per square foot at semimonthly intervals through the spring and fall, (Knapp, et al., 1958). Although the process has not as yet attained widespread practical use, some attention has been given to the control of House Fly larvae in poultry droppings by feeding the young chickens food that has been treated with an insecticide.

The application of insecticides to and around livestock, especially dairy cows, is regulated by the U.S. Department of Agriculture (USDA) and the Federal Food and Drug Administration (FDA) in the United States. The following recommendations on the selection of insecticides and their use are based on Handbook No. 331 (1968) and Addendum No. 1 for Agricultural Handbook No. 331 (June, 1969), that are prepared by the staff of the Agricultural Research Service of the United States Department of Agriculture. These recommendations are supplimented by contributions from other researchers and manufacturers of the various control agents.

Dairy Cattle

Cattle Grubs may be controlled by the application of five percent rotenone dust dispersed in water at the rate of twelve ounces of dust per gallon of water. One pint of the dispersion is applied to the back of the animal, and worked into the hair with either a sponge or a brush. Alternatively, four to five ounces of the one-and-one-half percent rotenone dust may be rubbed into the hair of the back of the animal. The five percent rotenone dust may be applied with a power sprayer to the animals one to three times every thirty days.

The **Face Fly** may be controlled by the application of either the oil solution or the emulsifiable concentrate of the pyrethrins in the form of a spray. Five-hundredths to one-tenth of a percent of active pyrethrins PLUS one-half to one percent of either piperonyl butoxide or piperonyl sulfoxide may daily be applied to the head and neck of the animal by the use of a hand sprayer. Dairy cattle may be treated with coumaphos by means of a self-treatment dust bag that is placed where the cattle will contact it periodically. Coumaphos should not be used in conjunction with oral drenches or other internal medication, such as allethrin, synergized pyrethrins, or phenothiozine or other organic phosphates. Animals that are younger than three months should not be treated with coumaphos. An animal that is treated with ronnel, trichlorfon, or Ruelene should not be treated with coumaphos.

The **Horn Fly** may be controlled by the application of one of the following: two ounces of four percent malathion dust, five hundredths of one percent of pyrethrins emulsifiable concentrate PLUS one-half percent piperonyl butoxide or sulfoxide, or coumaphos. The coumaphos is applied as described for treatment of the Face Fly. The malathion dust should be sprinkled on the back, neck, and flanks every three weeks, at least five hours before milking the animal. The pyrethrin emulsifiable concentrate is applied with water in the form of a spray of one to two quarts that is applied every three to seven days.

The **Horse Fly**, **Stable Fly**, and **Mosquitoes** may be controlled by the application of a spray that is prepared from five-hundredths to one-tenth percent of the pyrethrins PLUS one-half to one percent piperonyl butoxide or sulfoxide. One to two quarts of the spray may be applied per animal at two to three day intervals, the equivalent of one to two ounces per animal per day.

Lice are controlled by the application of 0.025 percent pyrethrins emulsifiable concentrate plus one-quarter of one percent piperonyl butoxide (sulfoxide), OR five percent rotenone wettable powder (one to two pounds per one hundred gallons of water) or one-half to one percent dust. The application of these treatments may be repeated every two to three weeks.

Screw-Worms are controlled by the application of either thirty-five percent diphenylamine (Smear 62) or three percent lindane (EQ 335). The insecticide is brushed or smeared on and around the wound, twice the first week and weekly until the wound is healed.

The control of **Ticks** may be effected by the same techniques that were used to control the Horse Fly with pyrethrins. Alternatively, two to four ounces of the five percent rotenone wettable powder may be sprayed on the animal as necessary.

Beef Cattle

The **Cattle Grub** may be controlled by the application of 0.26 percent ronnel (Trolene) in the feed at the rate of three-tenths of a pound per one hundred pounds of body weight. This agent is applied daily for fourteen days at the end of the adult Heel Fly season. This compound should not be fed to sick animals. Alternatively, one-and-one-tenth milligrams of Famphur per pound of body weight may be fed daily for thirty days, OR two and three-tenths milligrams per pound of body weight may be fed for ten days. Famphur must not be used within three days of treatment, either before or after, with any of the following: carbaryl (Sevin), malathion, dioxathion, Ruelene, ronnel, or thrichlorfon. Finally, coumaphos (Co-Ral) wettable powder, either at one-quarter of one percent as a dip or 0.375 to one-half of a percent as a spray, depending on the size of the body and the amount of hair on the animal, may be used to control the Cattle Grub. The entire body of the animal to be treated should be wet to the skin by this material at the end of the Heel or Bot Fly season. The material must not be used on sick animals or those that are less than three months old. Animals that are from three to six months old should be sprayed lightly, and the use of the pyrethrins, or allethrin should not be conducted in conjunction with this material. A final means of treatment is the use of 0.375 percent Ruelene in the form of the emulsifiable concentrate or the wettable powder. This spray should not be used on lactating cows or dry cows within twenty-eight days of freshening or at the same time as treatments with ronnel, coumaphos, or trichlorfon.

Sixty days must elapse between the last application of ronnel and the slaughter of the animal, while twenty-eight days are required in the case of the Ruelene.

The **Horn Fly** may be controlled by the application of either 0.15 percent Delnav (Dioxathion) emulsifiable concentrate or one-half percent methoxychlor emulsifiable concentrate or wettable powder. One to two quarts of the Delnav spray or two quarts of the methoxychlor spray may be applied to the backs of the animals every three weeks as necessary. Calves that are less than three months of age must not be dipped in Delnav.

Five percent of DDT or toxaphene in one gallon of oil may be used to treat the "black rubbers" strung on twenty feet of cable. (de Ong, 1960) Thirty days must elapse between the last application of the material and the slaughter of the animal. Finally, the Horn Fly may be controlled by the application of one to two quarts of dispersed 0.06 percent coumaphos (Co-Ral) that is applied as a spray to the backs of the animals every three weeks or as necessary.

The control of the **Lice** on the animal may be effected through the use of either of three agents. The wettable coumaphos (Co-Ral) powder (0.06 percent) may be applied as a spray or a dip to healthy animals and those over three months of age. Three to six month old animals should be treated lightly, while the treatment of older animals is dependent on their body weight and amount of hair. Co-Ral must not be used in conjunction with either the pyrethrins or allethrin. A second agent for the control of the Lice is the application of 0.03 percent lindane emulsifiable concentrate or wettable powder in the form of either a dip or a spray. Calves that are less than three months of age should not be dipped in lindane, and thirty days must elapse between the last application of this chemical and the slaughter of the animal. Finally, one-half percent of methoxychlor may be used as a spray to treat the animals. This spray may be formed from either the wettable powder or the methoxychlor containing emulsifiable concentrate.

The **Screw Worm** may be controlled by the application of a five percent coumaphos (Co-Ral) dust on the wounds and adjacent areas. The coumaphos may also be applied from the wettable powder to form a one-quarter percent spray or a 0.125 percent dip. Alternatively, one-half percent ronnel (Korlan) emulsifiable concentrate may be sprayed to control the Screw Worm. If ronnel is used, fifty-six days must elapse between the last application of the chemical and the slaughter of the animal.

Ticks are controlled by the application of Co-Ral wettable powder to form a 0.125 percent dip or spray. The Co-Ral dip must not be repeated within a forty-five day period. Sick animals and those that are less than three months of age must not be treated with Co-Ral. Animals from three to six months of age must be sprayed lightly with this compound. Co-Ral must not be applied in conjunction with the use of allethrin or the pyrethrins. Alternatively, dioxathion, lindane, malathion, or toxaphene may be used with excellent results in the control of Ticks. The dioxathion (Delnav) emulsifiable concentrate may be applied as a 0.15 percent dip or spray to the animals. The lindane emulsifiable concentrate or wettable powder may be applied as a 0.03 percent dip or spray, but the dip should not be used for animals of less than three months of age. Lindane must not be used on emaciated animals or those that are lactating.

The one-half percent spray formed from a malathion wettable powder or emulsifiable concentrate may be used on the animals that are at least one month of age. One-half percent dip or spray formed from the toxaphene emulsifiable concentrate or wettable powder may be used for the control of the Ticks.

The following minimum intervals between the last application of the chemical and the slaughter of the animal must be observed: toxaphene–twenty-eight days; lindane dip–sixty days; lindane spray–thirty days.

Sheep and Goats

Fleeceworms may be controlled by the application of a 0.125 percent spray composed of the coumaphos (Co-Ral) wettable powder. This spray must not be applied to sick animals. The dioxathion (Delnav) emulsifiable concentrate may be used for a 0.15 percent dip or spray. Neither the Co-Ral nor the Delnav should be used on milk goats. One part of three percent lindane (EQ 335) per nine parts of water may be used to wet infested areas and three inches around the area, but this treatment must not be used on emaciated or lactating animals. Finally, a two-and-one-half percent ronnel livestock aerosol or a one-half percent emulsifiable concentrate that forms a dip or spray may be used to wet the infested and surrounding areas.

The Delnav, malathion, ronnel and toxaphene must not be used on milk goats. Animals that are three to six months old should be treated lightly and animals of a more tender age than that should generally not be treated. These chemicals should not be used in conjunction with the use of allethrin or the pyrethrins.

Lice on Sheep and Goats may be controlled by the application of one of a number of chemicals, as shown in Table XI-1.

TABLE XI-1

AGENT	*FORM*	*MEANS OF APPLICATION*	*CONCENTRATION*
Co-Ral	WP	Spray, Dip	0.06%
DDT[1,2]	EC, WP	Spray	0.50%
		Dip	0.25%
malathion[2]	EC, WP	Spray, Dip	0.50%
		Dust	5.00%
methoxychlor	EC, WP	Spray, Dip	0.50%
		Dip	
rotenone	WP (5%)	Spray	1-2 lb/100 gallons water
		Dust	0.5-1.0%
toxaphene[2]	EC, WP	Spray	0.50%
		Dip	0.25%

[1]Use only once WP = wettable powder
[2]Do not use on milk goats EC = emulsifiable concentrate

The choice of the application method, whether spray, dip, or dust, and the dosage depend on the size of the animals and the amount of hair. The following minimum number of days must elapse between the last application of the chemical and the slaughter of the animal: Co-Ral–fifteen days; DDT–thirty days; toxaphene–twenty-eight days.

Keds (Sheep Tick) are controlled by the application of one of the agents shown in Table XI-2.

TABLE XI-2

AGENT	*FORM*	*MEANS OF APPLICATION*	*CO CONCENTRATION*
dieldrin[1,2,3]	dust	once after shearing	1.5%
DDT[2,3]	EC, WP	Dip	0.25%
		Spray	0.50%
malathion[2,4]	EC, WP, dust	Dip, Spray	0.50%
pyrethrins[5]		Spray	0.10%
rotenone	WP(5%)	Spray	8 oz./100 gallons water

[1]do not use on sick animals
[2]do not use on milk goats
[3]use only once; other chemical may be repeated as necessary
[4]do not use on animals younger than one month
[5]plus one percent piperonyl butoxide or sulfoxide

With the exception of the dieldrin and the DDT, these agents may be reapplied to the animals as necessary. The minimum days that must elapse between the last application of the dieldrin and the slaughter of the animal is ninety days; thirty days must elapse if DDT is used to control the Keds.

Screw-Worms may be controlled by the application of one of the following chemicals: coumaphos in the form of the wettable powder to form a one-quarter percent spray or dip or the five percent dust, OR three percent lindane OR a two-and-one-half percent ronnel livestock aerosol. The lindane is brushed or smeared on and around the wound, twice during the first week then weekly until the wound is healed. The minimum amount of lindane that is necessary to treat the wound should be used. Lindane should not be used on emaciated or lactating animals. The ronnel should not be applied to sick animals or milk goats. Ronnel must not be applied to animals within twenty-one days of their slaughter.

Ticks are controlled by the application of one of the agents listed in Table XI-3.

TABLE XI-3

AGENT	*FORM*	*MEANS OF APPLICATION*	*CONCENTRATION*
lindane[1]	EC, WP	Dip, Spray	0.025-0.030%
malathion[2,3]	EC, WP	Dip, Spray	0.5%
ronnel[2]	EC, WP	Dip, Spray	0.25-0.50%
toxaphene[2]	EC, WP	Dip, Spray	0.5%

[1]do not use on emaciated or lactating animals
[2]do not use on milk goats
[3]do not use on animals less than one month old

The following minimum number of days must elapse between the last application of the agent and the slaughter of the animal: lindane—thirty days if the spray was used, sixty days if the animals were dipped in the insecticide; ronnel—eighty-four days; toxaphene—twenty-eight days.

Poultry

Lice or **Mites** may be controlled by the application of either coumaphos, malathion, carbaryl, or nicotine sulfate. One-half percent of the coumaphos (Co-Ral) dust may be applied directly to the birds, but must not be used more frequently than weekly and must not be applied within ten days of the vaccination of the birds. A three percent malathion roost paint may be applied at the rate of one pint per one hundred and fifty feet of roost, or the malathion may be applied as a five percent dust or a one percent spray formed from either the emulsifiable concentrate or the wettable powder. One to two gallons of the spray per one thousand square feet or one pound of the dust per fifty to sixty square feet of litter is generally sufficient. Alternatively, one pound of a four to five percent dust may be applied directly to each one hundred birds by dipping only three to four inches of the tail of the bird into the dust. One pound of a five percent carbaryl (Sevin) dust may be applied per one hundred birds for the control of lice and mites, but the carbaryl spray should not be used on the birds. The Sevin dust may be reapplied only once, in four weeks. Seven days must be allowed to pass between the last application of the Sevin and the slaughter of the birds. Finally, one pint of a forty percent nicotine sulfate concentrate may be applied to each one hundred and fifty to two hundred feet of roosts, but must not be applied to the birds or while the birds are in the poultry house.

For control of only the **Lice**, a one percent rotenone dust may be applied at the rate of one pound per one hundred birds. This dust may also be applied to the roost and the nests.

The application of one to two ounces of wettable sulfur powder per gallon of water can be used to control the **Depluming Mite** only. The birds may be thoroughly immersed in this dip or a one hundred percent sulfur dust may be used to control this pest.

The **Fowl Tick** may be controlled by thoroughly covering the ceiling, floors, and walls of the poultry house with one of the following insecticides: one to two gallons of a three percent malathion spray that is formed from either the wettable powder or the emulsifiable concentrate, OR one to two gallons of a two percent carbaryl (Sevin) wettable powder-based spray, OR two gallons of a three-tenths of one percent naled emulsifiable concentrate-based spray. Each of the above sprays are based on the amount of active ingredient and should cover one thousand square feet of area.

Swine

Lice that infest Swine may be controlled by the application of one of the insecticides listed in Table XI-4.

TABLE XI-4

AGENT	*FORM*	*MEANS OF APPLICATION*	*CONCENTRATION*
DDT[1]	EC, WP	Dip, Spray	0.5%
		Dust	10.0%
Delnav[2]	EC	Dip, Spray	0.15%
malathion	EC, WP	Dip, Spray	0.5%
		Dust	5.0%
methoxychlor	EC, WP	Dip, Spray	0.5%
ronnel[3]	EC, WP	Spray	0.25%
toxaphene	EC, WP	Spray	0.5%

[1] Use only once
[2] do not dip animals that are less than three months of age
[3] do not apply to animals that are sick

EC = emulsifiable concentrate
WP = wettable powder

All of the above treatments may be repeated after two to three weeks as necessary with the exception of the DDT which must only be applied once. The following minimum number of days must elapse between the last application of the insecticide and the slaughter of the animal: DDT—thirty days; ronnel—forty-two days; toxaphene—twenty-eight days.

The **Common Mange**, caused by tiny **Mites**, results in scruffy areas around the eyes, neck, and back of the swine. This contagious disease can be combatted by a thorough application of lindane in the form of two quarts of the twenty percent active emulsifiable concentrate or four pounds of the twenty-five percent wettable powder per one hundred gallons of water. Thirty days must elapse between the last application of the insecticide and the slaughter of the animal.

Infested Areas

Note: The following treatments must not be allowed to contaminate animal feed, water, milk, or milking equipment.

House Flies and **Stable Flies** both inside and outside of the barns can be controlled by the application of one-tenth to one-quarter percent of pyrethrins plus one to two percent piperonyl butoxide or sulfoxide. This combination is used as a space spray. One-tenth percent Diazinon as a liquid bait or one percent as a dry bait may be sprinkled or broadcast daily as needed. One to two gallons per one thousand square feet of two-and-one-half to five percent methoxychlor can be prepared as a spray from either the emulsifiable concentrate or the wettable powder. This spray can be applied to the resting places of these insects. The Stable Flies will not be affected by the baits that are described above.

Imported Fire Ants may be controlled by one of the following insecticides. One-quarter percent Aldrin may be formed by dispersing either the wettable powder or the emulsifiable concentrate in water. This liquid can be used to drench the nest of the Ants, or the Aldrin can be used to dust the nest thoroughly. Either the five percent dust of chlordane, the one-and-one-half percent dieldrin dust, OR a two-and-one-half percent heptachlor dust may be used to dust the nest thoroughly. Alternatively, the nest may be drenched with one of the following insecticides: one-quarter percent of either chlordane, dieldrin, or helptachlor. All three insecticides are available as an emulsifiable concentrate and the chlordane and dieldrin are available in the form of the wettable powder.

GENERAL COMMENTS ON THE CONTROL OF INSECTS AROUND LIVESTOCK

Fly Control

The resistance of the House Fly to the effect of a number of insecticides is shown by Hansens (1960). In a review of the subject in the New Jersey area, he stated that "chlorinated hydrocarbons are of virtually no value. Organophosphorus materials are decreasing in usefulness and even space sprays are less effective than (they were) previously. Of all the materials available, diazinon is still giving best results in many locations and kills many flies even where resistance is present. . . . Emphasis on sanitation becomes more important than ever in fly control as resistance problems increase in complexity."

In a study of the fly control problem in dairy barns in Pennsylvania, Cheng, et al. (1961) found dimethoate to be a promising control agent. An application of one-half percent of dimethoate on the barns and pens controlled both the Horn Fly and the House Fly for five weeks; application of one percent dimethoate exercised control for more than nine weeks. Limited results in the control of the Stable Fly in the dairy barns was also noted by the investigators. Analysis of the samples of milk that were collected from lactating cows in the treated barns showed no detectable amounts of dimethoate in any of the samples reported. Further studies concerning this material appear to be warranted.

A three year study was made under varied farm conditions in different parts of Pennsylvania of the effects, if any, on milk production of fly control. The studies included "differences in herd management, degree of fly infestation, efficiency of treatment, spray formulations, and methods of application, and the absorption of insecticides by treated animals." The authors concluded "that among the factors involved in determining the affect of fly control on milk production, herd management with particular reference to the supply of pasture and supplements could play a dominant role," (Cheng and Kesler, 1961).

Chemical attractants as a means of bringing House Flies into baits or traps have demonstrated the superiority of malt extract for this purpose. "The most attractive material consisted of a combination, in aqueous solution, of malt extract five percent, ethyl alcohol one-half percent, skatole (beta-methyl indole) 0.02 percent, and acetal one percent. For field use, three hundred and eighty cubic centimeters of this attractant solution was made into a paste with two hundred and forty grams of fine-ground peat and forty grams of alfalfa meal". In this mixture, the attractiveness of fermenting malt extract was much greater than the non-fermenting malt, (Brown, et al., 1961).

Dimethilan, one percent in a dry sugar bait, controlled resistant house flies at some dairies and poultry farms in Florida, but failed at other farms. (Smith, et al., 1960, Weinberg, et al., 1961). Ribbons of a spongy material that were two meters long and impregnated with a sweet syrup and one gram of Dimethilan per meter gave effective control of flies for several weeks. Cotton fly cords (3/32 inches in diameter) that were impregnated with a twenty-five percent solution of ronnel in xylene also gave good results. The ribbons and cords were suspended variously from the rafters or ceiling as well as on racks nearer the floor.

The sterilization of House Flies by the addition of certain chemical compounds to the food is receiving much attention. Such compounds are non-toxic, but, by causing changes in the cell development of the fly, affect the reproductive capacity of both sexes. The development of a practical chemo-sterilant would enable the inducement of sterility in the population of flies without increasing the number of insects to do so. The effect on flies that feed on chemo-sterilant treated food would be similar to those that have been irradiated. Of the hundreds of compounds that have been evaluated, aphoxide, aphomide, and apholate have caused sterilization in the male and female flies. "When a mixed population of normal males and males treated with aphoxide or apholate were caged with normal females, some of the eggs laid were viable and some nonviable, but with aphomide the eggs were mostly viable . . . Normal females caged first with treated and then with untreated males laid only nonviable eggs, whereas females mated first to normal and then to treated males laid only viable eggs," (LaBrecque, 1961).

The compound Barthrin, combined with the synergist piperonyl butoxide was reported to be an effective agent for the control of the House Fly. A study of a number of other synergists including sulfoxide and sesamex showed that none of the tested group was superior to piperonyl butoxide, (Gersdorff, et al., 1960).

Face Fly Control

The history of the introduction and spread of the Face Fly in the northeastern quarter of the United States was recorded by Sabrosky (1961). The chemical control of the fly has been discussed by a number of authors, but, as yet, the results that are achieved are erratic and the control is temporary. The control of this pest is important since its favorite haunt is the face, near the eyes especially, but it also frequents the brisket and forelegs, providing great annoyance to the animal.

The residual treatment that is achieved by spraying barns and sheds with either diazinon, dimethoate, malathion, or ronnel is generally ineffective. Saturation spraying of the animals with DDT, malathion, ronnel, or toxaphene was also unsatisfactory. Poison baits that were applied to the forehead with a small paint brush gave some protection, with the favorite bait composed of seventy-five percent corn syrup and twenty-five percent water to which DDVP or dimethoate at a one percent level was added, (Bruce, et al., 1960).

Pyrethrins, combined with a synergist plus a repellent such as Crag Fly Repellent or one of the NGK repellents, gave a short period of protection to the animal, but would require daily application. Frequent use of the baits resulted in matted and sticky hair, (Granett and Hansens, 1961).

Applications of pyrethrins at the one-tenth percent level plus one percent piperonyl butoxide OR two percent barthrin plus one percent piperonyl butoxide were made on the head and forequarters of infested animals. Only the latter combination was effective into the second day, (Fales, et al., 1961).

Sprays, dusts, and smears that were applied to the faces of cattle were found to be generally ineffective by Dorsey, et al. However, one-tenth percent

Diazinon and one percent dimethoate "gave fair to good protection for seven to fourteen days". The smears were variable in their effectiveness. Formulations of one to two percent concentrations of Diazinon, Dylox, and Co-Ral plus Dylox gave protection to the animal for about seven days. Applications of Diazinon and Sevin and Co-Ral dusts gave good to fair control for seven to fourteen days. The authors concluded that control measures should be started early in the season and applied on a community-wide basis for maximum effectiveness and control. (Dorsey, et al., 1962)

Horn and Stable Fly Control

As was reported under the section dealing with the control of the House Flies, Cheng, et al., stated that applications of dimethoate at the level of one-half percent controlled the Horn Fly for five weeks while a one percent treatment of the dairy barn and pen yielded protection for nine weeks. Limited results in the control of the Stable Fly were also noted. Following these treatments, no detectable amounts of the dimethoate were found in the samples of milk taken from the treated barns (Cheng, et al., 1961).

Further experiments made with both spray and aerosol formulations that contained R-1207 and dimethoate for the control of flies on cattle were reported by Cheng, et al., (Cheng, et al., 1962). The applications were made with a sprayer or aerosol applicator that was installed either in a doorway of the barn or at a chute in the pasture. These sprayers were controlled by a photoelectric cell. One application per cow per day was made as they passed through the chute, generally in the morning. "Both R-1207 and dimethoate, used either separately or as mixtures, gave excellent control of the Horn Fly." A spray that contained two percent R-1207 was moderately efficient, but one percent dimethoate had little effect on the Stable Fly. A mixture of two percent R-1207 and one percent dimethoate gave effective control of all species of the fly for seven hours and retained a lasting, residual effect against the Horn Fly. "Aerosol treatments containing two percent R-1207 and 0.3 percent synergized pyrethrins gave excellent control of the Horn Fly and the Stable Fly." "No dimethoate residues were found in milk from cows which had been treated with one-half percent or one percent dimethoate sprays, or with an aerosol containing 1.8 percent dimethoate", (Cheng, et al., 1962).

"Compressed-air barn foggers utilizing a fogging solution containing one-tenth percent pyrethrins and one percent piperonyl butoxide gave adequate control of house flies in dairy barns and Horn Flies on cows." The mortality rate was inadequate if the content of the pyrethrins was lowered and repellents, such as R-1207 or R-326, were substituted in the formulations. Fogging was two to four times more costly than the methoxychlor powder treatment of cows for Horn Flies. Fogging did not give protection in the pasture against Stable Flies or Horse

Flies. Daily fogging of barns eighty to ninety feet long with twelve fluid ounces of solution requires 8.4 gallons for a season of ninety days, (DeFoliart and Eschle, 1961).

A comparative test was made between two methods of controlling the Horn Fly on grazing cattle: first, by pouring a concentrated form of the insecticide over the backs of the animals and second, applying a similar amount of the toxicant as a spray to the entire body of the animal. Two materials were used in these experiments, Ruelene and toxaphene. The Ruelene spray was a one-quarter percent mixture of eight-tenths of a gallon or 7.2 grams of actual toxicant per animal. The pour-on mix was twenty milliliters of the emulsifiable concentrate plus eighty milliliters of water per animal or 9.6 grams of toxicant each. The toxaphene spray was a one-half percent mix that was applied until the animal was thoroughly wetted. The pour-on, applied to the back of the animal, was twenty-five milliliters of the emulsifiable concentrate plus seventy-five milliliters of water per animal or eighteen grams of actual toxicant per animal.

The Ruelene test, both spray and pour-on, began July 1. By July 14 it was evident that the Ruelene spray treatment had failed. The pour-on method showed some effectiveness up to July 20. The toxaphene applications were made on August 1. Both treatments caused a reduction in the number of flies through August 11. The spray effects lasted until August 23. The effects of the pour-on application of toxaphene extended at least to August 31. The reports indicated that the pour-on method is superior to the standard spray procedure for the control of the Horn Fly on cattle, (Rogoff and Kohler, 1961).

Screw-Worm Fly Control

Since 1958 an active campaign to eradicate the Screw-worm from Florida and other southeastern states has been conducted by the U.S. Department of Agriculture in cooperation with the Florida Livestock Board. This program consisted in growing immense numbers of male Screw-Worm Flies, sterilizing them by radiation, and releasing the sterilized flies in districts where the fly occurred naturally to mate with the local female flies. As large areas are now free from the infestation of this pest, it is necessary to prevent livestock that are infested from being moved into these areas. Livestock-inspection stations are established along the principal east-west highways. Livestock that enter the eradication area through the insepction station are unloaded into chutes and examined for Screw-Worms. Infested wounds have been treated with a dressing of EQ-335 (Division of Insects Affecting Man and Animals, 1951). As an additional precaution, the animals were sprayed with either Bayer 22408 or Co-Ral (one quarter percent). The former was effective on sheep, but not on cattle (Wrich and Bushland, 1960). Because of the danger to wet animals that are moved by truck in cold weather, a five percent dust of Co-Ral was applied with a hand duster as a practical substitute for spraying, (Writch, 1961).

In addition to the sterilization of the insects by irradiation, the development of chemo-sterilants has proceeded. LaBrecque, et al. (1960) found that amethopterin was the most active of several chemo-sterilants that was investigated. The addition of amethopterin to the food of adult flies caused sterility in the females, but not in the males if a single feeding was used. This compound reduced subsequent oviposition when it was given in later meals. The discovery of the effects of apholate on House Flies led to experiments with Screw-Worms. Complete sexual sterility occurred in several instances. Sterilization resulted by treating the larval, pre-pupal, or adult stage, but pupal treatment resulted in only partial sterility. The adults were sterilized by their feeding on the apholate (Chamberlain, 1962).

Cattle Grub

A comparative test in Iowa was conducted to determine the effects of the use of feed additives and sprays under field conditions in the control of Cattle Grubs. Ronnel was administered orally to steers at three locations and two dosage levels: fifteen milligrams per kilogram per day for seven days, and twenty-five milligrams per kilogram per day for six days. Co-Ral was applied as a spray to steers at two locations either as a one-half percent suspension or one-half percent emulsion. A fourth lot of steers was used as a control group. "There was no difference in grub control, weight gain, feed efficiency, or carcass grade between treatments at each location" during the test period that ran 162 days. "Grub control at the three locations ranged from 76.3 to 98.6 percent", (Raun and Herrick, 1960).

The application to the back of the animal and by intramuscular injection of low volumes of the systemic insecticide Co-Ral were shown to give more effective control of the Cattle Grub than the usual application of one gallon of a one-half percent spray. The applications to the backs of the animals were made with a small paint brush. Oil suspensions gave better coverage than did the water-borne formulations. The volumes that were applied to the backs of the animals ranged from fifty to two hundred and fifty milligrams per kilogram of body weight. All of the treatments, with the exception of fifty milliliters of two percent Co-Ral in oil and eight ounces of the two percent Co-Ral in water, were ninety-seven to one hundred percent effective in the control of the Cattle Grubs. (Drummond and Graham, 1962)

The intramuscular injections were made with formulations of peanut oil and diethyl succinate, the latter being preferred. Both five and ten milligrams per kilogram were dosages that were effective in the control of the Cattle Grub, but the two-and-one-half milligram per kilogram dosage yielded only partial control, (Drummond and Graham, 1962).

A summary of the effectiveness of systematically active insecticides against the Cattle Grub was prepared. Details of the testing procedures and the preliminary results were given by Drummond in 1960. The most effective insecticide treatments against Grubs at dosages that were tolerated by the cattle are shown in Table XI-5.

TABLE XI-5

AGENT	*DOSAGE**	*MEANS OF APPLICATION*
Amer. Cyanamid CL-38023	25	intramuscular
Bayer 37341	25	drench
	0.5%	spray
Bayer 37342	25 & 100	capsule
	25	intramuscular
	0.5%	spray
	2%	pour-on
butonate	25	per day in feed for five days
Ruelene	10	intramuscular
	10	subcutaneous
Stauffer R-1504	50	capsule

*milligrams per kilogram of body weight

Poultry Lice

A comparison was made of egg production between louse-infested and non-infested laying Leghorn pullets. Four pullets and a rooster in each of eleven pens were infested with the Chicken Body Louse and eleven similar pens were kept louse-free as controls. The egg production was recorded for twenty-two weeks after the infestation. The birds were fed a standard sixteen percent mash. No significant difference in egg production was observed between the two groups of hens for the entire period of the experiment. The fertility and hatchability of the chickens were not adversely affected by the infestation of the Lice. The total number of Lice recovered from a single bird ranged as high as 8,646, (Stockdale and Raun, 1960).

Poultry Lice may be controlled by painting the roosts with a mixture of three percent malathion that is prepared by mixing the insecticide with either kerosene, diesel oil, or water, following the directions of the manufacturer. Alternatively, the roosts may be painted with a forty percent solution of nicotine sulfate, but the nicotine sulfate must not be used on other areas of the chickenhouse or on the birds, while the birds are in the house, (Staff, U.S.D.A. Entomology Research Division, 1960).

The species of Lice, commonly found infesting poultry in southern California, have been controlled by the use of dust-bath boxes that are placed in pens with litter on the floor. Alternatively, the natural dusting in wallows in rangeland type pens were treated. "One pound of a four percent malathion dust per wallow or per dust-bath box (three feet long by two feet wide by one foot deep) was used for each one hundred hens." The effectiveness of the treatment lasted from eighty-four to one hundred and eight days, (Rodriguez and Riehl, 1960).

Mites

The **Chicken Mite**, when present in large numbers, is a serious pest and may cause a decrease in the production of eggs. This pest usually leaves the body of its host during the day and hides on the roosts or their frames. Because of this habit, control measures are directed toward the pen and roosts rather than the birds. Dusts are preferable to sprays as they do not cause wet spots in the manure. One thorough application of a four percent malathion dust has been found sufficient to control the mite for at least five months, (Rodriguez and Riehl, 1960).

The **Northern Fowl Mite** was controlled on White Leghorns and Rhode Island Red chickens by the application of either Co-Ral (Bayer 21/199) at one-half percent, OR one percent Dipterex (Bayer L13/59), OR one percent ronnel dust. Three to four grams of the above dusts were applied to each bird. Each treatment yielded satisfactory control for up to twenty-eight days. No detectable residues were reported in the eggs from either the Co-Ral or ronnel treated hens, (Knapp, et al., 1960)

In tests on turkeys that were infested with the Northern Fowl Mite, excellent results were reported from treatment with both ronnel and malathion. Three experiments were made with ten to fifteen bird flocks of infested turkey hens. In one room, the birds, floor, and walls were treated with two gallons of a one-half percent malathion spray, while a second pen was treated with the same amount of ronnel. In a third pen, a four percent malathion dust was applied as a treatment of the litter at the rate of one pound per forty square feet of floor space. All three treatments gave excellent control for two months (Bigley, et al., 1960).

XII

Insects and Diseases of Trees and Forests

The loss to standing timber from the attack of insects and disease is equal to, if not greater than, that caused by fire. "On the 664 million acres of forest land in the United States, enough timber is destroyed each year by insects alone to build well over one million houses. Disease has almost destroyed the American Chestnut and is killing great numbers of the American Elm. White Pine is under severe attack from Blister Rust. Added to these insect and disease pests are the invasions from various fungi and bacteria and the weakening effects of crowding and other adverse conditions that now affect every species of tree. The enemies of the forest, serious though they are, are being identified and methods of combating them that have been, and are being, developed will do a great deal to reduce the heavy losses that were suffered in the past. (Staff, U.S.D.A., Forest Service, 1962)

The application of pesticides in forests involves the possible effects on wildlife, grazing livestock, and recreational practices. Seasonal plans for pest control are examined by the Federal Pest Control Board before the plans are implemented. The use of chemicals in the control of pests of the forest has been reviewed and approved by a committee of the National Academy of Sciences as the only effective method of fighting epidemics of certain pests under specific conditions. Such applications must be made with due consideration and care given to the protection of the wildlife. (Staff, U.S.D.A., Forest Service, 1962)

DISEASES

Diseases of Leaves

The **Leaf Spot** diseases are caused by either bacteria or fungi and often become apparent in July or August. A majority of the Leaf Spots are manifested as rather round, small, brown, dead portions of the tissue of the leaf. Later these spots enlarge and some produce the tiny, black fruiting bodies of their causal fungus. The margins of these spots vary from reddish-brown to purplish. The Leaf Spots may be controlled by the use of a spray that contains one of several fungicides that are available.

Figure XII-1. Blister Rust on white Pine. (Illinois Natural History Survey.)

Figure XII-2. Leaf Spot on Box Elder Leaves. (Illinois Natural History Survey.)

Powdery Mildew, a fungus infection, manifests itself as a white or gray powdery growth on the leaves. This powdery deposit often contains the tiny black fruiting bodies of the fungus. Although many varieties of trees are capable of being infected by Powdery Mildew, this disease rarely causes serious damage to the tree. Powdery Mildew usually is controlled by the application of a fungicide that contains a sulfur compound.

Figure XII-3. Chrysanthemum Powdery Mildew. (Illinois Natural History Survey.)

Diseases of Roots, Trunk, Branches and Twigs

Canker is the name that is given to areas of dead tissue on the tree, if the tissue is a portion of a twig, branch, or trunk. Cankers are elongated and vary in length from an inch to more than a foot. The Canker may encircle a branch and kill the portion of the tree that grew beyond the diseased area. Some Cankers are conspicuous and sunken, while others are barely noticeable. This tissue that is diseased may become cracked and discolored. One of the most effective methods that can be employed to control Cankers is the removal of the diseased tissue.

Gall diseases affect the twigs, branches, trunks, and roots of the tree. These diseases result form the infection of the tree by either bacteria, fungi, or viruses. The agents that cause the infection often enter the tree through wounds and the Galls that result from the infection can seriously stunt or kill an infected tree.

Vascular diseases can result from the infection of the tree by certain bacteria, viruses, or fungi. These invaders develop in the inner bark or the sapwood of the tree, causing the leaves on one or more branches to wilt and die. The tree is often killed if severely infected by one of the vascular diseases. The wilt-producing bacteria and fungi often discolor the young sapwood of an infected tree. Some of the vascular diseases are very serious pests of the trees and forests.

Figure XII-4. A branch of a poplar Tree with Dothichiza canker. (Illinois Natural History Survey.)

Figure XII-5. Chinese Chestnut branches with Branch Canker. (Illinois Natural History Survey.)

Abnormal growths on the roots of the tree is one of the common forms of **Root** disease. Roots that are diseased are unable to provide adequate water and food for the tree to maintain a normal pattern of growth. The death of the terminal areas of the branches often results from diseases of the roots, a condition known as **Terminal Staghead.**

INSECTS – GENERAL

Bark Beetles are the most destructive insects to infest the forests of the United States. These pest are capable of causing great epidemics in which thousands of trees are killed. During a severe attack, a large tree may be girdled and killed by the Beetle within a few weeks. The various, closely related species of the Bark Beetle are widely distributed throughout the western and south-eastern States in the coniferous forests of pine, fir, and spruce.

Figure XII-6. Western Yellow Pine Gall Rust. (Illinois Natural History Survey.)

The reddish-brown or black adult Bark Beetle is four to eight millimeters in length. The life history and feeding habits of the various species that are known as the Bark Beetle are quite similar. The adults appear in the spring from the inner layers of the bark and cut galleries in the surface of the wood as well as in the inner bark. The eggs of this pest are laid in the tunnels that are packed with sawdust. The emergent larvae cut large chambers as they feed in the inner bark. At this point in their development, some species create Pitch Tubes at the entrance holes, but other species leave no distinguishing marks to indicate their presence at this point of their development. A tree that has been killed by these pests possesses characteristically bleached-out foliage, especially at the top of the tree. The trees that are infested are logged as quickly as practical to prevent the release of another generation of beetles to infect healthy trees. If immediate logging is impractical, the tree may be felled and the broods of the beetles can be destroyed either by the application of chemicals or by burning. Rudinsky, et al. (1960) reported that newly felled Douglas Fir logs were effectively protected by the application of either endrin, lindane, Thiodan, Sevin, or heptachlor. The sprays were in the form of solution, suspension, or emulsion. Two hundred milligrams per square foot of timber was the dosage of Thiodan, endrin, and lindane used, while the heptachlor and Sevin were applied at dosages of four hundred milligrams per square foot of timber. During the twenty week testing program that Rudinsky reported, the emulsions and solutions were more effective than the application of the insecticides from the suspensions.

The group of insects known as **Borers** are less injurious to the forests than the Bark Beetles, since the larvae principally attack the sapwood and the heartwood of the tree, although they may feed for a short time within the inner bark. The eggs of this pest are laid on the bark of dead or dying trees or newly felled logs. Healthy trees appear to be attacked only in wounds and are generally not attacked by the Borers. Two principal groups of Beetle Borers are identifiable: The **Round-Headed**, or **Long-Horned Beetle** whose antennae are either as long or longer than its body, and the **Flat-Headed Beetles** who have short antennae and larvae that have large, flattened thoracic segments. The Round-Headed Beetle is the principal enemy of coniferous trees since their winding burrows degrade the quality of the lumber. Exceptions of all rules are legion as evidenced by the **Giant California Prionus** that breeds in stumps and decaying roots and limbs that belong to live, deciduous oaks. This pest may kill orchard and native trees. The adult Flat-Headed Beetles are beautifully colored with a metallic luster that extends their seven to twenty millimeter length. A few species of Flat-Headed Beetles attack the trees of the forest, but these pests are known principally for their invasion of sunburned or otherwise weak orchard and ornamental trees. Some species of this pest require two or more years to mature. The control of this pest in ornamental trees consists of digging the burrowing larvae from the tree or painting the tree with a penetrating oil or chemical that will be tolerated by the tree and not tolerated by the Beetle.

Insect damage to field-piled pulpwood of Balsam Fir and Spruce in Minnesota was found to be caused principally by the Round-Headed Borers and the wasp-like **Horntails**. The closely packed piles of pulpwood that were stacked in the shade of trees were damaged less severely than the loose piles that were placed in sunny clearings, (Wilson, 1962).

The **Ambrosia Beetles** are serious enemies of trees in the forest since they burrow both in the sapwood and the heartwood of unseasoned logs and poles. The winding holes that are caused by these Beetles are small, but surrounded by a dark brown stain. The small, brown Beetles have a cylindrical shape. These pests attack dying or newly felled trees, sawlogs, and green lumber. The Beetles do not feed on the wood, but throw it out of the tunnels in the form of a fine dust. To add to the destructive effect of this pest, the adult carries spores of certain fungi into the tunnels. Given suitable moisture, the spores begin to grow along the walls of the galleries and are used as the foodstuff of both the adults and the larvae. Without the proper moisture level, the fungi will not grow and the beetles starve. Thus, seasoned wood is immune to the attack of these Ambrosia Beetles.

To control infestations of the Ambrosia Beetles, logs that are cut in the summer or the fall should be removed from the timber within two weeks after cutting and either milled or placed in water. Logs that are cut through the late fall until the early spring are less subject to infestation. Freshly cut lumber should be piled in such a manner that it will dry as rapidly as possible.

The large, thick-waisted female **Horntail**, or **Wood Wasp**, is twenty to thirty-five millimeters long and sports a long ovipositor. These metallic blue or black females have smoky wings. The smaller males are decorated, in some species, with yellow-banded stripes on their abdomen. The yellowish white larvae reach a length of twenty-five to thirty-five millimeters at their maturity. The Horntails attack unseasoned or dead wood of almost all of the western coniferous trees and even attack cured Redwood lumber. (Keen, 1939)

Carpenter Moths attack the broad-leaved trees and especially the oak under climates that are similar to that in California. The burrows of this pest wind through the sapwood and the heartwood, degrading the quality of the lumber. The life span of the Carpenter Moth requires three years for its completion, (Keen, 1939).

Defoliating Insects

Injury to the foliage of the tree may affect the tree in several different ways. Defoliation prevents the production of food and thus checks the rate of growth of the tree. Injury to the foliage reduces the evaporation of moisture from the tree and may do so to the point that fermentation may occur in the tree. In either case, defoliation, if severe, may cause the death of the tree. Evergreens are more susceptible to injury from defoliation than are deciduous trees. Pine and spruce trees may be killed in a single year by severe defoliation, while oak and maple trees may tolerate a number of severe attacks.

The populations of defoliating insects vary greatly in number from outbreaks that may cover thousands of acres to the attack on a single tree. These variations in the population of the insects may be due to the attack of parasites, disease, inclement weather, or other factors that are not yet identified. Measures that may be used to control outbreaks of defoliating pests consist chiefly of the application of chemicals in the form of either dust or sprays from aircraft, (Keen, 1939).

The adult **Spruce Budworm** is a small, brown moth that has a wing expanse of approximately one inch. The eggs, laid on the upper branches of the pine and fir, hatch in midsummer. The young larvae wander about for a few days in search of a suitable place to spin their cocoons in which they will hibernate. Emerging in the spring as the buds of the tree open, they feed on the buds until they are hollow. As the new needles grow, these are eaten by the Spruce Budworm. The larvae mature in about four weeks to construct their loose cocoon of silk and dead needles. The moths emerge from the cocoon in about twelve days, (Keen, 1939).

Large areas of the spruce and fir forests in the inter-mountain areas of the United States, from Washington to Minnesota, have been infested during the past ten years by this pest, causing heavy losses. The attacks vary in severity due to the attacks of parasites of the Budworm and other conditions that are unfavorable to the development of the pest. Several million acres of forest were infested by this pest between 1951 and 1958. The common control procedure

Figure XII-7. Spraying for Spruce Budworm in Montana. (United-Heckathorn.)

consisted of the application of one pound of DDT dissolved in a gallon of petroleum solvent per acre of forest. This mixture was applied by aircraft. The persistence of the DDT residue on the foliage of the tree caused this treatment to be rated as the most successful of any insecticide that was tested. Certain differences in the policy of spraying have been developed by the United States and Canada. In the United States, sprays that are applied during the early stages of an infestation were recommended to prevent the mortality of the trees and reduce the danger of reinfestation from buffer zones that were more lightly attacked. In Canada, sprays generally were applied in areas of high infestation and a further year of attack would threaten the life of the trees. This policy was dictated in part by economic limitations and the difficulty of assembling enough planes to treat 6.5 million acres of forest land. Since 1953, the dosage has been reduced to one-half pound of DDT in one-half U.S. gallon of oil solvent per acre in the Canadian spraying program. If this dosage is carefully applied the tree is protected and the lesser concentration of DDT is less hazardous to other forms of life. In the Oregon-Washington areas, the heavier dosage was applied to five million acres and resulted in a remarkable recovery of the defoliated trees, restricting the losses of timber to ten thousand acres. No evidence was noted of serious damage to the natural control of the pest. Studies of the effects of the spray program on the parasites failed to show that any species were eliminated. In some of the Canadian provinces, severe concern is raised on the effects of such spray programs on the fish, and similar fears are being raised in the United States. Fisheries investigations in both eastern and western Canada showed drastic losses of the youngest stages of the salmon, even at the reduced dosage of one half pound of DDT per acre. As a result, reductions of dosages to one-quarter of a pound of DDT in one-half U.S. gallon of oil solvent per acre were begun in the early sixties, (Webb, 1960).

The **Jack-Pine Budworm** is closely related to the Spruce Budworm and is often found in the eastern United States extending into Canada. This pest feeds on a number of the pines, white spruce, and Eastern larch. The most consistant target of its attack is the Red Pine in addition to the Jack-Pine. The Jack-Pine Budworm feeds in the top of the tree and kills the crown, thus stunting the growth of the tree, (Kulman and Hodson, 1961).

The small, yellow **European Pine Shoot Moth** lays its eggs on the needles of the pine. The young caterpillars feed on the tips of the shoots and burrow into the buds and downward into the new growth. This feeding and burrowing deforms the tree, causing forks and stunting the growth of the tree. These pests are found mostly in the north central areas of the United States. This insect attacks principally the Red and Scotch Pine in the forest and in Christmas tree plantations.

Two applications of insecticides may be used to control this pest, one in the spring and the second application in mid-summer. Satisfactory control of the spring brood was obtained with the application by hydraulic machines of either dimethoate, DDT, or methoxychlor at the rate of two pounds of active ingredient per one hundred gallons of water. For the summer applications, two pounds of carbaryl (Sevin) per hundred gallons of water was superior to the application of DDT. Mist blower applications of ten pounds of active DDT ingredient in forty gallons of water per acre yielded ninety percent control of the pest in both the spring and summer treatments, (Butcher and Carlson; Miller and Haynes, 1961).

The **Sawflies** resemble flies, but have two pair of wings. These heavy-bodied pests vary in size from three to nine millimeters long. The head, thorax and the base of the abdomen are similar in width. The greenish-yellow, hairless larvae have six to eight pair of prolegs. A number of species attack the conifers by feeding on the needles. First these insects feed in clusters, then they scatter. When mature, the larvae drop to the ground to pupate.

The usual applications of DDT are sufficient to control outbreaks of Sawflies. In addition to the DDT or substitute chemical sprays, the aerial treatment of the pest with a polyhedral virus was made in Maryland. In the samples that were collected, a mortality of ninety-three percent was found, but no symptoms appeared in the larvae that were collected in the following year, (McIntyre and Dutky, 1961)

The **Gypsy Moth** was accidently introduced to the United States from Europe in 1869. By 1889, forest and residential trees were defoliated over a three hundred and sixty mile area. At the present time this pest has infested a large part of the New England states, eastern New York, northeastern Pennsylvania, and a small area of Michigan. Nearly all of the trees that are native to Pennsylvania are attacked by this pest, particularly the oak, popular, willow, birch, and apple. In addition, many valuable forest trees are subject to the attack of the older larvae. These trees include the beech, red cedar, hemlock, and all species of the pine and spruce. Other, less favored, trees that are attacked include the elm, cherry, and maple and the blueberry, barberry and cranberry bushes. When food is scarce, the Gypsy Moth is not above attacking grass and cultivated crops. In addition to the death of the trees that results from the recurring defoliation caused by the Gypsy Moth, many other trees are weakened and their growth is stunted making them susceptible to damage from the Bark Beetles, Borers, and fungus diseases. The defoliation of the trees increases the danger of erosion, reduces the water storage capacity of the area, and destroys the natural habitat of the wild life. The aesthetic and monetary value of the land is reduced. In addition to these losses, almost one hundred million dollars have been spent, and more must be spent before this pest is either eradicated or permanently held in check.

Native parasites and predators are of value in checking the moth, but none effectively reduces the level of the infestation. A great deal of effort has been expended in importing and colonizing parasites and predators that feed on the Gypsy Moth. Of the forty species that have been imported, nine parasites and two predators have become firmly established. The combined effort of the native and the introduced enemies of the Gypsy Moth are of value in reducing the populations of this pest, but they have not prevented either the spread of the moth or large scale defoliation.

Early methods of chemical control of the Gypsy Moth included the application of five pounds of lead arsenate plus one-and-one-half pints of fish oil. This mixture checked the outbreak of the pest, but did not act as an eradicant. The application of creosote to the clusters of eggs was also used in the early campaigns to control the Gypsy Moth. In 1945 DDT was initially used at dosages of five pounds of active ingredient per acre. This dosage was reduced to one pound per acre within three years and one hundred percent eradication of the moth was achieved with minimal losses to other forms of wildlife. The formulation that was used for many years was standardized around one pound of technical grade DDT, one pint of xylene and sufficient kerosene or No. 2 fuel oil to make one gallon. This gallon is sufficient for the treatment of one acre of land. In recent years, the search for alternative chemicals to replace the DDT has resulted in several promising compounds, including carbaryl (Sevin).

Comparative tests of several insecticides were conducted by Keller, et al., 1962. They reported that Sevin, Thiodan, Diazinon, and Bayer 22048, applied at dosages of one-half, one, and two pounds of active ingredient, respectively, per acre were "highly effective" against the Gypsy Moth, but were less effective than the standard application of DDT.

Stream and pond insects or fish-food organisms suffer the greatest loss from applications of DDT. Some research has indicated that these organisms recover from these losses so rapidly that sufficient food is available to maintain the fish population. Instances of mistakes in judgment or other errors have resulted in fish-kills in the Spruce Budworm campaigns, but even in these cases the recovery to normal fish populations were reported to occur in two to three years. "In all of the sprayed territory in Pennsylvania, consisting of over one million acres, the general insect fauna has suffered no permanent damage. As near as can be determined, the only insect exterminated has been the Gypsy Moth. The effect of DDT applied at the rate of two pounds per acre also was not prolonged and most of the species were present in considerable number two or three weeks later." Losses of other forms of wildlife have been practically non-existent in all of the areas of Pennsylvania that were sprayed with DDT at the rate of one pound of the insecticide per acre in the programs to eradicate the Gypsy Moth, (Nichols, 1962).

Because of the frequent spraying of the woodlands of Connecticut with DDT, the question has arisen concerning the effect of this program on the aquatic insects. Hitchcock (1960) states that only two types of insects disappeared from the streams that were treated. Losses ranged as high as 42.8 percent of the genera in one stream. However, insect drift from untreated portions of the stream would later colonize the main channels of the treated parts.

The value of the use of insecticides as a measure in the conservation of the forests is well established. What remains now to establish is the effect, if any, of the dosages of DDT and malathion that have been used. Hartenstein (1960) found that the usual dosages of DDT and slightly higher dosages of malathion were ineffective in reducing or increasing the beneficial Forest Soil Mites and the tiny insects that are known as Collembola.

The aerial application of DDT-oil solutions for the purpose of the eradication of the Gypsy Moth were found to be very persistent on foliage and little affected by weathering. "However, the growth of new grass, with its great increase in both mass and volume, was found to be a very important factor in residue "loss". It was concluded that DDT, as an eradicant of Gypsy Moth, "should not be used in dairy farm areas and cannot be used without (the resultant) contamination of the milk supply," (Huddleston, 1960)

SPECIFIC TREES

Insects and Diseases of the Ash

Anthracnose, resulting from the infection of the tree by the fungus *Gloeosporium aridum*, appears as large, uneven, brown areas on the leaves of the tree. The application of a spray that contains either a copper fungicide or an organic mercury compound will control this disease.

Weakened ash trees may be severely damaged by the invasion of one of the fungi that cause **Cankers**, specifically *Cytospora* sp., *Phomopsis* sp., and *Physalospora obtusa*. Fertilization of the infected trees accompanied by the removal and burning of the diseased portions of the tree are effective means of controlling this disease.

Figure XII-8. Green Ash with insect injury by Flat Headed Borers. (Illinois Natural History Survey.)

In July and August, the application of a spray that is composed of one-and-one-half pints per one hundred gallons of water of the thirty-and-one-half percent Dimethoate emulsifiable concentrate will kill all of the young **Ash Borers.**

Insects and Diseases of the Beech

Beech trees sometimes suffer from the **Nectria Beech Bark Canker,** a disease that results from the infection of the tree by *Nectria coccinea* var. *faginata*. The Beech Scale is the vector of this disease. Control of the Beech Scale is achieved by the application of the following spray during the dormant season to the trunk and lower limbs of the tree: twelve gallons of Lime Sulfur 33 per one hundred gallons of water.

Both the **Wooly Beech Aphid** and the **Giant Aphid** may be controlled by spraying the limbs and the foliage with one of the following formulations when the Aphids appear: four pounds of twenty-five percent malathion wettable powder or one quart of the fifty-seven percent malathion emulsifiable concentrate per one hundred gallons of water.

Insects and Diseases of the Birch

Birch Anthracnose, *Gloeosporium betularum,* and **Brown Leaf Spot,** *Cylindrosporium betulae,* can be controlled by the application of a spray that contains sixteen pounds of the 8-8-100 Bordeaux mixture per one hundred gallons of water. The spray should be applied to the foliage when the disease becomes noticeable and the treatment may be repeated in ten days if necessary.

Figure XII-9. Birch Twig showing mature fruiting bodies of *Melanconium betulinum.* (Illinois Natural History Survey.)

Cytospora ambiens (Valsa ambiens), Melanconium betulinum, Nectria cinnabarina, Nectria ditissima, Sphaeropsis alnicola, and *Valsa betulina* are several of the fungi that are able to cause **Cankers** on the branches and trunks of the birch tree. Some of these diseases can be controlled by removing the diseased areas and burning them. The **Melanconium Canker** is the most common form of the Canker and often kills branches and entire trees on occasion.

Lime Sulfur 33 that is sprayed on the tree in the spring before the buds have opened will help to control the **Birch Casebearer**. Twelve gallons of lime sulfur per one hundred gallons of water should be used in the spray. In July and August, the application of either four pounds of twenty-five percent malathion wettable powder or one quart of fifty-seven percent malathion emulsifiable concentrate per one hundred gallons of water is also effective in a program to control the Casebearer.

The application of a spray to the foliage about May 15 and again around July 12 (in the northeastern United States) will control the first and second broods, respectively, of the **Birch Leaf Miner.** The spray may be composed of one of the following: two pounds of carbaryl (Sevin) fifty percent wettable powder, OR one pint of thirty-and-one-half percent Dimethoate emulsifiable concentrate per one hundred gallons of water.

The **Bronze Birch Borer** may be controlled by the application to the foliage in May or June of ten pounds of eighty-five percent carbaryl (Sevin) wettable powder per one hundred gallons of water. Additional control procedures include the fertilization of the tree and wrapping the trunk of the tree with a form of tree wrap.

Figure XII-10. A Chinese Chestnut Tree Infected with Chestnut Blight. (Illinois Natural History Survey.)

Diseases of the Chestnut

The American Chestnut is extremely susceptible to the **Chestnut Blight** that is caused by *Endothia parasitica*. The major means of control of this disease is the use of blight-resistant hybrids or planting the Chinese Chestnut.

Figure XII-11. Branch of an American Chestnut Tree exhibiting Chestnut Blight. (Illinois Natural History Survey.)

Insects and Diseases of the Elm

The **Dutch Elm Disease**, resulting from the infection of the tree by *Ceratocystis ulmi*, has already defaced many American towns by destroying the large populations of the very susceptible American elm. This fungus grows in the water-conducting tissues of the tree and promotes the formation of gums by the tree. These gums plug the water-conducting tissues, causing the wilt of the tree and inevitable death. The vectors of this disease are the **Smaller European Elm Bark Beetle**, *Scolytus multistriatus*, and the **Native Elm Bark Beetle,** *Hylurgopinus rufipes*. The adult beetles feed on the crotches of twigs and the small branches of the elm trees. Their eggs are laid on weakened trees or old logs. In the spring when the adults emerge from these breeding places, they frequently are contaminated with the spores of the Dutch Elm Disease and proceed to infect the healthy trees upon which they feed. To attempt to control this disease, all diseased elm wood must be burned promptly to eliminate the breeding places of the Beetles. To kill the emergent adults, the application of a spray that contains four gallons of twenty-four percent methoxychlor emulsifiable concentrate per one hundred gallons of water is recommended. This spray should be applied in the early spring when the adults emerge. The cooperative efforts between members of the community, including the rural areas, are necessary to slow the ravages of this disease.

Figure XII-12. Elm Trees Killed by Dutch Elm Disease. (Illinois Natural History Survey.)

Figure XII-13. European Elm Bark Beetle feeding in crotch of Elm Shoot. (Illinois Natural History Survey.)

Insects of the Fir

The Bagworm can be controlled by the application of an insecticide containing spray to the foliage as the larvae of the pest appear in May or June. This spray may contain one of the following chemicals: two pounds of fifty percent carbaryl (Sevin) wettable powder, OR one quart of carbaryl 4F, OR four pounds of twenty-five percent malathion wettable powder, OR one quart of fifty-seven percent malathion emulsifiable concentrate per one hundred gallons of water.

Insects of the Hemlock

The control of the **Hemlock Scale** can be effected by the application of a spray while the insect is in the "crawler stage", in the latter part of May in the area of Washington, D.C.. One spray that is effective in the control of this pest is composed of one pint of thirty-and-one-half percent Dimethoate emulsifiable concentrate per one hundred gallons of water.

Dicofol may be sprayed on the foliage of the tree when the **Two-Spotted Spider Mites** appear. The spray is composed of either one quart of the 18.5 percent emulsifiable concentrate or two pounds of the 18.5 percent wettable powder per one hundred gallons of water. The treatment may be repeated in a month if it is necessary.

Insects of the Hickory

The **Gall Aphid** may be controlled by the application of a spray that contains two quarts of twenty-four percent methoxychlor emulsifiable concentrate plus one gallon of Summer Oil per one hundred gallons of water. This spray should be applied during the early summer when the young crawlers emerge from the galls on the tree.

Insects and Diseases of the Maple

Tar Spot is the common name for the infection of the Maple by the fungus *Rhytisma acerinum*. The common name arises from the thick, tar-like spots that appear on leaves that are infected with this disease. The disease overwinters on fallen leaves and infects new foliage in May and June. Old leaves should be burned and the new foliage should be sprayed in the spring. A suitable spray is composed of one-and-one-half pounds of seventy-six percent Ferbam wettable powder per one hundred gallons of water. The spray may be repeated in two to three weeks if necessary.

Several times during the summer months, the application of a spray that contains either four pounds of twenty-five percent malathion wettable powder or one quart of fifty-seven percent malathion emulsifiable concentrate per one hundred gallons of water will control the **Norway Maple Aphid**.

Figure XII-14. Maple Leaf Showing Tar Spot (Illinois Natural History Survey.)

Diseases of the Oak

Oak trees that are severely infected with either of the following diseases should be removed and burned: **Bleeding Canker,** *Phytopthora cactorum*, or **Chestnut Blight**, *Endothia parasitica*, or **Oak Wilt**, *Ceratocystis fagacearum*. The Oak Wilt can spread through natural root grafts and, making it advisable to sever the natural root grafts.

Avoiding injuries to the roots and trunk of the trees is advisable to control the **Shoestring Root Rot** of the Oak, *Armillaria mellea*. Infected areas of the tree can be removed by surgery if such removal is necessary.

Insects and Diseases of the Pine

The **Pine Tip Blight**, *Diplodia pinea*, primarily affects the Scotch and Austrian Pines. This disease may be controlled by the application of a spray that contains sixteen pounds of an 8-8-100 Bordeaux mixture per one hundred gallons of water. The spray should be applied when the buds open in the spring, with a second application two weeks after the first. Blighted twigs should be removed from the tree and burned.

Both the **Nantucket Pine Moth**, *Rhyacionia frustrana*, and the **European Pine Shoot Moth**, *Rhyacionia buoliana*, may be controlled by the application of either two pounds of fifty percent carbaryl (Sevin) wettable powder or one quart of carbaryl 4F per one hundred gallons of water. This spray should be applied to the new foliage of the tree.

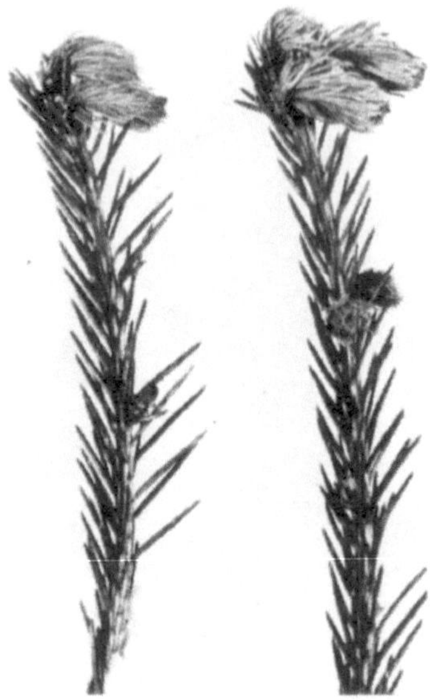

Figure XII-15. Spruce with Sphaeropsis Tip Blight.

In the late spring when the crawler stage of the **Pine Needle Scale,** *Phenacaspis pinifoliae*, appears, the application of a spray containing either four pounds of twenty-five percent malathion wettable powder or one quart of fifty-seven percent malathion emulsifiable concentrate per one hundred gallons of water will control this pest.

The application of two quarts of twenty-four percent methoxychlor emulsifiable concentrate plus one gallon of Summer Spray Oil per one hundred gallons of water will control the nymphs of the **Pine Bark Aphid.** This spray should be applied in the late spring.

Diseases of the Poplar

The removal and burning of trees is warranted if they are severely infected by any of the following diseases: **Poplar Anthracnose,** *Marssonina populea*, or **Poplar Bacterial Canker,** *Phytomonas rimaefaciens*, or **Poplar Canker and Leaf Spot,** *Dothichiza populae.*

BIBLIOGRAPHY

1. Carter, J.C., Illinois Trees: Their Diseases, Illinois Natural History Survey, Circular 46, 3rd printing with alterations, June 1964, Urbana, Ill.
2. Staff, U.S.D.A., Forest Service, (1962).
3. Wallner, W. E., and Hart, J. H., Dutch Elm Disease Control, Extension Bulletin 506, Cooperative Extension Service, Michigan State University, East Lansing, Mich. (1967).
4. Whitten, R. R., Elm Bark Beetles, U.S.D.A., Leaflet No. 185, Washington, D.C. (May, 1967).

BIBLIOGRAPHY
WEEDS

1. ———, Chemunique, Atlas Chemical Industries, Inc., Wilmington, Delaware, Vol. 17, No. 5 (October, 1968).
2. Dawson, J. H., Lee, W. O., and Timmons, F. L., Controlling Dodder in Alfalfa, U.S.D.A. Farmers' Bulletin No. 2211, Washington, D.C., January, 1965.
3. Day, B. E. and Russell, R. C., The Effect of Drying on Survival of Nutgrass Tubers, Bulletin 751, California Agricultural Experiment Station, August, 1955.
4. Gay, C. W. and Dwyer, D. D., Poisonous Range Plants, Cooperative Extension Service Circular 391, New Mexico State University, Las Cruces, N.M., June, 1967.
5. Harvey, W. A., Jones, L., and Barnes, R. M., Dodder or Dollars?, Agricultural Extension Service, College of Agriculture, University of California.
6. Hauser, E. W. and Arle, H. F., Johnson Grass as a Weed, Farmers' Bulletin No. 1537, U.S.D.A., Washington, D.C., August, 1958.
7. Knake, E. and Slife, F., Controlling Giant Foxtail in Illinois, Circular 828, Extension Service, College of Agriculture, University of Illinois, Urbana, Ill., May, 1963.
8. Major, J., McKell, C. M., and Berry, L. J., California Agricultural Experiment Station Extension Service, January, 196.
9. Miller, J. P., Johnsongrass Control, Circular 552, Cooperative Extension Service, University of Georgia, Athens, Ga., January, 1966.
10. Mitich, L. W., Field Control of Quackgrass, Circular A-293, Extension Service, North Dakota State University, Fargo, N.D., October, 1964.
11. Mitich, L. W., Leafy Spurge Control, Circular A-295, Extension Service, North Dakota State University, Fargo, N.D., November, 1965.
12. Mitich, L. W., Noxious Weeds, No. 1 (Leafy Spurge, April, 1967), No. 2 (Field Bindweed, May, 1967), No. 4 (Perennial Sowthistle, May, 1967), No. 5 (Russian Knapweed, May, 1967), Extension Service, North Dakota State University, Fargo, N.D.
13. Morse, M.D., Wild Blackberry Control, OSA #23, University of California Agricultural Extension Service, December, 1963.
14. Rosenberg, D. Y., Nichols, C. W., and Schlocker, A., Branched Broomrape *(Orobanche ramosa)* A Threat to California Crops, Extension Service, California Agricultural Experiment Station, June, 1965.
15. Simpson, C. F. and West, E., Coffe-Weed (Bagpod) Seed Poisoning of Cattle, Circular S-58, Agricultural Experiment Station, University of Florida, Gainesville, Florida, July, 1953.

16. Slife, F. and Knake, E., Controlled Johnsongrass in Illinois, Circular 827, College of Agriculture, University of Illinois, Urbana, Ill., May, 1963.
17. Sommers, C. E., Successful Farming's Guide to Poisonous Plants, Successful Farming, Vol. 65, No. 3 (March, 1967), pp. 77-82.
18. Sylwester, E. P., Identification and Control of White Snakeroot, Weed Control Series No. WC-69, Iowa State University, Ames, Iowa, October, 1962.
19. Sylwester, E. P., Black Nightshade, Weed Control Series No. WC-18, Iowa State University, Ames, Iowa, February, 1963.
20. Sylwester, E. P., Downy Brome Grass, Weed Control Series No. WC-36, Iowa State University, Ames, Iowa, April, 1964.
21. Sylwester, E. P., Pokeweed, Weed Control Series No. WC-11, Iowa State University, Ames, Iowa, July, 1964.
22. Sylwester, E. P., Control of Teasel, Weed Control Series No. WC-78, Iowa State University, Ames, Iowa, November, 1964.
23. Sylwester, E. P., Unicorn Plant or Devil's Claw, Weed Control Series, No. WC-79, Iowa State University, Ames, Iowa, December, 1964.
24. Sylwester, E. P., Noxious Weeds, Bulletin No. 82-C, Iowa Department of Agriculture, Des Moines, Iowa, 1965.
25. Sylwester, E. P., Cocklebur, Weed Control Series No. WC-10, Iowa State University, Ames, Iowa, April, 1965.
26. Sylwester, E. P., Control of Canada Thistle, Weed Control Series No. WC-71, Iowa State University, Ames, Iowa, January, 1966.
27. Sylwester, E. P., Horsetail, Weed Control Series No. WC-37, Iowa State University, Ames, Iowa, August, 1965.
28. Sylwester, E. P., Wild Four O'Clock, Weed Control Series No. WC-16, Iowa State University, Ames, Iowa, August, 1965.
29. Sylwester, E. P., Whorled Milkweed, Weed Control Series No. WC-30, Iowa State University, Ames, Iowa, September, 1965.
30. Sylwester, E. P., Dodder, Weed Control Series No. PM-339, Iowa State University, Ames, Iowa, September, 1966.
31. Sylwester, E. P., Wild Barley *(Hordeum Jubatum* L.*)* Weed Control Series, Iowa State University, Ames, Iowa, June, 1967.
32. Tucker, J. M., Cordy, L. J., Harvey, W. A., and Fuller, T. C., Nitrate Poisoning in Livestock, Circular 506, Extension Service, California Experimental Station, Berkeley, Cal., December, 1961.
33. Tucker, J. M., Fowler, M. E., Harvey, W. A., and Berry, L. J., Poisonous Hemlocks: Their Identification and Control, Circular 530, Extension Service, California Agricultural Experiment Station, Berkeley, Cal., October, 1964.
34. Tucker, J. M. and Kinball, M. H., Poisonous Plants in the Garden, Bulletin AXT-22, Agricultural Extension Service, University of California, December, 1966.

FIELD CROPS

1. Barnes, G. and Roberts, J. E., Rice Insect Control Recommendations, Leaflet No. 330, Agricultural Extension Service, University of Arkansas, January, 1966.
2. Bernard, R. L., Millis, D. E., Graffis, D. W., and Scott, W. O., Soybean Varieties in Illinois for 1967, Circular 964, College of Agriculture Cooperative Extension Service, University of Illinois, Urbana, Ill., March, 1967.
3. Bigger, J. H., Corn Rootworm Resistance to Chlorinated Hydrocarbon Insecticides in Illinois, Journal of Economic Entomology, Vol. 56, No. 1 (February, 1963), pp. 118-119.
4. Bigger, J. H. and Petty, H. B., Insect Infestation of Corn Roots in Illinois, Bulletin 704, Agricultural Experiment Station, University of Illinois, Urbana, Ill., January, 1965.
5. Barnes, G. and Roberts, J. E., Control Insects on Soybeans Leaflet 193, Agricultural Extension Service, University of Arkansas, May, 1967.
6. Boewe, G. H., Diseases of Wheat, Oats, Barley, and Rye, Circular 48, Illinois Natural History Survey, Urbana, Ill., June, 1960.
7. Clark, F. and Myers, J. M., Fumigation and Equipment for Nematode Control in Soils for Flue-Cured Tobacco, Circular S-27, Agricultural Experiment Station, University of Florida, January, 1951.
8. Eyer, J. R., Habits and Control of the Mexican Bean Beetle in Southern New Mexico, Bulletin 377, Agricultural Experiment Station, New Mexico College of Agriculture and Mechanical Arts, June, 1963.
9. French, J. C., Tobacco Insect Control, Circular 413, Cooperative Extension Service, University of Georgia, Athens, Ga., January, 1967.
10. Harris, E. D., Jr., The Cowpea Curculio, Leaflet No. 56, Cooperative Extension Service, University of Georgia, June, 1966.
11. Harris, E. D., Jr., Insect Control of Commercial Beans and Southern Peas, Circular 567, Cooperative Extension Service, University of Georgia, Athens, Ga., February, 1967.
12. Hooker, A. L., Corn Stalk Rot, Report on Plant Diseases No. 200, Cooperative Extension Service, University of Illinois, Urbana, Ill.
13. Janes, R. L., Cereal Leaf Beetle Control, Cooperative Extension Service, Michigan State University, Extension Bulletin #523, East Lansing, Mich., June, 1966.
14. Janes, R. L., Guyer, G., and Ruppel, R., Michigan Corn Production Insect Control, Extension Bulletin #439, Cooperative Extension Service, Michigan State University, East Lansing, Mich., March, 1966.
15. Janes, R. L. and Ruppel, R., Field Bean, Soybean and Sugar Beet Production, Insect and Nematode Control in Michigan, Extension Bulletin #499, Farm Science Series, Cooperative Extension Service, Michigan State University, East Lansing, Mich., December, 1965.

16. Janes, R. L. and Ruppel, R., Insect Control in Forages and Small Grains, Extension Bulletin #523, Cooperative Extension Service, Michigan State University, East Lansing, Mich., June, 1966.
17. Jordan, C. R., Soybean Insects and Their Control, Leaflet 29, Cooperative Extension Service, University of Georgia, Athens, Ga., February, 1967.
18. Kantack, B. H. and Berndt, W. L., Grasshopper Control in South Dakota-1967, F.S. 321, Cooperative Extension Service, South Dakota State University.
19. Khare, M. N., Shurtleff, M. C., and Hooker, A. L., Sorghum Smuts, Report on Plant Diseases No. 208, Cooperative Extension Service, University of Illinois, Urbana, Ill.
20. King, C. L., Chemical Controls of Plant Diseases, Extension Service, Kansas State University.
21. Koehlor, B., Cornstalk Rots in Illinois, Bulletin #658, Agricultural Experiment Station, University of Illinois, Urbana, Ill., April, 1960.
22. Luckmann, W. H., and Petty, H. B., Controlling Corn Borers in Field Corn with Insecticides, Circular No. 768, Cooperative Extension Service, University of Illinois, Urbana, Ill., January, 1965.
23. Luckmann, W. H., Rhodes, A. M., and Wann, E. V., Silk Balling and Other Factors Associated with Resistance of Corn to Corn Earworm, Journal of Economic Entomology, Vol. 57, No. 5 (October, 1964), pp. 778-779.
24. Maier, C. R., Effects of Crop Residues on the Pinto Bean Rot Complex, Bulletin #491, Agricultural Experiment Station, New Mexico State University, February, 1965.
25. Malm, N. R. and Hsi, D. C., New Mexico 31- A Charcoal Rot Resistant Grain Sorghum Line, Research Report #93, New Mexico State University, May, 1964.
26. Malm, N. R., Williams, D. H., and Jones, H. D., Performance of Sorghum Hybrids and Varieties in New Mexico, 1961, Research Report #68, New Mexico State University, March, 1962.
27. Matteson, J. W. and Decker, G. C., Development of the European Corn Borer at Controlled Constant and Variable Temperatures, Journal of Economic Entomology, Vol. 58, No. 2 (April, 1965), pp. 344-349.
28. McLaughlin, R. L., The Role of Certain Gram-negative Bacteria and Temperature in Larval Mortality of the Armyworm, *Pseudaletia unipuncta* (Haworth) Journal of Insect Pathology, Vol. 4, No. 3 (September, 1962).
29. Owen, J. H., Clark, F., and Brothers, S. L., Angular Leafspot of Tobacco, Circular 190, Agricultural Extension Service, University of Florida, January, 1959.
30. Pady, S. M., et al., Diseases of Wheat in Kansas, Bulletin 368, Agricultural Experiment Station, Kansas State College of Agriculture and Applied Science, Manhattan, Kan., January, 1955.

31. Petty, H. B. and Moore, S., III, Profitable Control of Underground Corn Insects, Circular 805, Cooperative Extension Service, University of Illinois, Urbana, Ill., April, 1959.
32. Petty, H. B. and White, C. E., Control of Spittlebugs on Legumes, Circular 689, University of Illinois College of Agriculture in cooperation with Illinois Natural History Survey, 1953.
33. Roberts, J. E. and Barnes, G., Control of Insects on Growing Corn, Leaflet No. 274, Agricultural Extension Service, University of Arkansas, July, 1966.
34. Roselle, R. E. and Weekman, Corn Rootworm Control, E.C. 66-1596, Extension Service, University of Nebraska, January, 1965.
35. Safir, G. R. and Shurtleff, M. C., Blue Dwarf of Oats, Report on Plant Diseases No. 118, Cooperative Extension Service, University of Illinois, Urbana, Ill.
36. Shannon, E. L., and Hsi, D. C. H., Control of Peanut Diseases, Plant Service Guide 400 A-502, Cooperative Extension Service, New Mexico State University.
37. Shurtleff, M. C., Fungicide Seed Treatment for Small Grains, Report on Plant Diseases No. 1001 (Revised), Cooperative Extension Service, University of Illinois, Urbana, Ill.
38. Shurtleff, M. C. and Bever, W. M., Smuts of Barley, Report on Plant Diseases No. 100 (Revised), Cooperative Extension Service, University of Illinois, Urbana, Ill.
39. Shurtleff, M. C., Bever, W. M., and Quebral, F. C., Loose Smut of Wheat, Report on Plant Diseases No. 112, Cooperative Extension Service, University of Illinois, Urbana, Ill.
40. Shurtleff, M. C. and Britton, M. D., Helminthosporium Leaf Spots and Blotches of Forage Grasses, Report on Plant Diseases No. 309, Cooperative Extension Service, University of Illinois, Urbana, Ill.
41. Shurtleff, M. C. and Britton, M. P., Brown Stripe or Leaf Streak of Forage Grasses, Report on Plant Diseases No. 310, Cooperative Extension Service, University of Illinois, Urbana, Ill.
42. Shurtleff, M. C. and Britton, M. P., Selenophoma Leaf Spot or Speckle of Forage Grasses, Report on Plant Diseases No. 311, Cooperative Extension Service, University of Illinois, Urbana, Ill.
43. Shurtleff, M. C. and Britton, M. P., Stagonospora Leaf Spot or Blotch of Forage Grasses, Report on Plant Diseases No. 312, Cooperative Extension Service, University of Illinois, Urbana, Ill.
44. Shurtleff, M. C. and Britton, M. P., Anthracnose of Forage Grasses and Cereals, Report on Plant Diseases No. 313, Cooperative Extension Service, University of Illinois, Urbana, Ill.

45. Shurtleff, M. C. and Britton, M. P., Stripe Smut of Forage Grasses, Report on Plant Diseases No. 314, Cooperative Extension Service, University of Illinois, Urbana, Ill.
46. Shurtleff, M. C. and Jedlinski, H., Oat Smuts, Report on Plant Diseases No. 114, Cooperative Extension Service, University of Illinois, Urbana, Ill.
47. Shurtleff, M. C. and Hooker, A. L., Corn Ear Rot, Report on Plant Diseases No. 205, Cooperative Extension Service, University of Illinois, Urbana, Ill.
48. Shurtleff, M. C. and Hooker, A. L., Storage Rots of Corn, Report on Plant Diseases No. 206, Cooperative Extension Service, University of Illinois, Urbana, Ill.
49. Shurtleff, M. C., et al., Diseases of Corn in the Midwest, North Central Regional Extension Publication No. 21, Cooperative Extension Service, University of Illinois, Circular 967, Urbana, Ill., May, 1967.
50. Sill, W. H., Jr., Wheat Streak Mosaic, Reprint No. 1, Kansas Agricultural Experiment station, August, 1959.
51. Smith, N. A., Michigan Corn Production Disease Control, Extension Bulletin 440, Cooperative Extension Service, Michigan State University, East Lansing, Mich., September, 1965.
52. Staff, Insect Control Guide for Small Grains, Field Beans, Soybeans, Sugar Beets - Nebraska, E.C. 66-1504 (compiled by extension entomologists) Extension Service, University of Nebraska.
53. Staff, 1967 Suggested Insecticide Guides - Insect Control for Field Crops, Circular 899, Cooperative Extension Service of the University of Illinois in cooperation with the Illinois Natural History Survey, Urbana, Ill,. January, 1967.
54. Staff, The Alfalfa Weevil: How to Control It, Leaflet No. 368, U.S.D.A., Washington, D.C., November, 1967.
55. Staff, Leaflets Nos. 475, 479, U.S.D.A., Washington, D.C., 1960.
56. Staff, The Pea Aphid on Alfalfa: How to Control It, Leaflet No. 529, U.S.D.A., Washington, D.C., February, 1967.
57. Staff, Compendium of Plant Diseases, Rohm and Haas Company, Philadelphia, Pa., 1959.
58. Strayer, J. R., Sugar Cane Insect Control, Extension Entomology Mimeograph #40, Florida Agricultural Extension Service, May 2, 1966.
59. Sylwester, E. P., Ergot of Rye, Weed Control Series WC-39, Cooperative Extension Service, Iowa State University, January, 1966.
60. Tappan, W. B. and Strayer, J., Shade Grown Tobacco Insect Control for 1967, Extension Entomology Mimeograph #43, Florida Agricultural Extension Service, December 17, 1-66.
61. Thompson, S. S., Granville Wilt, Leaflet No. 6, Cooperative Extension Service, University of Georgia, Athens, Georgia, August, 1964.

62. Thompson, S. S., Angular Leafspot of Tobacco, Leaflet No. 73, Cooperative Extension Service, University of Georgia, Athens, Ga., August, 1967.
63. Thompson, S. S., Southern Stem Rot and Sore Shin of Flue Cured Tobacco, Leaflet No. 37, Cooperative Extension Service, University of Georgia, Athens, Ga., October, 1965.
64. Thompson, S. S., Tobacco Mosaic and its Control, Leaflet No. 53, Cooperative Extension Service, University of Georgia, Athens, Ga., May, 1966.
65. White, L. D., Plant Feeding Mites of South Dakota, Technical Bulletin #27, Agricultural Experiment Station, South Dakota State University, May, 1966.
66. Wilson, M. L. and Melton, B. A., Breeding for Resistance to Bacterial and Fusarium Wilt in Alfalfa, Bulletin No. 468, Agricultural Experiment Station, New Mexico State University, July, 1962.

INDEX

A

B

D

E

Q

R

T

U

www.ingramcontent.com/pod-product-compliance
Ingram Content Group UK Ltd.
Pitfield, Milton Keynes, MK11 3LW, UK
UKHW041131071026
15587UKWH00009B/423

9 780820 600062